EAI/Springer Innovations in Communication and Computing

Series editor
Imrich Chlamtac, CreateNet, Trento, Italy

Editor's Note

The impact of information technologies is creating a new world yet not fully understood. The extent and speed of economic, life style and social changes already perceived in everyday life is hard to estimate without understanding the technological driving forces behind it. This series presents contributed volumes featuring the latest research and development in the various information engineering technologies that play a key role in this process.

The range of topics, focusing primarily on communications and computing engineering include, but hardly limited to, wireless networks; mobile communication; design and learning; gaming; interaction; e-health and pervasive healthcare; energy management; smart grids; internet of things; cognitive radio networks; computation; cloud computing; ubiquitous connectivity, and in mode general smart living, smart cities, Internet of Things and more. The series publishes a combination of expanded papers selected from hosted and sponsored European Alliance for Innovation (EAI) conferences that present cutting edge, global research as well as provide new perspectives on traditional related engineering fields. This content, complemented with open calls for contribution of book titles and individual chapters, together maintain Springer's and EAI's high standards of academic excellence. The audience for the books consists of researchers, industry professionals, advanced level students as well as practitioners in related fields of activity include information and communication specialists, security experts, economists, urban planners, doctors, and in general representatives in all those walks of life affected ad contributing to the information revolution.

About EAI

EAI is a grassroots member organization initiated through cooperation between businesses, public, private and government organizations to address the global challenges of Europe's future competitiveness and link the European Research community with its counterparts around the globe. EAI tens of thousands of members on all continents together with its institutional members base consisting of some of the largest companies in the world, government organizations, educational institutions, strive to provide a research and innovation platform which recognizes excellence and links top ideas with markets through its innovation programs.

Throughs its open free membership model EAI promotes a new research and innovation culture based on collaboration, connectivity and excellent recognition by community.

More information about this series at http://www.springer.com/series/15427

Antonio Puliafito • Kishor S. Trivedi
Editors

Systems Modeling: Methodologies and Tools

 Springer

Editors
Antonio Puliafito (iD)
Department of Engineering
University of Messina
Messina, Italy

Kishor S. Trivedi
ECE Department School of Engineering
Duke University
Durham, NC, USA

ISSN 2522-8595 ISSN 2522-8609 (electronic)
EAI/Springer Innovations in Communication and Computing
ISBN 978-3-030-06421-1 ISBN 978-3-319-92378-9 (eBook)
https://doi.org/10.1007/978-3-319-92378-9

This Springer imprint is published by the registered company Springer Nature Switzerland AG
The registered company address is: Gewerbestrasse 11, 6330 Cham, Switzerland

Contents

Chapter 1
Systems Modelling: Methodologies and Tools

Antonio Puliafito ⓘ and Kishor S. Trivedi

Modern systems implement multiple and complex operations to manage the user demand, thereby ensuring adequate quality levels. They are usually made of a collection of interconnected (autonomous) subsystems, with a common goal to be pursued, that are perceived as a whole, single, integrated facility.

Several heterogeneous technologies and processes are usually combined (computing, networking, manufacturing, marketing, mechanical, economical, biological, etc.) that involve complex interactions, interferences, and dependencies. Basic functionalities have to be provided through adequate mechanisms, but also advanced ones implementing specific quality-driven policies have to be delivered. For these reasons, functional and non-functional properties are key issues to be addressed during the whole system life cycle, both at design time and at run time, as well as during maintenance stages, thus requiring adequate methods and techniques for their evaluation. Simply speaking, functional requirements describe what the system should do, while non-functional requirements describe how the system works [1]. Typical functional requirements include Administrative functions, Business Rules, Transaction corrections, Authorization and Authentication levels, Reporting Requirements and Historical Data, External Interfaces, Legal and Regulatory Requirements. Non-functional requirements can be seen as quality attributes of a system, i.e., criteria that judge the operation of a system, rather than specific behaviors [2]. Some typical non-functional requirements are Performance (such as Response Time, Throughput, and Utilization), Scalability, Availability, Reliability, Recoverability, Maintainability,

A. Puliafito (✉)
Department of Engineering, University of Messina, Messina, Italy
e-mail: apuliafito@unime.it

K. S. Trivedi
ECE Department School of Engineering, Duke University, Durham, NC, USA
e-mail: ktrivedi@duke.edu

© Springer International Publishing AG, part of Springer Nature 2019
A. Puliafito, K. S. Trivedi (eds.), *Systems Modeling: Methodologies and Tools*,
EAI/Springer Innovations in Communication and Computing,
https://doi.org/10.1007/978-3-319-92378-9_1

1

and Security. Systems modelling specifically focuses on non-functional parameters, as the intention is to quantitatively evaluate the behavior of a system.

Measurements are derived from a real system running under real operating conditions. They report the actual system performance in the condition in which the system is working. For this reason, measurements are very specific since they heavily depend on the characteristics of the measured system, and on the particular workload the system is experiencing during the measurement itself [3].

There are several situations in which relying upon measurements alone is not sufficient. For instance, when the performances of two systems have to be compared, it is difficult to ensure that the operating conditions under which measurements are performed are equivalent, thus yielding a fair comparison. To overcome such problem, *benchmarks* [4] feed the system with an artificial workload, so to perform observations in equivalent working conditions, and meaningful comparisons can be made.

Measurements and benchmarks need a system to be observed. When the performance analysis regards a system that is not yet operational (i.e., it is not available), its representative and detailed approximation (usually indicated as a *prototype*) has to be developed, either in hardware or in software. Prototypes are used to make observations, possibly using benchmarks as artificial workloads.

In measurements, benchmarks, and prototypes the performance of the system is evaluated through observation of the system's behavior, in working conditions, i.e., when it processes the actual user requests or the benchmark. However, evaluating the performance of a system is an important task not only during and after the system implementation, but also at the early design stage to compare possible alternative choices. When the design and production of a new system is the consequence of ever-increasing performance, like in the computer and telecommunications fields, early performance analysis is absolutely mandatory.

During the design process, measurements on real systems are obviously not possible, and also prototype implementations present difficulties due to the necessity of specifying many details that are far from being decided. A more interesting alternative to solve these problems is the use of models to define characteristics of the systems under study, and to investigate their best configuration by modifying the parameters of different components. *System modelling* is the art of developing an abstract representation of the real system to derive and analyze its behavior in terms of performance and dependability under different functioning conditions, without resorting to measurements on the real system as a whole or its prototype [5–8]. We note here that measurements of system components (or subsystems) do provide input parameter values for the overall system model [7].

System models can be analyzed/solved through three different approaches:

- simulation
- analytical (closed-form or numerical)
- hybrid (combining simulation and analytic methods).

In all the three cases, the study of the system is carried out using a description that includes only some of its main characteristics. In simulative solutions, the description is embedded into a computer program that mimics the system dynamics, whereas analytical methods consist in developing and solving sets of mathematical equations governing system dynamics [5, 8, 9]. The most attractive advantage of simulation is that the system under study can be represented in a very detailed way, without imposing many restrictions on the model, whereas in analytical models many simplifying assumptions are often introduced, so that the underlying mathematical equations are tractable.

In simulative solutions the accuracy of details is limited only by the time employed to obtain the final measures; more detailed the description of the model, longer is the time required for its solution. It is very common that a very detailed model requires days or even weeks to be evaluated. "Time" is the critical parameter in simulation modelling. Techniques for parallel and distributed simulation do exist to somewhat alleviate this problem [10].

Usually, analytical models have a higher level of abstraction, and they require shorter time to solve. This advantage becomes crucial when, for instance, sensitivity analysis [11] or optimization [12] is to be carried out. In fact, analytical models are described by sets of equations, and by taking formal derivatives, new equations can be derived that enable the computation of derivatives of the measures of interest with respect to the parameters [11]. Whereas a new simulation run, possibly very time consuming, has to be executed with a new set of parameter values with respect to which the sensitivity analysis is to be performed [6]. We note here significant research on perturbation theory that reduces some of these difficulties with simulation [13]. Benefits of simulation and analytic models can be combined via hybrid model solution techniques; however, this is not as common and we expect it to develop further in the future [14].

Modern systems are inherently distributed and aim to implement and provide services that are able to meet ever-increasing quality standards, while minimizing costs. Systems being part of critical infrastructures have to meet tight dependability, timeliness, and performance requirements and specifications. The inherently unpredictable nature of such systems requires a quantitative evaluation of deterministic and probabilistic timed models for their design and maintenance. Techniques for checking and verifying if and how a distributed system satisfies the requirements (validation) are specifically required; a proper evaluation of non-functional aspects (evaluation) is often mandatory; the optimization of the overall behavior of the system (optimization) is crucial to reduce costs and deliver high quality solutions. Validation is part of quality management and imposes that a product, service, or system is checked, inspected, and/or tested to verify that the requirements are satisfactory [7, 15]. Evaluation analyzes the system's non-functional properties such as performance, reliability, and availability [1]. Optimization [12, 16] is instead related to the identification and selection of the best configuration for the distributed system according to some given (usually multiple) parameters in order to meet high-level requirements such as overall costs and sustainability, i.e., the ability to continue at a particular level for a period of time.

Validation, evaluation, and optimization techniques and methodologies are sometimes overlapped, i.e., often validation techniques include evaluation and/or optimization and vice versa. In particular, evaluation and optimization often overlap, although the former usually investigates a single non-functional aspect of the system, while optimization problems usually evaluate the system looking at multiple, complex, and/or composed properties such as dependability, performability [17], and sustainability, often also including costs. In any case, all of them often rely on models to provide their useful insights. All such considerations and needs have given birth to an abundance of literature devoted to formal modelling languages combined with analytical and simulative solution techniques.

The aim of this book is to provide an overview of techniques and methods dealing with such specific issues in the context of systems modelling and to cover aspects such as correctness, validity, performance, reliability, availability, energy efficiency, and sustainability. This book collects some of the papers presented at the 10th EAI International Conference on Performance Evaluation Methodologies and Tools Conference (Valuetools 2016), held in Taormina (Italy) from the 25th to the 28th of October 2016, which have been significantly extended an improved.

Following this path, the book has been organized into four parts dealing with modelling theory (Part I), applications to communication systems and infrastructures (Part II), optimization and quantitative evaluation techniques applied to Cloud computing and the Internet of Things (Part III), and tools development for the analysis of specific areas of interests (Part IV).

The chapters have been selected to provide a good, although not exhaustive, coverage of issues, models, and techniques related to validation, evaluation, and optimization of complex systems, hoping that this will be useful in guiding students, researchers, and practitioners when approaching the quantitative assessment of distributed systems. Indeed, a key objective of this book is to help bridge the gap between modelling theory and practice through specific examples.

Specifically, included in Part I are seven contributions that cover theoretical aspects of systems modelling. Chapter 2 by Lei Zhang and Douglas G. Down introduces a numerically Stable MVA (SMVA) algorithm for closed product-form queueing networks that allows for load-dependent queues and offers a numerically stable, efficient, and accurate approximate solution. Chapter 3 by Esa Hyytiä, Rhonda Righter, Olivier Bilenne, and Xiaohu Wu studies the M/D/1 queue and its generalization, the M/iD/1 queue, when jobs have firm deadlines on waiting (or sojourn) times. Explicit value functions are derived for these M/D/1-type of queues that enable the development of efficient cost-aware dispatching policies to parallel servers. Chapter 4 by Elvio Gilberto Amparore and Susanna Donatelli defines X-PPN, an extension of the Phased Petri Net formalism that provides the modeller more freedom in the structure and in the stochastic distribution of the phases (from deterministic to general) and in the definition of the dependencies among the system and the phase net. Chapter 5 by Steffen Bondorf and Fabien Geyer tackles the problem of analyzing multicast flows with deterministic network calculus without accommodating for it by pessimistic changes to the network model and thus allowing for the derivation of more accurate performance bounds than existing approaches. Chapter 6 by D. Cerotti, M. Gribaudo, R. Pinciroli, and G. Serazzi focuses on Pool Depletion Systems (PDS), i.e., systems where a large number of tasks must be executed by one or more subsystems with a finite capacity, to find the allocation policy of all the tasks in the pool that minimizes the time required to execute all the tasks; Markov models, simulation, and fluid approximation techniques are considered for this purpose. Chapter 7 by Raymond Marie shows the importance of intermittent servers in order to reduce the response times without increasing significantly the idle times of servers, producing a closed form solution for the steady-state probability distribution and for different metrics such as expected response times for customers or expectation of busy periods. Chapter 8 by Zdravko Botev and Pierre L'Ecuyer studies and compares various methods to generate a random variate or vector from the univariate or multivariate normal distribution truncated to some finite or semi-infinite region, with special attention to the situation where the regions are far in the tail.

Part II focuses on specific applications to the telecommunication field and is composed of four chapters. Chapter 9 by Francesco Bianchi and Francesco Lo Presti compares two approaches to the Virtual Network Embedding problem based on Markov Reward Processes, to achieve a good trade-off between resource utilization and QoS (e.g., latency). Chapter 10 by Aditi Gupta, Dharmaraja Selvamuthu, and Subrat Kar proposes a load balancing Component Carrier selection scheme which can be optimized for the Quality of Service (QoS) required by users. Feedback fluid queue model is developed to analyze and optimize the performance of the proposed scheme. Chapter 11 by Ramona Kühn, Andreas Fischer, and Hermann de Meer discusses security requirements of Virtual Networks (VNs) and shows how they can be modelled to be mapped into the provided security mechanisms in the physical network. Chapter 12 by Philippe Olivier, Florian Simatos, and Alain Simonian analyzes the impact of inter-cell mobility on data traffic performance in dense networks with small cells.

Part III deals with Cloud computing and Internet of Things, proposing different techniques in five chapters. Chapter 13 by Nicola Bicocchi, Claudia Canali, and Riccardo Lancellotti proposes a technique to infer VMs communication patterns starting from input/output network traffic time series of each VM. They discuss both the theoretical aspect of such technique and the design challenges for its implementation. Chapter 14 by Emiliano Casalicchio surveys the state-of-the-art solutions and discusses research challenges in autonomic orchestration of containers. A reference architecture of an autonomic container orchestrator is also proposed. Chapter 15 by Dario Bruneo, Salvatore Distefano, Francesco Longo, Giovanni Merlino, and Antonio Puliafito describes an approach to network virtualization based on popular off-the-shelf tools and protocols in place of application-specific logic, acting as a blueprint in the design of the Stack4Things architecture, an OpenStack-derived framework to provide IaaS-like services from a pool of IoT devices. They quantitatively evaluate the underlying mechanisms demonstrating that the proposed approach exhibits mostly comparable performance with respect to standard technologies for virtual private networks.

Part IV presents some tools for systems modelling. Chapter 16 by Gábor Horváth and Miklós Telek introduces BuTools, a collection of computational methods that are useful for Markovian and non-Markovian matrix analytic performance analysis.

Chapter 17 by Riccardo Di Pietro, Maurizio Giacobbe, Carlo Puliafito, and Marco Scarpa presents J2CBROKER, a tool that simulates a Cloud Brokerage ecosystem, i.e., an environment where a software broker acts as an intermediary between service customers and providers in order to allow the former to discover and select the services that best suit their needs. Chapter 18 by Sushma Nagaraj and Armin Zimmermann compares several state-of-the-art transient removal algorithms and proposes a software framework for a systematic comparison of such algorithms. Finally, Chap. 19 by Abdalkarim Awad, Peter Bazan, and Reinhard German presents a co-simulation framework that captures two important worlds of the smart grid, namely the communication world and power world. Real data as well as simulation models are used to simulate several home appliances.

The chapters have been written by leading experts in distributed systems, modelling formalisms, and evaluation techniques, from both academia and industry. We wish to thank all of them for their contributions and cooperation. Special thanks go to the Springer staff for the support and valuable advice they have always provided. We hope that practitioners will find this book useful when looking for solutions to practical problems, and that researchers can consider it as a first-aid reference when dealing with systems modelling from a qualitative and quantitative perspective.

References

1. S. Lauesen, *Software Requirements: Styles and Techniques* (Addison-Wesley, Harlow, 2002)
2. H. Kaur, A. Sharma, Non-functional requirements research: survey. Int. J. Sci. Eng. Appl. **3**(6) (2014). ISSN-2319-7560
3. A. Neely, The evolution of performance measurement research: developments in the last decade and a research agenda for the next. Int. J. Oper. Prod. Manag. **25**(12), 1264–1277 (2005)
4. R. Dattakumar, R. Jagadeesh, A review of literature on benchmarking. Benchmarking: Int. J. **10**(3), 176–209 (2003)
5. K.S. Trivedi, *Probability and Statistics with Reliability, Queuing, and Computer Science Applications*, 2nd edn. (Wiley, Hoboken, 2001); revised paperback, 2016
6. R.A. Sahner, K.S. Trivedi, A. Puliafito, *Performance and Reliability Analysis of Computer Systems: An Example-Based Approach Using the SHARPE Software Package* (Kluwer Academic Publishers, Dordrecht, 1996)
7. K.S. Trivedi, A. Bobbio, *Reliability and Availability Engineering: Modeling, Analysis, and Applications* (Cambridge University Press, Cambridge, 2017)
8. G. Bolch, S. Greiner, H. de Meer, K. Trivedi, *Queueing Networks and Markov Chains*, 2nd edn. (Wiley, Hoboken, 2006)
9. J. Banks (ed.), *Handbook of Simulation: Principles, Methodology, Advances, Applications, and Practice* (Wiley, Hoboken, 1998)
10. R.M. Fujimoto, *Parallel and Distributed Simulation Systems* (Wiley, Hoboken, 2000)
11. J.T. Blake, A.L. Reibman, K.S. Trivedi, Sensitivity analysis of reliability and performability measures for multiprocessor systems, in *Proceedings of the 1988 ACM SIGMETRICS Conference on Measurement and Modeling of Computer Systems*, (ACM, New York, 1988), pp. 177–186
12. R. Ghosh, F. Longo, R. Xia, V. Naik, K. Trivedi, Stochastic model driven capacity planning for an infrastructure-as-a-service cloud. IEEE Trans. Serv. Comput. **7**, 667–680 (2014)

13. R. Suri, M. Zazanis, Perturbation analysis gives strongly consistent sensitivity estimates for the M/G/1 queue. Manag. Sci. **34**, 39–64 (1988)
14. S.-J. Hsieh, Hybrid analytic and simulation models for assembly line design and production planning. Simul. Model. Pract. Theory **10**, 87–108 (2002)
15. J.P.C. Kleijnen, Validation of models: statistical techniques and data availability, in *Proceedings of the IEEE 1999 Winter Simulation Conference*, Phoenix, AZ, USA, 5–8 Dec 1999
16. C.B. Gupta, *Optimization Techniques in Operation Research* (I K International Publishing House, New Delhi, 2008)
17. J.F. Meyer, Closed-form solutions of performability. IEEE Trans. Comput. **31**, 648–657 (1982)

Part I
Modelling Theory

Chapter 2
SMVA: A Stable Mean Value Analysis Algorithm for Closed Systems with Load-Dependent Queues

Lei Zhang and Douglas G. Down

2.1 Introduction

The Mean Value Analysis (MVA) algorithm [16] is an efficient solution for steady-state analysis of queueing networks. However, it relies on product-form assumptions, which can be violated by common features introduced in modern computer systems, e.g., simultaneous resource possession, locking behaviours, priority scheduling, high service demand variability, and process synchronization (see Chapter 15 in [14]). An approximate solution is to reduce a non-product-form network by using Flow-Equivalent Servers (FESs) [7]. An FES is load dependent, whose service rate with n jobs present is equal to the observed throughput of the original network with n jobs. The performance model can then be analysed using the load-dependent MVA algorithm [15].

Unfortunately, the load-dependent MVA algorithm suffers from numerical instability issues [15, 16]. The underlying reason is that the computation of state probabilities can yield negative results when the utilization is close to one. Consequently, negative values of mean performance measures (i.e., response times and throughputs) can be produced. Static and dynamic scaling techniques are potential approaches to cope with precision limits, but they are complicated to implement. In addition, Casale and Serazzi [4] show that they do not work in general, as the mean queue length computations are not affected. To the best of our knowledge,

L. Zhang
Department of Computer Science, Ryerson University, Toronto, ON, Canada
e-mail: leizhang@ryerson.ca

D. G. Down (✉)
Department of Computing and Software, McMaster University, Hamilton, ON, Canada
e-mail: downd@mcmaster.ca

the literature is lacking efficient solutions for the numerical instability of load-dependent MVA.

In this paper, we propose a Stable MVA (SMVA) algorithm for closed networks with load-dependent queues (the initial idea was presented in [21]). The main contributions of this paper include: (1) the SMVA algorithm, which is an efficient approximate solution for closed networks with load-dependent queues, and (2) an extended multi-class model used to determine class-level performance metrics.

This paper is structured as follows. Section 2.2 introduces the required background. Section 2.3 provides a review of solutions proposed in the literature for the numerical instability. We then present SMVA in Sect. 2.4. In Sect. 2.5, the results from SMVA are compared with other MVA algorithms in two case studies. Section 2.6 gives the multi-class SMVA algorithm. This paper ends with Sect. 2.7, in which we give a summary of the pros and cons of SMVA.

2.2   Background

The exact MVA algorithm for closed networks with load-dependent queues has numerical instability issues. It may exhibit numerical difficulties under heavy load conditions which eventually result in unreasonable results, such as negative throughputs, response times, and queue lengths. The numerical problem is that the probability of a resource being idle is calculated in every iteration of the load-dependent MVA algorithm. The calculation is as follows:

$$P_m(0|n) = 1 - \sum_{i=1}^{n} P_m(i|n), \qquad (2.1)$$

where $P_m(i|n)$ is the probability that i jobs are at the mth resource when a total of n jobs are in the system. When the utilization is close to one, (2.1) can yield negative values, and those errors propagate as the MVA algorithm iterates. Subsequently, other calculations which have direct or indirect dependence on (2.1) may result in negative values, such as mean response times and throughputs, which do not make any physical sense.

2.3   Related Work

Chandy and Sauer [6] provide the initial reports of the numerical instability of the MVA algorithm with load-dependent queues. Reiser [15] confirms this issue. To replace (2.1) in a single-class model, he proposes a new calculation of $P_m(0|n)$ evaluated by $P_m(0|n-1)$ and the throughput $X^{[m]}(n)$ in an m-complement system, which is defined as a queueing system without the mth queue and all other parameters remaining the same. However, the expense of the evaluation grows exponentially as the number of load-dependent queues increases.

Tucci and Sauer [20], and Hoyme et al. [11] independently propose two similar tree-structured MVA algorithms, which are invulnerable to numerical instability. The main idea is to build a tree data structure, where queues are leaves. The internal nodes are intermediate functions, resulting from convolving all queue functions in the subtree with the internal node as the root. For dense queueing networks, tree MVA algorithms can give even worse performance than the original MVA algorithms whose complexities grow linearly, but they are efficient when customers visit only a small number of queues.

Casale et al. [5] suggest an approximate MVA algorithm (QD-MVA) for queue-dependent stations in a multi-class setting. Its computational cost is $O(MC)$ for a model with M queues and C classes. However, it may not converge in some instances. Moreover, it relies on queue-dependent functions to analyse queue-dependent service times, which introduces excessive computational requirements. They show that the QD-MVA algorithm has very good accuracy for the estimation of mean queue lengths, but the results from QD-MVA on other performance metrics, such as mean response times and system throughput, are not provided.

In the literature, Seidmann's approximation [18] is also widely used to address MVA's numerical issues [8, 9, 13]. The basic idea is to replace a multi-server queue with k servers by two tandem servers. The first one is a single server queue with service demand D/k, where D is one server's service demand. The second one is a pure delay server with service demand $D \cdot (k-1)/k$. In practice, Seidmann's approximation can yield noticeable errors under intermediate loads, but it has the same time and space complexities as the original MVA algorithm. However, Seidmann's approximation assumes that the servers in the multi-server queue are load independent. Such an approximation may not be realistic when the FES technique is employed.

To address numerical issues, Casale [3] introduces the Conditional MVA (CMVA) algorithm, which avoids the computations of the state probabilities, and as a consequence, overcomes the limitation. Although the CMVA algorithm is an exact solution, its time and space complexities grow much faster than the original MVA algorithm. Given M is the number of queues, and N is the number of jobs, the time and space complexities for the original MVA algorithm only grow as $O(MN)$, while those for the CMVA algorithm grow as $O(MN^{L+1})$, where L is the number of load-dependent queues in the system . This may cause significant time and memory issues for the computation when N or L is large, which is very common in performance evaluation for stress tests.

2.4 Stable Mean Value Analysis

We study a closed queueing network with N jobs and M queues, and we focus on a generic load-dependent queue with service demand $D_m(n_m)$, where m is the index of the queue, and n_m is the number of jobs at the queue (with $n = \sum_{m=1}^{M} n_m$ where

$m = 1, \ldots, M$ and $n = 1, \ldots, N$). Here, we assume that the service demand of the load-dependent queue becomes a constant beyond some $\bar{N}_m$, i.e., there exists a finite $\bar{N}_m$ such that $D_m(n_m) = D_m(\bar{N}_m)$ for all $n_m \geq \bar{N}_m$. This assumption is reasonable for many systems, in particular when $D_m(n_m)$ becomes sufficiently close to $D_m(\bar{N}_m)$.

The basic idea of the SMVA algorithm is inspired by Seidmann's approximation, replacing the load-dependent queue with two tandem servers. The first is a load-independent (LI) queue with service demand $D_m^q = D_m(\bar{N}_m)$. The second is a load-dependent (LD) delay centre with service demand

$$
D_m^d(n_m) = \begin{cases} n_m D_m(n_m) - D_m(\bar{N}_m), & \text{if } n_m < \bar{N}_m \\ (\bar{N}_m - 1) D_m(\bar{N}_m), & \text{if } n_m \geq \bar{N}_m. \end{cases} \tag{2.2}
$$

To make sure the service demands in (2.2) are positive, we assume that $n_m D(n_m) \geq D(\bar{N}_m)$, for $n_m < \bar{N}_m$. In multi-core computer systems, it is a common assumption that $D_m(n_m)$ decreases as n_m increases, so $D_m(n_m) > D_m(\bar{N}_m)$ when $n_m < \bar{N}_m$ and $n_m D_m(n_m) \geq D_m(\bar{N}_m)$ holds. Although the delay centre is load dependent, there is no need to calculate its state probabilities because it does not have a queue. As a result, the SMVA algorithm is numerically stable.

Under light load, the two tandem servers behave as a server which has service demand $D_m(n_m)$. If n_m jobs are being served and no jobs are waiting in the queue, the time spent by a job in the approximating node is $D_m(\bar{N}_m) + n_m D_m(n_m) - D_m(\bar{N}_m) = n_m D_m(n_m)$. If there are jobs waiting in the first queue, the time spent by a job in the approximating node is dominated by the time spent at the first queue. The node behaves as a server which has service demand $D_m(\bar{N}_m)$. As a result, this approximation should perform well for both light and heavy loads. Note that SMVA is identical to Seidmann's approximation when $n_m D_m(n_m) = D_m(1)$, for $n_m \leq \bar{N}_m$.

Once we finish the service demand parameterization, the mean response times at the load-independent queue in the approximating network with n jobs can be computed by the arrival theorem [12, 19]:

$$
R_m^q(n) = D_m^q[1 + Q_m(n - 1)], \tag{2.3}
$$

where $Q_m(n - 1)$ is the mean queue length at the mth queue with $n - 1$ jobs in the network.

To compute the mean response time at the delay centre, we need to estimate the mean number of jobs, because its service demand is load dependent. We employ the Bard-Schweitzer approximation [1, 17] to estimate the mean number of jobs at the delay centre. The Bard-Schweitzer approximation is based on the following idea: The number of jobs at each queue increases proportionately as the total number of jobs increases in the network. Mathematically:

$$
\frac{Q_m(n - 1)}{Q_m(n)} = \frac{n - 1}{n}. \tag{2.4}
$$

There are two things that we need to clarify here: (1) We use the term "mean number of jobs" rather than "mean queue length", because it is a pure delay centre, and it has no jobs waiting for service. (2) When we mention the mean number of jobs, we refer to the actual mean number of jobs of the original network instead of those of the approximating network. Let $Q^o_m(n-1)$ be the mean number of jobs at the mth queue when there are $n-1$ jobs in the network, and $Q^e_m(n)$ be the estimated mean number of jobs at the mth queue when there are n jobs in the network. We can rewrite (2.4) as:

$$Q^e_m(n) = \begin{cases} 1, & \text{if } n = 1, \\ \dfrac{n}{n-1} Q^o_m(n-1), & \text{if } n > 1. \end{cases}$$

Then, we can compute the mean response times at the delay centre as $R^d_m(n) = D^d_m(\lceil Q^e_m(n) \rceil)$. The ceiling function ensures that the index of the load-dependent service demands starts from one, rather than zero.

The system throughput is calculated using Little's Law:

$$X(n) = n / \left\{ Z + \sum_{m=1}^{M} [R^q_m(n) + R^d_m(n)] \right\}, \tag{2.5}$$

where Z is the mean think time. To compute the mean queue length at the mth queue in the approximating network, we just continue applying Little's Law: $Q^a_m(n) = X(n) \cdot R^q_m(n)$. The mean queue length at the mth queue in the original network is:

$$Q^o_m(n) = X(n) \cdot [R^q_m(n) + R^d_m(n)].$$

Algorithm 1 illustrates the single-class SMVA algorithm in detail. SMVA has two features: (1) SMVA is numerically stable, because it avoids the calculation of stationary probabilities at load-dependent queues; (2) SMVA is efficient, because its time and space complexities are both $O(MN)$.

There are two things that we would like to highlight in Algorithm 1. Firstly, we assume that all queues are load dependent in Algorithm 1. If the mth queue is load independent, we can simply set $D^q_m = D_m$ and $D^d_m = 0$, and Algorithm 1 is still applicable. Secondly, we do not check whether $Q^e_m(n)$ and $Q^o_m(n)$ converge to each other in SMVA, because we are iterating over n and we do not guess the initial values of $Q^e_m(n)$ (the Bard-Schweitzer approximation has both of them). In the Appendix, we propose an alternative SMVA algorithm which has a comparison between $Q^e_m(n)$ and $Q^o_m(n)$.

Algorithm 1 The single-class SMVA algorithm

Input:
$Z, M, N, D_m(n), \bar{N}_m$
Output:
$Q_m^o(N), X(N), R(N)$
Condition:
$n D_m(n) \geq D_m(\bar{N}_m), \forall n, m$
Initialization:
$Q_m^a(0) = 0, \forall m = 1, \ldots, M$
Iteration:

    for $m = 1 \to M$ **do**
        for $n = 1 \to N$ **do**
            $D_m^q = D_m(\bar{N}_m)$

$$D_m^d(n) = \begin{cases} n D_m(n) - D_m(\bar{N}_m), & \text{if } n < \bar{N}_m \\ (\bar{N}_m - 1) D_m(\bar{N}_m), & \text{if } n \geq \bar{N}_m \end{cases}$$

        end for
    end for
    for $n = 1 \to N$ **do**
        for $m = 1 \to M$ **do**
            if $n = 1$ **then**
                $Q_m^e(n) = 1$
            else

$$Q_m^e(n) = \frac{n}{n-1} Q_m^o(n-1)$$

            end if
            $R_m^q(n) = D_m^q[1 + Q_m^a(n-1)]$
            $R_m^d(n) = D_m^d(\lceil Q_m^e(n) \rceil)$
        end for
        $X(n) = n / \{Z + \sum_{m=1}^M [R_m^q(n) + R_m^d(n)]\}$
        for $m = 1 \to M$ **do**
            $Q_m^a(n) = X(n) \cdot R_m^q(n)$
            $Q_m^o(n) = X(n) \cdot [R_m^q(n) + R_m^d(n)]$
        end for
    end for
    $R(N) = \sum_{m=1}^M [R_m^q(N) + R_m^d(N)]$

2.5 Experimental Results

In order to verify the accuracy and the efficiency of the SMVA algorithm, we compare the results of the SMVA algorithm, the CMVA algorithm, and Seidmann's approximation in two different closed queueing networks. The first one is a closed network with one generic load-dependent queue (FES), and the second one is a closed network with two FESs. To generate the input parameters—service demands—for these two queueing networks, we set up a testbed on an Intel i7-2600 quad-core computer with 8 GB memory, 1 TB hard drive, and Ubuntu 12.04.3 LTS. We employ JBoss 3.2.7 as the application server, MySQL 5.1.70 as the database server, and TPC-W [10] to generate the workload. TPC-W can simulate three

Fig. 2.1 Service demands for browsing

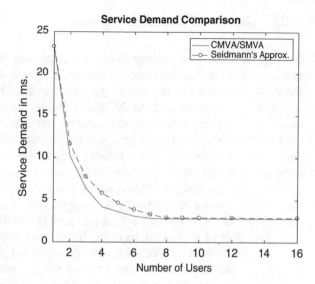

Fig. 2.2 Service demands for shopping

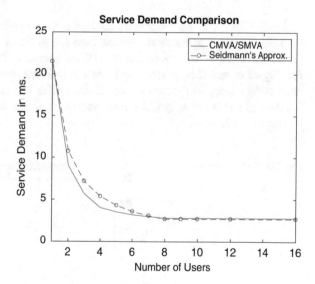

workloads for an e-commerce environment—browsing, shopping, and ordering. We choose the first two workloads in our tests, and plot their service demands in Figs. 2.1 and 2.2.

2.5.1   One Load-Dependent Queue

For the first network, we aggregate and model the computer system by an FES. We then run the browsing workloads to obtain the system throughputs, and calculate the service demands of the FES to parameterize the MVA algorithms (as shown in Fig. 2.1). As can be seen in the figure, the service demand curve adopted by Seidmann's approximation can only address the ideal case of a load-dependent server, where $D(n) = D(1)/n$ for $n \leq 8$. By overestimating the service demands in such a case, the outcomes of Seidmann's approximation are conservative in terms of performance metrics, but this may not be true in general.

To test the accuracy of SMVA under different loads, we vary the number of users and the mean think time in the system. Both the mean response time and the throughput are compared for the three candidate MVA algorithms. Three sets of results are presented. The results of the first set are presented in Figs. 2.3 and 2.4, where N ranges from 1 to 30, and $Z = 0$. The results of the second set are presented in Figs. 2.5 and 2.6, where N ranges from 1 to 300, and $Z = 0.7$ s. The results of the third set are presented in Figs. 2.7 and 2.8, where N ranges from 1 to 1300, and $Z = 3.5$ s.

Using CMVA as the benchmark (as it is an exact solution), SMVA works better than Seidmann's approximation in all three cases. However, we also observe some errors for both of the approximate MVA algorithms from those figures, except for Figs. 2.6 and 2.8, where the largest errors are only -1.05% and -0.23%, respectively (negative means underestimate). The reason is that for those figures, the throughput is given by (2.5). As Z increases, the error in R has a smaller effect on the accuracy of X.

Fig. 2.3 Response time with $Z = 0$

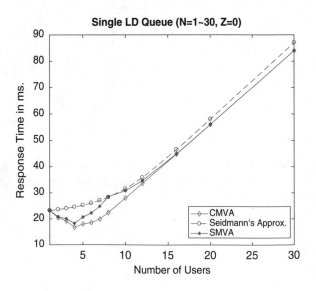

Fig. 2.4 Throughput with
$Z = 0$

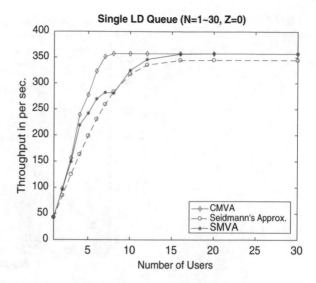

Fig. 2.5 Response time with
$Z = 0.7\,\mathrm{s}$

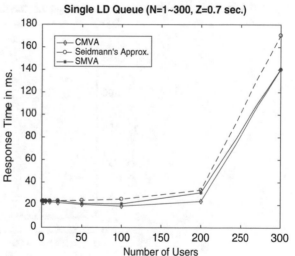

To verify the observations from Figs. 2.3, 2.4, 2.5, 2.6, 2.7 and 2.8, we calculate the Root-Mean-Square Percentage Error (RMSPE) for both SMVA and Seidmann's approximation in the three test sets. Here, RMSPE $= \sqrt{\sum_{i=1}^{T} E_i^2 / T}$, where E_i is the percentage error of the ith estimate, and T is the total number of estimates. The results can be found in Table 2.1, and they verify the two observations that we have in the figures:

- In terms of accuracy, SMVA works better than Seidmann's approximation in all cases.
- Errors in throughput are minor when Z is relatively larger than R.

Fig. 2.6 Throughput with
$Z = 0.7\,\mathrm{s}$

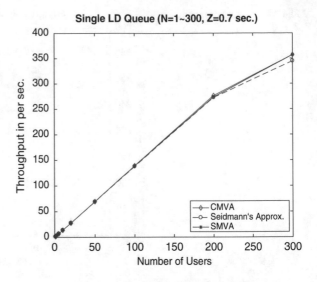

Fig. 2.7 Response time with
$Z = 3.5\,\mathrm{s}$

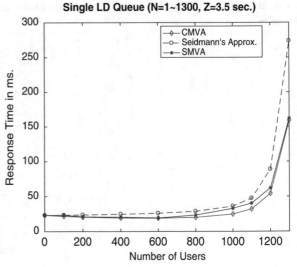

We also would like to quantify some large errors from SMVA and Seidmann's approximation. In Figs. 2.3 and 2.4, the largest error for the mean response time of SMVA is 27.16%, and the largest error for the throughput of SMVA is -21.36% when $N = 8$. In contrast, the errors for both the mean response time and the throughput of Seidmann's approximation are the largest when $N = 4$ (46.78% and -31.87%, respectively). We have similar observations from Figs. 2.5, 2.6, 2.7 and 2.8. In Fig. 2.5, the largest error for the mean response time of SMVA is 33.17% when $N = 200$. In Fig. 2.7, the largest error for the mean response time of SMVA is 32.48% when $N = 1000$. Seidmann's approximation has its worst case when $N = 1300$, the error for the mean response time is 71.91%. As a conclusion,

Fig. 2.8 Throughput with
$Z = 3.5$ s

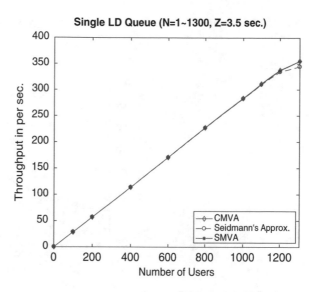

Table 2.1 RMSPEs in one
LD queue

Test case		SMVA	Seidmann
$Z = 0.0$ s	R	12.93%	25.54%
	X	10.63%	18.72%
$Z = 0.7$ s	R	12.02%	20.47%
	X	0.38%	1.29%
$Z = 3.5$ s	R	15.24%	42.42%
	X	0.13%	1.03%

the SMVA algorithm works well when the system is under light or heavy loads. However, some errors are significant when the system is under intermediate loads.

2.5.2 Two Load-Dependent Queues

For the second queueing network, we add one more FES to the previous network. We derive the service demands of the second FES from the shopping web interaction workloads of TPC-W (as shown in Fig. 2.2). We choose the shopping workloads, because the service demand curve is close to (but not the same as) the one derived from the browsing workloads in the first FES (in Fig. 2.1), so that no single queue can dominate the performance in the network.

As discussed in Sect. 2.5.1, we vary the number of users and the mean think time in the system, and compare the mean response time and the throughput among the three candidate MVA algorithms. Since the results of throughputs are almost identical when Z is larger than zero (similar to Figs. 2.6 and 2.8), those results are not shown. The results of the first set are presented in Figs. 2.9 and 2.10, where N

Fig. 2.9 Response time with
$Z = 0$

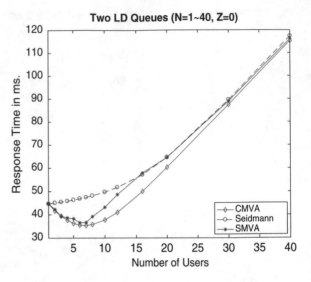

Fig. 2.10 Throughput with
$Z = 0$

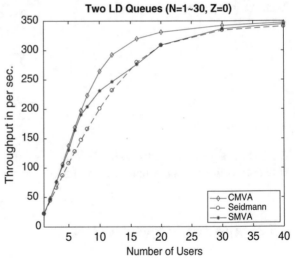

ranges from 1 to 40, and $Z = 0$. The results of the second set are presented in
Fig. 2.11, where N ranges from 1 to 400, and $Z = 0.7$ s. The results of the third set
are presented in Fig. 2.12, where N ranges from 1 to 1200, and $Z = 3.5$ s. Unlike
the test set of $Z = 3.5$ s in Sect. 2.5.1, where we could have maximum 1300 jobs
in the system, we cannot have 1300 jobs in this test case, because the calculation
space requirement of the CMVA algorithm grows exponentially as the number of
load-dependent queues increases. In this case, it is $O(MN^3)$, and it exceeds the
maximum capacity of the memory on our machine. For instance, the initialization
of the service demands for a single queue requires $N^3 \times 8$ bytes $= 16.37$ GB.

Fig. 2.11 Response time
with $Z = 0.7\,$s

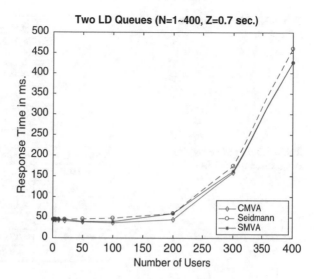

Fig. 2.12 Response time
with $Z = 3.5\,$s

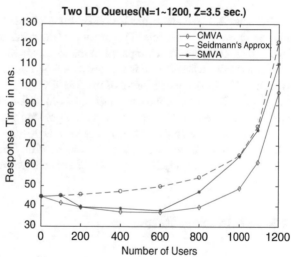

As can be seen from Figs. 2.9, 2.10, 2.11 and 2.12, the results are quite consistent with those from Figs. 2.3, 2.4, 2.5, 2.6, 2.7 and 2.8, respectively. Similarly, we have three observations as follows:

- Compared to CMVA, SMVA works better than Seidmann's approximation in all three cases.
- As the value of the mean think time grows, the errors in estimated throughputs from SMVA decrease.
- Compared to the results under light and heavy workloads, larger errors are observed for SMVA under intermediate workloads.

Table 2.2 RMSPEs in two
LD queues

Test case		SMVA	Seidmann
$Z = 0.0$ s	R	8.47%	22.15%
	X	7.40%	17.18%
$Z = 0.7$ s	R	11.23%	16.21%
	X	0.66%	1.39%
$Z = 3.5$ s	R	16.34%	26.27%
	X	0.26%	0.39%

Table 2.3 Largest error
comparison for SMVA

Test case		Single LD queue	Two LD queues
$Z = 0.0$ s	R	27.16%	18.60%
	X	−21.36%	−15.68%
$Z = 0.7$ s	R	33.17%	33.33%
	X	−1.05%	−1.96%
$Z = 3.5$ s	R	32.48%	32.40%
	X	−0.23%	−0.44%

In Table 2.2, we show the RMSPEs for both SMVA and Seidmann's approximation in the three test sets. The results in Table 2.2 can be seen to verify the first two observations above. Compared to the results in Table 2.1, the SMVA algorithm performs better when $Z = 0$ in the network with two LD queues than in the network with a single LD queue in terms of the RMSPE.

In Table 2.3, we compare the largest errors of the SMVA algorithm in these two case studies. As can be seen, the SMVA algorithm performs better when Z is zero in the second case study, but has very similar results when Z is larger than zero. This observation is consistent with the observation from the RMSPEs in Tables 2.1 and 2.2. The underlying reason is not clear and is worth future study.

2.6 Multi-class Extension

For completeness, we extend the SMVA algorithm to the case of multi-class closed networks. As in single-class networks, the scheduling discipline in multi-class networks is also constrained to preserve the product-form nature of the steady-state distribution, for example the scheduling disciplines considered in the BCMP theorem [2] may be employed. We consider that there are C classes of transactions, where the job population vector is given by $\mathbf{n} = (n_1, \ldots, n_c, \ldots, n_C)$, where $0 \leq n_c \leq N_c$ and $1 \leq c \leq C$. The service demand of class c at the mth load-independent queue is given by $D_{m,c}^q = D_{m,c}(\bar{N}_{m,c})$. The service demand at the delay centre becomes

$$D_{m,c}^d(n_c) = \begin{cases} n_c D_{m,c}(n_c) - D_{m,c}(\bar{N}_{m,c}), & \text{if } n_c < \bar{N}_{m,c}, \\ (\bar{N}_{m,c} - 1) D_{m,c}(\bar{N}_{m,c}), & \text{if } n_c \geq \bar{N}_{m,c}. \end{cases}$$

Then, the multi-class SMVA iterates over all feasible $\mathbf{n} = (n_1, \ldots, n_C)$ such that $\sum_{c=1}^{C} n_c = n$ and $1 \le n \le N$ to compute the mean response times at load-independent queues:

$$R_{m,c}^q(\mathbf{n}) = D_{m,c}^q[1 + Q_m(\mathbf{n} - 1_c)].$$

Here, $\mathbf{n} - 1_c = (n_1, \ldots, n_c - 1, \ldots, n_C)$ is the job population vector with one less class c job in the system. The mean response time at a pure delay centre is $R_{m,c}^d(\mathbf{n}) = D_{m,c}^d(\lceil Q_m^e(\mathbf{n}) \rceil)$, where

$$Q_m^e(\mathbf{n}) = \begin{cases} 1, & \text{if } n_c = 1, \\ \dfrac{n_c}{n_c - 1} Q_m^o(\mathbf{n} - 1_c), & \text{if } n_c > 1. \end{cases}$$

The system throughput of class c is calculated by

$$X_c(\mathbf{n}) = n_c / \left\{ Z_c + \sum_{m=1}^{M} [R_{m,c}^q(\mathbf{n}) + R_{m,c}^d(\mathbf{n})] \right\}.$$

The mean queue length at the mth load-independent queue is

$$Q_m^a(\mathbf{n}) = \sum_{c=1}^{C} X_c(\mathbf{n}) \cdot R_{m,c}^q(\mathbf{n}).$$

Finally, the mean queue length at the original load-dependent queue is

$$Q_m^o(\mathbf{n}) = \sum_{c=1}^{C} X_c(\mathbf{n}) \cdot [R_{m,c}^q(\mathbf{n}) + R_{m,c}^d(\mathbf{n})].$$

Both the time and space complexities of the multi-class SMVA algorithm are $O(MN^C)$. Due to the complexities, we have not evaluated the accuracy of the multi-class model in a case study.

2.7 Conclusions

In this paper, we present the SMVA algorithm in both a single-class model and a multi-class model. Compared to the CMVA algorithm and Seidmann's approximation, the SMVA algorithm has two advantages:

- The time and space complexities of SMVA are a significant improvement over CMVA, especially when the number of jobs in the system is very large, or when the number of load-dependent queues is larger than one.
- The SMVA algorithm is better able to handle cases when the service demands of a load-dependent node do not have a linear relationship.

In terms of accuracy, we also have two additional observations about the SMVA algorithm:

- The SMVA algorithm works as well as the CMVA algorithm when the system is under light or heavy loads. However, the errors of the SMVA algorithm increase when the system is under intermediate loads (but it still performs better than Seidmann's approximation).
- When the mean think time increases, the SMVA algorithm might produce less accurate estimates of the mean response times under intermediate load. In contrast, the estimated throughput becomes more accurate.

The accuracy of SMVA under intermediate loads is closely linked to the accuracy of the underlying approximations. It is inspired by Seidmann's approximation. Consequently, it behaves as Seidmann's approximation under intermediate workloads. In addition, it employs the assumption in the Bard-Schweitzer approximation to estimate the mean number of jobs at delay centres, which may also add errors to the results.

Acknowledgements The work reported here was supported by the Natural Sciences and Engineering Research Council of Canada.

Appendix

In Algorithm 1, we estimate the mean number of jobs at a delay centre, $Q_m^e(n)$. In the same iteration, new values are calculated as $Q_m^o(n)$. A natural thought would be to add a comparison between these two values, similar to a technique in the Bard-Schweitzer approximation. To accomplish this, we propose an alternative SMVA algorithm (A-SMVA) for a single-class system. The details of A-SMVA can be seen in Algorithm 2. Compared to SMVA, A-SMVA differs as follows:

- Iterations over $n = 1 \rightarrow N$ are removed. Instead, we focus only on the performance metrics with N jobs in the system.
- Initialize $Q_m^e(N)$ with estimated values, e.g., $N/(M + 1)$.
- Employ (2.4) to replace $Q_m^a(N - 1)$ by $Q_m^q(N)$ in (2.3), which is $Q_m^q(N) \times (N - 1)/N$.
- Choose an error criterion—ϵ, e.g., 0.01.
- Compare the difference between $Q_m^e(N)$ and $Q_m^o(N)$, and compare the difference between $Q_m^q(N)$ and $Q_m^a(N)$. If the maximum difference is larger than ϵ, replace $Q_m^e(N)$ by $Q_m^o(N)$, and $Q_m^q(N)$ by $Q_m^a(N)$. Otherwise, stop the iteration.

Algorithm 2 The single-class A-SMVA algorithm

Input:
$Z, M, N, D_m(n), \bar{N}_m, \epsilon$
Output:
$Q_m^o(N), X(N), R(N)$
Condition:
$N \cdot D_m(N) \geq D_m(\bar{N}_m), \forall m$
Initialization:
$Q_m^a(N) = 0, \forall m = 1, \ldots, M$
$Q_m^o(N) = N/(M+1), \forall m = 1, \ldots, M$
Iteration:

  for $m = 1 \rightarrow M$ **do**
    $D_m^q = D_m(\bar{N}_m)$

$$D_m^d(N) = \begin{cases} N \cdot D_m(N) - D_m(\bar{N}_m), & \text{if } N < \bar{N}_m \\ (\bar{N}_m - 1) D_m(\bar{N}_m), & \text{if } N \geq \bar{N}_m \end{cases}$$

  end for
  while $\max_i \{|Q_m^e(N) - Q_m^o(N)|\} > \epsilon$ or $\max_i \{|Q_m^q(N) - Q_m^a(N)|\} > \epsilon$ **do**
    for $m = 1 \rightarrow M$ **do**
      $Q_m^e(N) = Q_m^o(N)$
      $Q_m^q(N) = Q_m^a(N)$
      $R_m^q(N) = D_m^q [1 + \dfrac{N-1}{N} Q_m^q(N)]$
      $R_m^d(N) = D_m^d(\lceil Q_m^e(N) \rceil)$
    end for
    $X(N) = N/\{Z + \sum_{m=1}^{M} [R_m^q(N) + R_m^d(N)]\}$
    for $m = 1 \rightarrow M$ **do**
      $Q_m^a(N) = X(N) \cdot R_m^q(N)$
      $Q_m^o(N) = X(N) \cdot [R_m^q(N) + R_m^d(N)]$
    end for
  end while
  $R(N) = \sum_{m=1}^{M} [R_m^q(N) + R_m^d(N)]$

We set $\epsilon = 0.01$, and compare the results of A-SMVA with the results of SMVA with the same input parameters in Sect. 2.5.1. The estimated mean response times from A-SMVA are slightly larger than the results from SMVA, but they have very similar trends.

When N is very large, A-SMVA can be efficient, because it avoids the iteration over $n = 1 \rightarrow N$. However, we have two concerns about A-SMVA:

- The initial values of $Q_m^e(N)$ may significantly affect the outputs. For example, if they are too close to zero, the whole iteration will be skipped.
- The chosen value of ϵ may have a significant effect on the outputs. If ϵ is too big, we have less iterations, but sacrifice accuracy. If ϵ is too small, it may not converge in some instances (although we have not observed this).

As a summary, we provide one more numerically stable approach to determine the performance metrics for closed queueing networks with load-dependent queues. One can adopt SMVA or A-SMVA depending on the requirements.

References

1. Y. Bard, Some extensions to multiclass queuing network analysis, in *Proceedings of the 3rd International Symposium on Modelling and Performance Evaluation of Computer Systems: Performance of Computer Systems* (North-Holland Publishing Co., New York, 1979), pp. 51–62
2. F. Baskett, K.M. Chandy, R.R. Muntz, F.G. Palacios, Open, closed, and mixed networks of queues with different classes of customers. J. ACM 22(2), 248–260 (1975)
3. G. Casale, A note on stable flow-equivalent aggregation in closed networks. Queueing Syst. 60(3–4), 193–202 (2008)
4. G. Casale, G. Serazzi, Stabilization techniques for load-dependent queueing networks algorithms, in *Communication Networks and Computer Systems: A Tribute to Professor Erol Gelenbe*, Chap 8, ed. by J.A. Barria (Imperial College Press, London, 2006), pp. 127–141
5. G. Casale, J.F. Pérez, W. Wang, QD-AMVA: evaluating systems with queue-dependent service requirements. Perform. Eval. 91, 80–98 (2015)
6. K.M. Chandy, C.H. Sauer, Computational algorithms for product form queueing networks. Commun. ACM 23(10), 573–583 (1980)
7. K.M. Chandy, U. Herzog, L. Woo, Parametric analysis of queuing networks. IBM J. Res. Dev. 19(1), 36–42 (1975)
8. Y. Chen, S. Iyer, X. Liu, D. Milojicic, A. Sahai, SLA decomposition: translating service level objectives to system level thresholds, in *Proceedings of the 4th International Conference on Autonomic Computing* (IEEE, 2007), pp. 3–13
9. Y. Chen, S. Iyer, X. Liu, D. Milojicic, A. Sahai, Translating service level objectives to lower level policies for multi-tier services. Clust. Comput. 11(3), 299–311 (2008)
10. T. Horvath, TPC-W J2EE implementation (2008). http://www.cs.virginia.edu/~th8k. Last accessed 27 Sept 2017
11. K. Hoyme, S.C. Bruell, P. Afshari, R.Y. Kain, A tree-structured mean value analysis algorithm. ACM Trans. Comput. Syst. 4(2), 178–185 (1986)
12. S.S. Lavenberg, M. Reiser, Stationary state probabilities at arrival instants for closed queueing networks with multiple types of customers. J. Appl. Probab. 17(4), 1048–1061 (1980)
13. X. Liu, J. Heo, L. Sha, Modeling 3-tiered web applications, in *Proceedings of the 13th IEEE International Symposium on Modeling, Analysis, and Simulation of Computer and Telecommunication Systems, 2005* (IEEE, 2005), pp. 307–310
14. D.A. Menascé, V.A. Almeida, L.W. Dowdy, L. Dowdy, *Performance by Design: Computer Capacity Planning by Example* (Prentice Hall PTR, Upper Saddle River, 2004)
15. M. Reiser, Mean-value analysis and convolution method for queue-dependent servers in closed queueing networks. Perform. Eval. 1(1), 7–18 (1981)
16. M. Reiser, S.S. Lavenberg, Mean-value analysis of closed multichain queuing networks. J. ACM 27(2), 313–322 (1980)
17. P. Schweitzer, Approximate analysis of multiclass closed networks of queues, in *Proceedings of International Conference on Stochastic Control and Optimization* (Free University, Amsterdam, 1979), pp. 25–29
18. A. Seidmann, J. Paul, S. Shalev-Oren, Computerized closed queueing network models of flexible manufacturing systems: a comparative evaluation. Large Scale Syst. 12, 91–107 (1987)
19. K.C. Sevcik, I. Mitrani, The distribution of queuing network states at input and output instants. J. ACM 28(2), 358–371 (1981)
20. S. Tucci, C.H. Sauer, The tree MVA algorithm. Perform. Eval. 5(3), 187–196 (1985)
21. L. Zhang, D.G. Down, A stable mean value analysis algorithm for closed systems with load-dependent queues, in *Proceedings of the 10th EAI International Conference on Performance Evaluation Methodologies and Tools* (ACM, New York, 2016), pp. 178–181

Chapter 3
Dispatching Discrete-Size Jobs with Multiple Deadlines to Parallel Heterogeneous Servers

Esa Hyytiä, Rhonda Righter, Olivier Bilenne, and Xiaohu Wu

3.1 Introduction

In the dispatching problem, each arriving job is routed to one of the available servers immediately upon arrival. Even though a single fast server would often be preferred, the parallel servers are needed to match increasing capacity demands. Moreover, short latency, in the absence of preemptive scheduling, requires parallel servers.

In this chapter, we consider a cost structure based on (firm) deadlines. Each job has a certain deadline for the maximum waiting time it can tolerate. If this waiting time is exceeded, a deadline violation cost is incurred, but the job must still be served. This cost structure stems from quality-of-experience metrics, where customers observe a good service level whenever the waiting time is "short," but as soon as a given customer-specific threshold is exceeded, the observed service quality drops, cf. video conferencing and other interactive systems. That is, the tail of the response time distribution is one of the most crucial performance measures [1]. For this reason, service level agreements (SLAs) are often defined in terms of acceptable waiting times [2].

Our basic setting has been studied recently in [3] in the context of M/G/1 queues, and in [4] for the standard M/D/1 queue. The results in [3] are either asymptotic or in the form of differential equations. In contrast, [4] gives exact closed-form

E. Hyytiä (✉)
Department of Computer Science, University of Iceland, Reykjavík, Iceland
e-mail: esa@hi.is

R. Righter
Department of Industrial Engineering and Operations Research, UC Berkeley, Berkeley, CA, USA

O. Bilenne · X. Wu
Department of Communications and Networking, Aalto University, Espoo, Finland

© Springer International Publishing AG, part of Springer Nature 2019

A. Puliafito, K. S. Trivedi (eds.), *Systems Modeling: Methodologies and Tools*,
EAI/Springer Innovations in Communication and Computing,
https://doi.org/10.1007/978-3-319-92378-9_3

expressions (that satisfy the aforementioned differential equations and asymptotic behavior) for the M/D/1 queue. In this chapter, we give exact results for the value function and admission cost for the M/iD/1 queue subject to a general deadline-based cost structure, where the service times are assumed to be random multiples of a fixed size d, and the deadlines and their violation costs can vary according to some probability distributions. Moreover, there can be multiple deadlines with added cost for each deadline that is violated, and the job arrival process can include batches. In summary, the model considered here is much more general than those of [3] and [4].

The approach of first deriving a value function for a single server queue, and then applying it to develop efficient dispatching rules for a system of parallel servers, is general. Traditionally the objective is the minimization of the mean sojourn time (see, e.g., [5–7]), possibly combined with minimizing energy consumption (see, e.g., [8, 9]). The value function for M/G/1-FCFS then enjoys elementary closed-form expressions. In contrast, other disciplines, such as processor sharing (PS), make the situation more complex and exact results are available only for M/D/1-PS and M/M/1-PS [5, 10]. Our approach also lends itself to minimization of blocked jobs in loss systems [11].

3.2   M/G/1 FCFS Queue with Deadlines

The basic model for a single M/G/1-FCFS queue with deadlines is as follows [3]. We let λ denote the arrival rate and X the service time of a job so that the offered load is $\rho = \lambda \, \mathrm{E}[X]$. Jobs whose waiting time in queue, W, exceeds time τ, referred to as the deadline, incur a unit cost. We assume that $\rho < 1$ for stability. The deadline must be non-negative, $\tau \geq 0$. Thus, in the special case when $\tau = 0$ all jobs that have to wait incur the unit cost. The mean cost rate is

$$r = \lambda \, \mathrm{P}\{W > \tau\}. \tag{3.1}$$

In the general case, we have multiple classes of jobs, each with its own arrival rate λ_i, target deadline τ_i and i.i.d. deadline violation cost H_i. The total arrival rate is $\lambda = \sum_i \lambda_i$, and the stability requirement is that $\lambda d = \rho < 1$. The mean cost rate in this case is

$$r = \sum_i \lambda_i \, \mathrm{E}[H_i] \, \mathrm{P}\{W > \tau_i\}.$$

Our first task is to derive the so-called *value function* with respect to the deadline cost structure. Formally, the value function is defined as

$$v(u) \triangleq \lim_{t \to \infty} \mathrm{E}[V(u, t) - rt],$$

where u is the current backlog (unfinished work) in the queue, and the random variable $V(u, t)$ denotes the deadline violation costs during time $(0, t)$ when the system is initially in state u. Given $\rho < 1$, the M/G/1 queue is stable, the system is ergodic, and the above limit is well-defined. (In fact, the limit is finite and well-defined also when $\rho \geq 1$ and the system is unstable).

We can use two complementary approaches to determine the value function. First, the value function $v(u)$ for the M/G/1 queue satisfies the integro-differential equation

$$v'(u) = \lambda \left(c(u) - \bar{c} + \mathrm{E}[v(u + X) - v(u)] \right), \tag{3.2}$$

where X denotes the random i.i.d. service time, $c(u)$ is the (mean) cost when a job arrives at state u, and $\bar{c}$ is the mean cost of a job. Additionally, we have the boundary condition,

$$v'(0) = 0. \tag{3.3}$$

In [3], it is shown that for the deadline cost structure, the value function is a linear function of u for $u > \tau$. The mean time before the system returns to state τ is $t = (u-\tau)/(1-\rho)$, during which on average λt jobs arrive (PASTA), each incurring the deadline violation cost, and rt is the mean costs incurred during the same time interval in equilibrium. Thus,

$$v(u) - v(\tau) = \frac{\lambda - r}{1 - \rho}(u - \tau), \qquad u > \tau. \tag{3.4}$$

For $0 \leq u \leq \tau$, (3.2) reduces to

$$v'(u) = -r + \lambda \mathbf{1}(u > \tau) + \lambda \, \mathrm{E}[v(u + X) - v(u)], \quad u \geq 0, \tag{3.5}$$

where $\mathbf{1}(u > \tau)$ is 1 if $u > \tau$ and zero otherwise. Given (3.5) expresses $v'(u)$ as a function of $v(u + t)$ with $t \geq 0$, and since $v(u)$ is known for $u \geq \tau$, $v(u)$ can be solved *numerically* backwards starting from $u = \tau$ as discussed in [3]. Moreover, explicit results are given for M/G/1 when (1) $\tau < X$ and the load $\rho < 1$, and when (2) $\tau \gg X$ and $\rho \to 1$ (the heavy-traffic regime).

The second approach is more general and gives the value function for the $\mathrm{M}^X/\mathrm{G}/1$ queue with an arbitrary cost function $c(w)$. In particular, the value function satisfies [12, Proposition 1]

$$v(u) - v(0) = \frac{\lambda u}{1 - \rho} \mathrm{E}[c(W + Y) - c(W)]. \tag{3.6}$$

where λ is the *job arrival rate*, $c(w)$ is the admission cost of a job with waiting time w, W is the waiting time in steady state, and $Y \sim \mathrm{U}(0, u)$.

Example 1 Suppose that $\tau = 0$, i.e., jobs that have to wait incur a unit cost. Then $r = \lambda \rho$, and as the linear regime starts immediately, (3.4) gives the value function for all $u > 0$, $v(u) - v(0) = \lambda(u - \tau)$, and $a(u) = \mathbf{1}(u > 0) + \rho$.

Unfortunately, the value function for the M/G/1 queue with respect to deadline $\tau > 0$ cannot be expressed in closed-form using elementary functions. Therefore, next we analyze the M/D/1 queue, where the service time is d, and the M/iD/1 queue where the service time is some random multiple of d. For both cases, we obtain explicit closed-form expressions for the corresponding value functions. Moreover, we give the value function for a general multi-class system, and discuss how the results can be applied to also analyze batch arrivals.

3.3 M/D/1 FCFS Queue with Deadlines

In this section, we consider the M/D/1 FCFS queue. First, in Sect. 3.3.1, we assume a single deadline τ that applies to all jobs and a unit deadline violation cost, $h = 1$. The obtained results are generalized in Sect. 3.3.2 to multiple job classes with distinct deadlines and deadline violation costs.

3.3.1 M/D/1 FCFS with a Single Deadline

Let us start with the M/D/1 FCFS queue with a single deadline τ for the waiting time. Note that this is equivalent to having a deadline $\tau + d$ for the sojourn time.

In general, the distribution of the waiting time cannot be expressed in simple terms, but instead is given in the form of the Laplace-Stieltjes Transform (LST) [13] or an infinite sum involving convolutions [14]. However, for the M/D/1 queue the waiting time distribution is available[1] [15]

$$P\{W \leq \tau\} = (1 - \rho) \sum_{i=0}^{\lfloor \tau/d \rfloor} \frac{(\lambda(id - \tau))^i}{i!} e^{-\lambda(id-\tau)}. \tag{3.7}$$

Similarly, for the M/D/1 queue, the integro-differential equation (3.5) simplifies [4]:

$$v'(u) + \lambda v(u) = -r + \lambda \mathbf{1}(u > \tau) + \lambda v(u + d), \quad u > 0. \tag{3.8}$$

In general, the mean cost rate r follows from the boundary condition (3.3). However, in our case $r = \lambda(1 - P\{W \leq \tau\})$, and $P\{W \leq \tau\}$ is given by (3.7). Substituting r into (3.4) thus yields the value function $v(u)$ for the tail $u \geq \tau$. For $u \leq \tau$, $v(u)$ can

---

[1]We use the convention that $0^0 = 1$ so that (3.7) holds also when $\tau = 0$.

be determined by solving (3.8). The following theorem, which is a special case of Theorem 2, was shown in [4].

Theorem 1 *The value function for an M/D/1 FCFS queue with respect to deadline at time τ with a unit violation cost is*

$$v(u) - v(\tau) = \lambda \left(\sum_{i=0}^{\lfloor \frac{\tau}{d} \rfloor} \frac{(\lambda(id - \tau))^i}{i!} e^{-\lambda(id-\tau)} \right) (u - \tau)$$

$$+ \sum_{i=0}^{\lfloor \frac{\tau-u}{d} \rfloor} \left(e^{-\lambda(id+u-\tau)} \sum_{j=0}^{i} \frac{(\lambda(id+u-\tau))^j}{j!} - 1 \right). \qquad (3.9)$$

Note that, in accordance with (3.4), $v(u)$ in (3.9) reduces to a linear function when $u \geq \tau$. Given the value function, we can write down the (marginal) *admission cost*, $a(u) = v(u + d) - v(u) + \mathbf{1}(u > \tau)$.

Corollary 1 *The admission cost of a job with deadline τ to the M/D/1 queue with backlog u is*

$$a(u) = \rho \sum_{i=0}^{\lfloor \frac{\tau}{d} \rfloor} \frac{(\lambda(id - \tau))^i}{i!} e^{-\lambda(id-\tau)} + 1 - \sum_{i=0}^{\lfloor \frac{\tau-u}{d} \rfloor} \frac{(\lambda(id+u-\tau))^i}{i!} e^{-\lambda(id+u-\tau)}.$$

$$\qquad (3.10)$$

Note that the first summation in (3.10) is a constant (cf. the linear term). Recall that $v(u_2) - v(u_1)$ corresponds to the expected difference in the number of deadline violations between a system that has an initial backlog of u_2 and a system that is initially in state u_1. Similarly, the admission cost $a(u)$ tells us the expected increase in the number of deadline violations if a job is admitted to the system currently in state u, including the cost for the job itself.

In the general case, for the M/G/1 queue with several reasonable cost structures, including deadline violations and latency, it holds that

$$v(0) = -\frac{r}{\lambda} + \mathrm{E}[v(X)],$$

where r denotes the corresponding mean cost rate (e.g., $\lambda \mathrm{E}[W > \tau]$ or $\lambda \mathrm{E}[T]$). The above yields a simple identity for the mean admission cost to an empty system,

$$\mathrm{E}[a(0, X)] = \frac{r}{\lambda},$$

We can verify that this holds also for the M/D/1 queue with the deadline cost structure, i.e., (3.10) at $u = 0$ reduces to

$$a(0) = \mathrm{P}\{W > \tau\}, \qquad \forall \, \tau \geq 0.$$

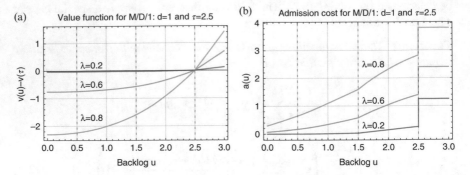

Fig. 3.1 Value function and the corresponding admission costs for an M/D/1 queue. (**a**) Value function. (**b**) Admission cost

As $v(u)$ for the M/D/1 queue, given in (3.9), is strictly increasing and convex, $a(u)$ is an increasing function of u. Moreover, for $u \leq \tau$, $a(u) = v(u + d) - v(u)$, and hence

$$a(u) \leq v(\tau + d) - v(\tau) = \frac{\rho}{1 - \rho} \, P\{W \leq \tau\}.$$

Therefore, the following bounds hold for $a(u)$,

$$P\{W > \tau\} \leq a(u) \leq \frac{\rho}{1 - \rho} \, P\{W \leq \tau\}, \qquad u \leq \tau,$$

which for $\tau = 0$ reduces to $\rho \leq a(0) \leq \rho$, in accordance with Example 1.

Example 2 Let $d = 1$ and $\tau = 2.5$. The corresponding value function and admission cost are illustrated in Fig. 3.1 for $\lambda \in \{0.2, 0.6, 0.8\}$. The value function is smooth (except at $u = \tau$), whereas the admission cost behaves quite differently. For example, the unit cost due to the immediate cost of a deadline violation when $u > \tau$ shows clearly.

3.3.2 M/D/1 FCFS with Multiple Job Classes

In this section, we extend the system model and consider the multi-class scenario, where all jobs have the same fixed service time d, but their deadlines and deadline violation costs can vary. More specifically, we assume k job classes such that class i jobs have deadline τ_i (from the arrival time) and each violation for class i jobs costs H_i. The corresponding (Poisson) arrival rates are $\lambda_1, \ldots, \lambda_k$, and are such that $\lambda d = \rho < 1$, where $\lambda = \sum_i \lambda_i$ (i.e., a stable system). For convenience, we further define $p_i = \lambda_i / \lambda$.

Let $v_i(u)$ denote the value function of a system with arrival rate λ and deadline τ_i. As $v_i(u) - v_i(0)$ corresponds to the number of extra jobs on average that exceed the deadline τ_i if the initial backlog is u instead of zero, then on average $p_i(v_i(u) - v_i(0))$ of them belong to class i (superposition of Poisson arrival processes), and, as class i violations cost H_i, we have

$$v(u) - v(0) = \sum_i p_i \, \mathrm{E}[H_i](v_i(u) - v_i(0)), \tag{3.11}$$

where each $v_i(u) - v_i(0)$ is given by (3.9) with (λ, τ_i). Note that this is valid because all job classes have the same service time d and are treated the same way under FCFS, and we can also assume that costs are paid upon arrival.

Similarly, the admission cost to the system can be determined, where the immediate cost is included only for the class of the arriving job. That is, if (3.10) is used, the admission cost of a class j job with violation cost h to an M/D/1 queue in state u is

$$a(u, j) = h \cdot \mathbf{1}(u > \tau_j) + \sum_i p_i \mathrm{E}[H_i] \left(a_i(u) - \mathbf{1}(u > \tau_i) \right).$$

Note that h can be replaced with $\mathrm{E}[H_j]$ if the violation cost of the given job is unknown.

We note that *without any technical difficulties*, we can extend the model so that each job class can also have several deadlines with arbitrary violation costs. That is, a penalty is paid for each deadline that is violated for the same job. Then we can approximate any cost structure based on the waiting and/or sojourn times. For example, a cost structure for a single class with unit violation costs $h_i = h$ and deadlines $\tau_i = ih$, where $i = 0, 1, \ldots$, converges to the cost structure where each job incurs a cost equal to its waiting time as $h \to 0$. This is equivalent to the (mean) waiting time. However, for clarity of presentation we omit such examples.

3.4 M/iD/1 FCFS Queue with Deadlines

In this section, we first consider a more general queueing model, the M/iD/1 queue, where the service time of a job is a random multiple of d, and derive its value function. Then we also demonstrate how it can be applied to deduce value functions for systems with batch arrivals.

3.4.1 Value Function for the M/iD/1 FCFS Queue

Consider the M/iD/1 queue, where job sizes are some multiple of d. More specifically, the service time of a job is id with probability p_i, and the size-specific arrival rates are $\lambda_i = p_i \lambda$.

We apply two results from previous work when deriving the corresponding value function. First, the waiting time distribution is available also for the M/iD/1 queue [16, Theorem 1]:

$$P\{W \leq \tau\} = (1 - \rho) \sum_{i=0}^{\lfloor \tau/d \rfloor} e^{-\lambda(id-\tau)} \sum_{L \in \mathscr{P}(i)} \frac{(id - \tau)^{|L|}}{H(L)} \prod_{j \in L} \lambda_j, \qquad (3.12)$$

where

$$\sum_{L \in \mathscr{P}(0)} \frac{(id - \tau)^{|L|}}{H(L)} \prod_{j \in L} \lambda_j \equiv 1.$$

Set $\mathscr{P}(i)$ is the set of partitions of i, i.e., the set of positive integers that sum to i, and $H(L)$ is

$$H(L) \equiv n_1! n_2! \cdots n_k!,$$

where the n_j denote the multiplicity of number j in partition L, i.e., $|L| = n_1 + \cdots + n_k$. For example, with size d and $2d$ jobs, when $\lambda_i = 0$ for $i \geq 3$, $\mathscr{P}(3) = \{(1, 1, 1), (1, 2)\}$, which corresponds to $(n_1, n_2) \in \{(3, 0), (1, 1)\}$. Second, given the waiting time distribution is available, we apply (3.6) from [12].

The value function for the standard M/D/1 queue with respect to a deadline at time τ, given in (3.9), resembles the corresponding expression for the waiting time distribution. Similarly, it turns out that value function for the M/iD/1 queue with respect to deadline resembles the corresponding expression (3.12) for the waiting time.

Theorem 2 *The value function for an M/iD/1 queue with respect to deadline at time τ with a unit violation cost is*

$$v(u) - v(\tau) = \frac{\lambda P\{W \leq \tau\}}{1 - \rho}(u - \tau)$$

$$+ \left(\sum_{i=0}^{\lfloor (\tau-u)/d \rfloor} \sum_{L \in \mathscr{P}(i)} \frac{|L|! \prod_{j \in L} p_j}{H(L)} \left[e^{-\lambda(id+u-\tau)} \sum_{j=0}^{|L|} \frac{(\lambda(id + u - \tau))^j}{j!} - 1 \right] \right),$$

$$\hspace{10cm} (3.13)$$

where

$$\sum_{L \in \mathscr{P}(0)} \frac{|L|! \prod_{j \in L} p_j}{H(L)} \left[e^{-\lambda(u-\tau)} \sum_{j=0}^{|L|} \frac{(\lambda(u-\tau))^j}{j!} - 1 \right] \equiv e^{-\lambda(u-\tau)} - 1. \quad (3.14)$$

Proof Suppose first that $u > \tau$. Then, as with the M/D/1 queue, all arriving jobs are late until the backlog returns to τ. Hence,

$$v(u) - v(\tau) = \frac{u - \tau}{1 - \rho} (\lambda - \lambda \, \mathrm{P}\{W > \tau\}) = \frac{\lambda \, \mathrm{P}\{W \leq \tau\}}{1 - \rho} (u - \tau). \quad (3.15)$$

Next we assume that $0 \leq u \leq \tau$. With the deadline cost structure, $c(w) = \mathbf{1}(w > \tau)$, the general result (3.6) yields

$$v(u) - v(0) = \frac{\lambda u}{1 - \rho} \left(\frac{1}{u} \int_0^u \mathrm{P}\{W + x > \tau\} \, dx - \mathrm{P}\{W > \tau\} \right)$$

$$= \frac{\lambda}{1 - \rho} \int_0^u 1 - \mathrm{P}\{W \leq \tau - x\} \, dx - \frac{\lambda u}{1 - \rho} (1 - \mathrm{P}\{W \leq \tau\})$$

$$= \frac{\lambda u}{1 - \rho} \mathrm{P}\{W \leq \tau\} - \frac{\lambda}{1 - \rho} \int_{\tau - u}^{\tau} \mathrm{P}\{W \leq x\} \, dx,$$

and thus,

$$v(u) - v(\tau) = \frac{\lambda(u - \tau)}{1 - \rho} \mathrm{P}\{W \leq \tau\}$$

$$- \frac{\lambda}{1 - \rho} \left(\int_{\tau - u}^{\tau} \mathrm{P}\{W \leq x\} \, dx - \int_0^{\tau} \mathrm{P}\{W \leq x\} \, dx \right),$$

$$= -\frac{\lambda \mathrm{P}\{W \leq \tau\}}{1 - \rho} (\tau - u) + \frac{\lambda}{1 - \rho} \int_0^{\tau - u} \mathrm{P}\{W \leq x\} \, dx. \quad (3.16)$$

The first term is the same linear function of u that holds also for the tail $u > \tau$, as given in (3.15). The second term is essentially an integral of the waiting time distribution, which is given by (3.12). When integrating (3.12) from 0 to $\tau - u$, where $\tau - u < \tau$, more terms appear in the sum, one for every interval of length d. In particular, for every i, the corresponding term appears when $x = id$, and thus the corresponding term can be evaluated by integrating from id to $\tau - u$. For $i = 0$, we have

$$\int_0^{\tau - u} e^{-\lambda(0-x)} \, dx = \frac{e^{-\lambda(u-\tau)} - 1}{\lambda}, \quad (3.17)$$

and for $i > 0$,

$$\int_{id}^{\tau-u} e^{-\lambda(id-x)}(id-x)^k \, dx = \frac{k!}{\lambda^{k+1}} \left[e^{-\lambda(id+u-\tau)} \sum_{j=0}^{k} \frac{(\lambda(id+u-\tau))^j}{j!} - 1 \right].$$

where $k = |L|$ as in (3.12). Therefore, the integral of (3.12) gives

$$\int_0^{\tau-u} P\{W \le x\} \, dx$$

$$= \frac{1-\rho}{\lambda} \left(\sum_{i=0}^{\lfloor(\tau-u)/d\rfloor} \sum_{L \in \mathscr{P}(i)} \frac{|L|! \prod_{j \in L} p_j}{H(L)} \left[e^{-\lambda(id+u-\tau)} \sum_{j=0}^{|L|} \frac{(\lambda(id+u-\tau))^j}{j!} - 1 \right] \right),$$

where $p_j = \lambda_j/\lambda$ and, similarly as with (3.12), (3.14) ensures that the first term, with $i = 0$, is according to (3.17). Substituting the above into (3.16) then gives (3.13). □

Note that the factor,

$$\frac{|L|! \prod_{j \in L} p_j}{H(L)} = \frac{(n_1 + \cdots + n_k)!}{n_1! n_2! \cdots n_k!} p_1^{n_1} p_2^{n_2} \cdots p_k^{n_k},$$

in (3.13) is the pmf of the multinomial distribution.

Example 3 For the M/D/1 queue, $L = \{1, 1, \ldots, 1\}$ and $|L| = i$, and the above reduce to the results discussed earlier: $P\{W \le \tau\}$ given in (3.7) and $v(u)$ given in (3.9).

3.4.2 Special Case: Systems J2 and B2 with Two Sizes

For simplicity, next we limit ourselves to the case where the service time is d or $2d$, with probabilities $1-q$ and q, respectively. Hence, the offered load is $\rho = \lambda(1+q)d$. We refer to this model as J2 (J for the job size). By varying q we can introduce a moderate variability in the job sizes.

In the second model, we consider batch arrivals. More specifically, batches arrive with rate λ and with probability of q the arriving batch has size $B = 2$, with the total service time of $2d$, and otherwise the batch consists of a single job, $B = 1$, and the service time is d. Again, the offered load is $\rho = \lambda(1 + q)d$. We call this model B2, and it allows us to increase variability in the job interarrival times moderately.

With batch arrivals, some jobs experience also waiting time due to the jobs arriving in the same batch. Hence, instead of W, we use U to denote the unfinished work in the queue. We observe that the unfinished work, U, will be the same for both models B2 and J2. Moreover, as the size of a job or a batch is either d or $2d$, the partitions $\mathscr{P}(i)$ are explicitly of the form,

$$\{\overbrace{1, 1, \ldots, 1}^{=n_1}, \overbrace{2, 2, \ldots, 2}^{=n_2}\}.$$

i.e., for $i = 1, 2, \ldots$, we can write

$$n_2 = 0, \ldots, \lfloor i/2 \rfloor,$$

$$n_1 = i - 2n_2,$$

$$|L| = n_1 + n_2 = i - n_2.$$

Finally, the size-specific arrival rates are $\lambda_1 = (1 - q)\lambda$ and $\lambda_2 = q\lambda$. Letting $n = n_2$, (3.12) reduces to

$$P\{U \le x\} = (1 - \rho) \sum_{i=0}^{\lfloor x/d \rfloor} e^{-\lambda(id-x)} \begin{cases} 1, & i = 0, \\ \sum_{n=0}^{\lfloor i/2 \rfloor} \frac{(\lambda(id-x))^{i-n}}{(i-2n)!n!} (1-q)^{i-2n} q^n, & i > 0. \end{cases}$$

$$(3.18)$$

3.4.3   Steady State Performance with J2 and B2

The mean cost rate with J2 follows immediately from (3.18),

$$r_{J2} = \lambda P\{U > \tau\}.$$

The mean cost rate with batch arrivals, i.e., with B2, follows similarly. If a batch of two jobs arrives when the backlog U is in $(\tau - d, \tau]$, the second job of the batch incurs a unit cost, while the first job receives service in time. Once the backlog increases beyond $U = \tau$, all arriving jobs incur a unit cost. This is illustrated in Fig. 3.2. Thus, the cost rate for $\tau - d < u \le \tau$ is λq, and for $u > \tau$ it is $(1 + q)\lambda$. Therefore, the mean cost rate with the B2 system is

$$r_{B2} = \lambda q\, P\{U > \tau - d\} + \lambda\, P\{U > \tau\}.$$

Note that with B2 each arriving batch incurs a cost of 0, 1, or 2, depending on how many jobs receive service late. These results generalize to arbitrary general job- and batch-size distributions by appropriate conditioning.

Fig. 3.2 Batch arrivals experience deadline violations also below τ

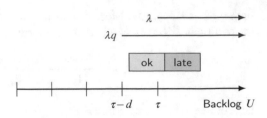

3.4.4 Value Functions for J2 and B2

Let us first consider the system J2, where the job sizes vary. From (3.13), one immediately obtains the value function for system J2:

Corollary 2 (J2) *The value function for an M/2D/1 queue with respect to deadline at time τ with a unit violation cost is*

$$v(u) - v(\tau)$$

$$= \lambda \left(e^{\lambda \tau} + \sum_{i=1}^{\lfloor \frac{\tau}{d} \rfloor} e^{-\lambda(id-\tau)} \sum_{n=0}^{\lfloor \frac{i}{2} \rfloor} \frac{q^n (1-q)^{i-2n} (\lambda(id-\tau))^{i-n}}{n!(i-2n)!} \right) (u-\tau)$$

$$- \mathbf{1}(u < \tau) \left(1 - e^{-\lambda(u-\tau)} \right) \tag{3.19}$$

$$- \sum_{i=1}^{\lfloor \frac{\tau-u}{d} \rfloor} \sum_{n=0}^{\lfloor \frac{i}{2} \rfloor} \frac{(i-n)! q^n (1-q)^{i-2n}}{n!(i-2n)!} \left(1 - e^{-\lambda(id-\tau+u)} \sum_{j=0}^{i-n} \frac{(\lambda(id-\tau+u))^j}{j!} \right),$$

where the first part corresponds to the linear component, and the second part (the last two rows) adjusts the value function for $0 \le u \le \tau$.

It is easy to see that as $q \to 0$, (3.19) reduces to (3.9). It is also worth noting that alternatively, as we did with the standard M/D/1 queue, it is possible to derive the value function directly by solving a set of the differential equations. That is, (3.4) holds for the tail $u \ge \tau$, and for $u \ge 0$ the value function $v(u)$ satisfies the integro-differential equation (3.5). Thus, as with the M/D/1 queue, we can solve the differential equation backwards one interval at a time starting from $(\tau - d, \tau]$ until $u = 0$ is included. The constants are solved by requiring continuity at points $u = kd$, $k = 1, 2, \ldots, \lfloor \tau/d \rfloor$, and the boundary condition $v'(0) = 0$. This straightforward procedure gives both the value function and the mean cost $\bar{c} = P\{W > \tau\}$, and thus does not require the use of (3.9).

Let us next consider batch arrivals. In particular, now we assume that each job has a unit service time d, but jobs arrive in batches of k jobs with probability of p_k, $k = 1, 2, \ldots$, and the batch arrival process is Poisson with rate $\lambda = \lambda_b$. As with varying job sizes, for simplicity, we limit ourselves to the case where a batch consists of one or two jobs; $p_1 = 1 - q$ and $p_2 = q$.

As with the mean performance, the value function can be deduced from the corresponding value function of the job size case $J2$.

Corollary 3 (Batch Arrivals) *The value function for an M/2D/1 queue B2 with respect to deadline at time τ with a unit violation cost and batch arrivals is*

$$v(u) - v(\tau) = \tilde{v}(u; \tau) - \tilde{v}(\tau; \tau) + q \left(\tilde{v}(u; \tau - d) - \tilde{v}(\tau; \tau - d) \right),$$

where $\tilde{v}(u; \tau)$ denotes the corresponding value function of system J2 with deadline at time τ and the job arrival rate λ_b.

Proof The first job of each batch experiences a deadline violation similar to that of the jobs in model J2. Let $v_1(u)$ denote their contribution to the total value function, for which we immediately have

$$v_1(u) - v_1(\tau) = \tilde{v}(u; \tau) - \tilde{v}(\tau; \tau),$$

where $\tilde{v}(u; \tau)$ denotes the corresponding value function of system J2 with deadline at time τ and the job arrival rate λ_b.

For the second job of a batch, which exists only in batches with two jobs (or more in the general case), we have

$$v_2(u) - v_2(\tau) = q \left(\tilde{v}(u; \tau - d) - \tilde{v}(\tau; \tau - d) \right),$$

as a fraction q of the batches have two jobs, and in those cases the second job is late whenever $u < \tau - d$. □

Even though here we have assumed a Bernoulli distribution for batch sizes, the same steps can be taken when batch sizes are i.i.d. random variables with an arbitrary distribution. In fact, it is possible to analyze a general class of models where the total service time of a batch is a random multiple of d with different kinds of internal structures of a batch. For example, each batch could start with a fixed size job, followed by jobs of varying size. Similarly as discussed in Sect. 3.3.2, different batches and jobs can have their own individual deadlines and deadline violation costs. In all these cases, the total value function follows by straightforward integration and superposition, as illustrated above.

Example 4 Let us consider three systems with $d = 1$, $\tau = 2$ and $\rho = 0.75$:

1. The basic M/D/1 queue, where $\lambda = 0.75$,
2. The J2 system with $q = 0.5$ and $\lambda = 0.5$,
3. The B2 system with $q = 0.5$ and $\lambda = 0.5$.

Fig. 3.3 Value functions
when $\rho = 0.75$ with three
different types of job- and
batch-size distributions

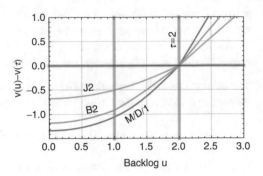

Figure 3.3 depicts the corresponding value functions. We can see that, as expected, they have a similar shape, but both the slope for $u > \tau$ and the convex portion for $0 \leq u \leq \tau$ are at different levels. In all cases, an appropriate quadratic function for $0 \leq u \leq \tau$ can be assumed to be an adequate approximation for the value function.

3.5 Parallel Servers

In this section, we consider a dispatching system with parallel heterogeneous servers, as illustrated in Fig. 3.4. In particular, we show how efficient dispatching policies can be derived based on the new results given earlier. We consider the following model for a multi-server system:

1. Jobs arrive according to Poisson process with rate λ.
2. Job sizes are i.i.d. and obey a discrete uniform distribution, X is 1, 2, 3 or 4.
3. All jobs have the same deadline τ for the waiting time.
4. Jobs are processed by m parallel servers with nominal service times d_i, $i = 1, \ldots, m$, where d_i denotes the time to serve one unit of a job's size.

The offered load to the system is $\rho_{\text{tot}} = \sum_i \lambda_i / c$, where $c = \sum_i 1/d_i$, and, for stability, $\rho_{\text{tot}} < 1$. We consider the following heuristic dispatching policies:

Definition 1 (RND) *Random split* routes a job to Server i with probability $p_i \propto 1/d_i$, so that the offered load ρ_{tot} is balanced among the m servers.

Definition 2 (JSQ) *Join-the-shortest-queue* routes a job to the server with fewest jobs. Ties are resolved in favor of the server with a higher index.

Definition 3 (LWL) *Least-work-left* routes a job to the server with the shortest backlog. Ties are again resolved in favor of the server with a higher index.

Note that RND is a *static* policy, i.e., its actions are independent of the system's state. Similarly, all policies based on job- or class-specific information are also static. For example, SITA chooses the server according to job's size [17, 18], and

Fig. 3.4 Dispatching system with $m = 4$ parallel servers

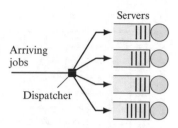

CIQ according to job's class [4]. Next we develop a new policy based on the value function by carrying out one policy improvement step. More specifically, the standard procedure (see, e.g., [3, 6, 11, 19] and [20, Section 11.5]) is as follows:

Definition 4 (FPI) The First Policy Iteration step yields an improved policy:

1. Choose a *static* policy α_0, e.g., RND.

   – With α_0, the system decomposes into m independent M/G/1 queues.

2. Compute the value functions and admission costs $a_i(u, x)$ for each queue.
3. The policy improvement step yields a new (dynamic) policy,

$$\alpha_{FPI}(u_1, \ldots, u_m, x) \in \arg\min_i a_i(u_i, x).$$

where u_i is the current backlog in queue i, and x is the size of the new job.

In general, it is difficult to go beyond the first iteration as a value function for a dynamic policy would be needed for the second iteration. One option is to carry out a *lookahead analysis* [21] that considers two consecutive decisions instead of one.

Example 5 (Two Identical Servers) Suppose we have two identical servers, $d_1 = d_2 = 1$. The mean service time is then $E[X] = 2.5$, the offered load (to the system) $\rho_{tot} = \lambda E[X]/c = (5/4)\lambda$, and the system is stable (with an appropriate routing) when $\lambda < 4/5$. The deadline for the waiting time is set at $\tau = 4$.

The simulation results are depicted in Fig. 3.5 (left). On the x-axis is the offered load ρ_{tot}, and on the y-axis the relative performance defined as the ratio of deadline violation rates between the given dispatching policy α and JSQ, $\alpha \in$ {RND, LWL, FPI}. We can observe that RND has very poor performance except in the heavy traffic regime where $\rho \approx 1$. The performance of LWL is significantly better than with JSQ when ρ is small, but under a high load also they become equal. In contrast, FPI, based on the value function (3.13) and RND, yields a clear improvement over other heuristic dispatching policies at all loads. The FPI policy, in contrast to LWL, may route a jobs to a longer queue if the backlogs are such that it will miss its deadline regardless of which queue it goes to. (For $u > \tau$, the admission cost is constant as illustrated in Fig. 3.1 (right)). The "notch" that appears under a very high load is due to the stability issues the FPI policy introduces. Indeed, it may

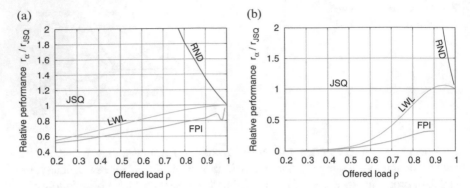

Fig. 3.5 Simulation results with two dispatching systems, when the job sizes obey a discrete uniform distribution on $\{1, 2, 3, 4\}$. The simulation run for each (policy,ρ)-pair included about 2×10^8 jobs. (**a**) Two identical servers. (**b**) Four heterogeneous servers

be beneficial to overload one server in order to minimize the deadline violation rate! This phenomenon, observed also in [3], gets more pronounced in the next example.

Example 6 (Four Heterogeneous Servers) Next we consider a heterogeneous four server system with service rates $1, 2, 3, 4$, i.e., $d_i = 1/i$, $i = 1, \dots, 4$. JSQ and LWL now choose a faster server in case of ties. Due to the system having faster servers, the target deadline is reduced to $\tau = 2$.

The simulation results are depicted in Fig. 3.5 (right). We can observe that RND has a very poor performance, as expected. The deadline violation rate with LWL is significantly smaller than with JSQ when ρ is small, but under a very high load LWL is actually worse than JSQ. LWL, balancing the backlogs, ensures that backlog in every queue is longer than τ, whereas JSQ, observing only the number of jobs, sometimes imbalances the backlogs a bit and avoids a deadline violation. The performance with FPI is now significantly better than with the other three dispatching policies. Moreover, under a sufficiently high load, FPI overloads the slow server. This is an artifact of the simple cost structure that does not penalize for the excessive (infinite) waiting times. The obvious fix would be to include, e.g., the mean waiting or response time to the objective function, as discussed in [3].

3.6 Summary

Past work has given explicit forms for the value function with respect to the deadline cost structure only for specific cases: (1) in the heavy-traffic regime as $\rho \uparrow 1$, (2) when all service times are larger than the (single) deadline, and (3) when the service times are constant. In the heavy-traffic regime, the value function for M/G/1 (with large deadline) is quadratic. When the deadline is smaller than the service time, the value function includes an exponential term.

In this chapter, we gave exact expressions for the value function with respect to (possibly multiple) deadlines for a single server queue under arbitrary load when service times are random multiples of a fixed size d. The model enables us to consider both varying job sizes and batch arrival processes, or some combination of them. The standard M/D/1 queue follows as a special case.

The basic results take the form of a double or triple sums with a finite number of terms. These results can be generalized for queues with multiple job classes having different target deadlines and violation costs. The availability of the value function enables policy iteration for developing cost-aware dispatching strategies for parallel servers, making these results immediately useful.

Acknowledgements This work was supported by the Academy of Finland in the FQ4BD project (grant nos. 296206).

References

1. J. Dean, L.A. Barroso, The tail at scale. Commun. ACM **56**(2), 74–80 (2013)
2. Z. Liu, M.S. Squillante, J.L. Wolf, On maximizing service-level-agreement profits, in *Proceedings of the 3rd ACM Conference on Electronic Commerce*, ser. EC '01 (ACM, New York, 2001), pp. 213–223
3. E. Hyytiä, R. Righter, Routing jobs with deadlines to heterogeneous parallel servers. Oper. Res. Lett. **44**(4), 507–513 (2016)
4. E. Hyytiä, R. Righter, O. Bilenne, X. Wu, Dispatching fixed-sized jobs with multiple deadlines to parallel heterogeneous servers. Perform. Eval. **114**, 32–44 (2017)
5. E. Hyytiä, A. Penttinen, S. Aalto, J. Virtamo, Dispatching problem with fixed size jobs and processor sharing discipline, in *23rd International Teletraffic Congress (ITC'23)*, San Fransisco (2011), pp. 190–197
6. E. Hyytiä, A. Penttinen, S. Aalto, Size- and state-aware dispatching problem with queue-specific job sizes. Eur. J. Oper. Res. **217**(2), 357–370 (2012)
7. K.R. Krishnan, Joining the right queue: a state-dependent decision rule. IEEE Trans. Autom. Control **35**(1), 104–108 (1990)
8. E. Hyytiä, R. Righter, S. Aalto, Task assignment in a heterogeneous server farm with switching delays and general energy-aware cost structure. Perform. Eval. **75–76**(0), 17–35 (2014)
9. A. Penttinen, E. Hyytiä, S. Aalto, Energy-aware dispatching in parallel queues with on-off energy consumption, in *30th IEEE International Performance Computing and Communications Conference (IPCCC)*, Orlando, 2011
10. E. Hyytiä, J. Virtamo, S. Aalto, A. Penttinen, M/M/1-PS queue and size-aware task assignment. Perform. Eval. **68**(11), 1136–1148 (2011)
11. K.R. Krishnan, T.J. Ott, State-dependent routing for telephone traffic: theory and results, in *IEEE Conference on Decision and Control*, vol. 25 (1986), pp. 2124–2128
12. E. Hyytiä, R. Righter, J. Virtamo, L. Viitasaari, Value (generating) functions for the $M^X/G/1$ queue, in *29th International Teletraffic Congress (ITC'29)*, Genoa, 2017
13. L. Takács, A single-server queue with Poisson input. Oper. Res. **10**(3), 388–394 (1962)
14. L. Kleinrock, *Queueing Systems, Volume I: Theory* (Wiley Interscience, New York, 1975)
15. A. Erlang, Sandsynlighedsberegning og telefonsamtaler. Nyt tidsskrift for Matematik B **20**, 33–39 (1909)
16. J.F. Shortle, P.H. Brill, Analytical distribution of waiting time in the M/{iD}/1 queue. Queueing Syst. **50**(2), 185–197 (2005)

17. M. Harchol-Balter, M.E. Crovella, C.D. Murta, On choosing a task assignment policy for a distributed server system. J. Parallel Distrib. Comput. **59**, 204–228 (1999)
18. M. Harchol-Balter, *Performance Modeling and Design of Computer Systems: Queueing Theory in Action* (Cambridge University Press, Cambridge, 2013)
19. H. Wu, K. Wolter, Tradeoff analysis for mobile cloud offloading based on an additive energy-performance metric, in *Proceedings of Valuetools'14* (2014), pp. 90–97
20. P. Whittle, *Optimal Control: Basics and Beyond* (Wiley, New York, 1996)
21. E. Hyytiä, Lookahead actions in dispatching to parallel queues. Perform. Eval. **70**(10), 859–872 (2013)

Chapter 4
Modelling and Efficient Solution of Multiple-Phased Systems

Elvio Gilberto Amparore and Susanna Donatelli

4.1 Introduction and Paper Contribution

Multiple-Phased Systems (MPS) [13] are systems whose behaviour can be split in a set of successive periods, called phases. MPS are also called Phased Mission System in other works [20], since they can easily describe systems in which the behaviour is described as a mission structured into multiple *phases*. Each phase is described by a different duration, system configuration, desired task, etc. MPS have been shown to be useful in many contexts, like modelling systems with scheduled maintenance [12] (Scheduled Maintenance Systems—SMS). In a MPS, the standard question is to compute the reliability of the system, i.e. the probability that the system survives the mission, but optimization also plays a role [22], especially for SMS to determine the best maintenance policy, as well as sensitivity analysis [12, 16] to allow to reason about the structure of the mission and their parameters.

There has been a significant amount of work on MPS, especially in the late nineties and also more recently. Different techniques have been used for modelling and solving these systems, from combinatorial methods like reliability blocks and fault trees to state-based techniques. Combinatorial approaches based on reliability blocks and fault trees have been reported, for example, in [24, 28]. State-based approaches build Markovian systems either through an ad-hoc language as in EHARP [24] or through a high level modelling formalism as the Stochastic Petri Nets [21]. It is well known what are the relative advantages and disadvantages of combinatorial methods over state-based ones, but when the objective of the MPS analysis goes beyond the computation of the reliability of the system at the end of the

E. G. Amparore · S. Donatelli (✉)
Dipartimento di Informatica, Università di Torino, Torino, Italy
e-mail: amparore@unito.it; donatelli@unito.it; susi@di.unito.it

© Springer International Publishing AG, part of Springer Nature 2019   47
A. Puliafito, K. S. Trivedi (eds.), *Systems Modeling: Methodologies and Tools*,
EAI/Springer Innovations in Communication and Computing,
https://doi.org/10.1007/978-3-319-92378-9_4

mission, state-based approaches can provide a plus, like computing the probability of the system states at time t, or associate a reward structure to the system states.

Popular formalisms for state based models in performance evaluation and reliability analysis include various forms of Stochastic Petri nets: *Generalized Stochastic Petri Nets* (GSPN) [2], that allows exponentially distributed transitions and immediate ones (transitions that fire in zero time); *Deterministic and Stochastic Petri Net* (DSPN) [1, 18], an extension of GSPN to include also transitions with a deterministic duration, subject to the constraints that at most one deterministic transition is enabled in any one state; *Markov Regenerative Stochastic Petri Net* (MRSPN) [17], where transitions can have a generally distributed delay, again subject to the single enabling constraints as DSPN. The steady state solution of a GSPN requires the steady state solution of a continuous time Markov chain (CTMC), while the stochastic process underlying a DSPN and a MRSPN is a *Markov Regenerative Process* (MRgP) [23], which has attracted a significant attention from the performance community, since an MRgP can describe more complex behaviours than a CTMC, while still allowing an analytical solution.

The work in [21] examines various modelling approaches for MPS and concludes that a high level formalism based on DSPN shows the best trade-off between the modelling power (class of systems that can be specified) and the amount of human intervention in the definition of the model. In particular the authors identify a specific Petri net structure. The phases are described as a Petri net (the *Phase Net*—PhN) with a directed acyclic graph (DAG) structure, which only includes deterministic transitions and immediate ones, with exactly one deterministic transition enabled in each phase. The system behaviour is modeled using a GSPN (the *System Net*—SN). Transition enabling and firing rates in the SN may depend on the marking of the PhN and this permits a description of the system that is compact (a single net) but that is able to describe a behaviour that may differ from phase to phase. This results in a class of nets called Phased Petri Nets (PPN). The limitations imposed on the Phase net limit the modelling power of PPN, but allowed Mura et al. to show in [21] that the solution of the MPS can be decomposed in a sequence of transient solutions of Markov chains. The DEEM tool [13] implements this method.

An important contribution to the analysis of MPS is the work in [20]. The authors present a description of the MRgP underlying a PPN in which many of the limitations of the work in [21] are lifted. The paper presents a complete theoretical framework and practical indications on how to solve the resulting models for various classes of PPN, including the case of general distributions in the PhN. But no implementation is provided (apart from what is implemented in DEEM) and the practical indications still assume that the marking graph of the PhN is a DAG and that a SN event cannot interrupt a PhN event.

Contribution In this paper we define *eXtended PPN* (X-PPN), an extension of the PPN of [21] and [20] to allow a more general definition of PhN (to include a mix of general and exponential events and cyclic, ergodic behaviour) and a more flexible definition of dependencies (in particular a SN event can interrupt a phase).

When the resulting MRgP is ergodic the solution is computed through the so-called *matrix-free solution* of MRgP, provided by [18] and enhanced in [5]; when the MRgP is non-ergodic the solution is demanded to the *Component Method* algorithm for MRgPs [6]. The proposed solutions are implemented inside the GreatSPN [11] framework. The solvers of GreatSPN allows to solve X-PPN with more than a million states. We shall show through a scheduled maintenance system example the practical relevance of the extension introduced by X-PPN and of the associated tool. We shall also show that, when a X-PPN reduces to a PPN, the Component Method does exactly the same computations as the ad-hoc efficient solution of [21]. We shall not compare the two tools performances since DEEM uses older technologies and it is very inefficient in time. A comparison can be found in [7].

The paper develops as follows: Sect. 4.2 defines the X-PPN formalism, Sect. 4.3 the proposed solution techniques for X-PPN, Sect. 4.4 experiments the available solution technique on some X-PPN examples. Section 4.5 compares the behaviour of GreatSPN and DEEM on PPN models, while Sect. 4.6 concludes the paper and outlines future possible extensions and integration activities.

4.2 Extended Phased Petri Nets

In this paper we use the definitions and notations of MPS given in [21]: a mission is divided into m phases of deterministic duration δ_i, and the system behaviour in each phase is described by a CTMC with state space $\mathscr{S}_i$ and rate matrix $\mathbf{Q}_i$. Models are expressed as GSPN or DSPN where we indicate with P the set of places, T the set of transitions, I, O and H for the input, output and inhibitor arcs, respectively, $W(T)$ for the function that defines the timing aspects of the transitions. A function $W(t)$ can depend on the state (marking) of the net and $m_0 : P \rightarrow \mathbb{N}_{\geq 0}$ the initial marking.

The MPS is defined using a restricted class of DSPNs called *Phased Petri nets* (PPN) [20, 21] that model the system using two Petri nets [2]: a *Phase Net* (PhN) that defines the phase structure, and a *System Net* (SN) that describes the stochastic behaviour of the system components during each phase of the mission. Marking dependent rates and guards can be used to make the SN behaviour dependent from a specific phase, and the probability of immediate transitions in the PhN can depend from the marking of the SN: this allows the choice of the next phase to depend upon the state of the SN.

Definition 1 (Adapted from [13, 21]) A PPN is a DSPN resulting from the union of two Petri nets with disjoint set of places and transitions, the *phase net PhN* (a DSPN) and the *system net SN* (a GSPN), with the additional restrictions:

1. Transitions in *PhN* are either immediate or deterministic (no exponential);
2. The marking graph of *PhN* must be a directed acyclic graph (DAG);
3. At most a single deterministic transition is enabled in any state;
4. The firing of a transition in *SN* cannot disable a transition in *PhN*.

The above limitations are imposed to provide a guideline to the modeller and to allow for a decomposable solution [13, 20]. Places and transitions are disjoint but the two nets are dependent, due to the presence of arcs encompassing the two nets and to a definition of marking dependent rates of transitions of one net that can depend upon the places of the other net, which gives to the modeller the ability to express a system behaviour that differs from phase to phase. Note that requirement 1 and 3 imply that each phase is identified by a single marking of the PhN and condition 2 implies that the whole stochastic process is non-ergodic.

Definition 2 (From [17]) A GSPN extended to include deterministic and generally distributed durations of event is a Markov Regenerative Stochastic Petri Nets (MRSPN) if its marking process is an MRgP.

A Petri net with immediate, exponential and generally distributed transitions in which at most one general transition is enabled in a marking is a MRSPN, as the marking right after the firing of the single general transition enabled identifies a regeneration point. Point 3 of Definition 1 ensures that each PPN is a MRSPN, as in [20].

Definition 3 A X-PPN is a Petri net resulting from the union of two Petri nets with disjoint set of places and transitions, the *phase net PhN* (a MRSPN) and the *system net SN* (a GSPN), with the additional restriction:

1. At most a single general transition is enabled in any state;

Given the limitation of at most one general transition enabled in any marking, an X-PPN is a MRSPN. Note that the limitation is required since, due to the presence of marking dependencies, it is possible that a change of state in the SN enables a new general transition in the PhN. X-PPN represents an extension of PPN in which PhN may include general distributions, the marking graph of the PhN is not required to be acyclic, and the firing of a transition in the SN may disable a transition in the PhN, while marking dependencies allows the PhN behaviour to influence the SN one, and vice versa. In a X-PPN the behaviour of a phase is not strictly identified by a single marking, phases can have a duration which is exponential or generally distributed, and a system event may interrupt a phase. Note that since the marking graph of a PhN is not required to be acyclic, repetitive tasks can be modelled and it is possible that the marking process of a X-PPN is ergodic.

X-PPN can be defined through the graphical user interface of GreatSPN, with the rich definition of general distributions supported by the alphaFactory [9] library, which includes, among the others, deterministic, uniform, Erlang, Pareto, expolynomial, triangular distributions, as well as any linear combination of them. The usefulness and flexibility of X-PPN for the modelling of MPS is discussed in Sect. 4.4 through four examples of a scheduled maintenance system.

4.3   X-PPN Solution

An MRgP is a stochastic process defined by a sequence of time instants called *renewal times* in which the process loses its memory, i.e. the age of non-exponential (general) events is 0. A regeneration point is a renewal time associated to the state at that time. The process behaviour among regeneration points is a discrete-time Markov chain, the *embedded Markov chain* (EMC), while the behaviour between two regeneration points is described by a continuous-time process, the *subordinated process*, that we require to be a time-homogeneous CTMC. MRgPs have been studied extensively in the past [18, 25, 26], and many solid analysis techniques exist.

An MRgP can be represented as a discrete event system [14] with a finite state space, where in each state a general event g is taken from a set G. As the time flows, the age of g being enabled is kept, until either g fires (Δ event), or a Markovian transition, concurrent with g, fires. Markovian events may disable g (preemptive event, or $\bar{Q}$ event), clearing its age, or keep g running with its accumulated age (non-preemptive event, or Q event). Matrix Q accounts for the rates of the exponential transitions whose firings do not disable any general (deterministic) transition; Matrix $\bar{Q}$ accounts for the rates of the exponential transitions whose firings disable a general (or deterministic) one; Matrix Δ has, for each entry $\Delta_{i,j}$, the probability of ending in state j when the general transition fires in state i. MRgP are normally solved in steady state, since the solution at time t is significantly more complex.

The EMC matrix P is defined in terms of Q, $\bar{Q}$, and Δ in a standard manner (see [18] for example): the EMC is solved for steady state and the steady state distribution of the MRgP states is derived from the steady state solution of the EMC states. This method is what we term **Explicit**, and it can be used only for small MRgP as indeed, even if Q, $\bar{Q}$, and Δ are sparse, which is usually the case, matrix P is typically dense and expensive to compute, due to the matrix exponential terms. This problem has been solved in [18] with a technique based on the idea that P can be substituted by a function of the Q, $\bar{Q}$, and Δ matrices of the MRgP. These functions are used in vector×matrix products with P so that they can be computed without the need of constructing and storing P. This method is what we term **matrix-free** technique (actually P-free). Both methods (explicit and matrix-free) have been extended in [5] to deal with non-ergodic MRgP and to use a richer set of numerical solvers (like GMRES and the alike).

If the EMC is non-ergodic, its states can be classified as transient or recurrent states, and specialized solution methods can be devised to compute the probability distribution of recurrent states. The work in [3, 6] introduces an efficient steady-state solution for non-ergodic MRgPs, called **Component Method**, which computes the outgoing probability flow from transient to recurrent subsets of states, called *components*. The method is given in matrix-free terms. The basic idea of the Component Method is readily explained assuming that matrix P is available (in reality, since the method is matrix-free, the technique is more complicated). If the MRgP is non-ergodic it is indeed possible to rearrange the order of its states so that the EMC matrix P is in upper-triangular form (the *reducible normal form*, or RNF):

$$\mathbf{P} = \begin{bmatrix} \mathbf{T}_1 & \boxed{\quad\quad \mathbf{F}_1 \quad\quad} & \\ & \ddots & \ddots & \\ & & \mathbf{T}_k & \boxed{\mathbf{F}_k} & \\ & & & \mathbf{R}_{k+1} & \\ & & & & \ddots & \\ & & & & & \mathbf{R}_m \end{bmatrix} \qquad (4.1)$$

Matrix $\mathbf{P}$ has $k \geq 0$ transient subsets and $(m - k)$ recurrent subsets of states, with $m > k$. Let $\mathscr{S}_i \subseteq \mathscr{S}$ be the set of states in subset i, hereafter called the *component* i. The upper triangular form of $\mathbf{P}$ allows to interpret the $\mathscr{S}_i$ subsets as a DAG of components, and a way to build an RNF is to identify the set of *strongly connected components* (SCC) of the graph built by considering $\mathbf{P}$ as an adjacency matrix. The computation of an adequate set of components is the topic of [8]. Transient SCCs are the $\mathscr{S}_i$ for $i \leq k$, and bottom SCC (BSCC) are the $\mathscr{S}_i$ for the recurrent classes ($k < i \leq m$). When $\mathbf{P}$ is in RNF, the steady-state probability of the recurrent states can be computed using the *outgoing probability vectors* $\boldsymbol{\mu}_i$. For each state $s \in (\mathscr{S} \setminus \mathscr{S}_i)$, $\boldsymbol{\mu}_i(s)$ is the probability of reaching s in one jump when leaving $\mathscr{S}_i$. Vector $\boldsymbol{\mu}_i$ is given by:

$$\boldsymbol{\mu}_i = \left(\boldsymbol{\alpha}_i + \sum_{j<i} (\mathbf{I}_i \cdot \boldsymbol{\mu}_j) \right) \cdot (\mathbf{I} - \mathbf{T}_i)^{-1} \cdot \mathbf{F}_i, \quad i \leq k \qquad (4.2)$$

where $\boldsymbol{\alpha}_i$ is the vector of initial probabilities for the states in $\mathscr{S}_i$ and $\mathbf{I}_i$ are appropriate filter matrices (diagonal is 1 for states in $\mathscr{S}_i$, 0 otherwise). Since matrix inversion is usually expensive, a product of a generic vector $\mathbf{u}$ with $(\mathbf{I} - \mathbf{T}_i)^{-1}$ can be reformulated as a linear equations system $\mathbf{x} \cdot (\mathbf{I} - \mathbf{T}_i) = \mathbf{u}$ that is computed iteratively using vector×matrix products with $\mathbf{T}_i$. Each vector $\boldsymbol{\mu}_i$ may depend on the previous $(i-1)$ outgoing probability vectors, implicitly defining a computational order. Given the $\boldsymbol{\mu}_i$ vectors, the steady state probability of the recurrent subsets $\mathscr{S}_i$ is:

$$\boldsymbol{\pi}_i = \left(\boldsymbol{\alpha}_i + \sum_{j=1}^{k} (\mathbf{I}_i \cdot \boldsymbol{\mu}_j) \right) \cdot \lim_{n \to \infty} (\mathbf{R}_i)^n, \quad k < i \leq m \qquad (4.3)$$

The Component Method computes first Eq. (4.2) for all transient components, taken in an order that respects the condition $j < i$ of the formula, and then computes the probability for the recurrent subsets based on Eq. (4.3). The work in [6] explain how to compute (4.2) and (4.3) in matrix-free form. The computation of the products of a generic vector $\mathbf{u}$ with the matrix-free form of $\mathbf{T}_i$ and $\mathbf{F}_i$ is given by:

$$\mathbf{u}\mathbf{T}_i = \mathbf{I}_i \cdot (\mathbf{a}_i(\mathbf{u}) + \mathbf{b}_i(\mathbf{u}) + \mathbf{c}_i(\mathbf{u})) \qquad (4.4)$$

$$\mathbf{u}\mathbf{F}_i = (\mathbf{I} - \mathbf{I}_i) \cdot (\mathbf{a}_i(\mathbf{u}) + \mathbf{b}_i(\mathbf{u}) + \mathbf{c}_i(\mathbf{u})) \qquad (4.5)$$

Actually the above computation are performed using an *augmented set* $\widehat{\mathscr{S}_i}$ of component i, as we have to include in component i all the states that are reachable before the next regeneration point. For simplicity we do not use the augmented notation. The vector terms $\mathbf{a}_i(\mathbf{u})$, $\mathbf{b}_i(\mathbf{u})$ and $\mathbf{c}_i(\mathbf{u})$ are:

$$\mathbf{a}_i(\mathbf{u}) = \mathbf{u} \cdot \left(\sum_{g \in G} \mathbf{I}_i^g \cdot \int_0^\infty e^{\mathbf{Q}_i \, x} \cdot f_g(x) \, dx \right) \cdot \mathbf{\Delta}_i$$

$$\mathbf{b}_i(\mathbf{u}) = \mathbf{u} \cdot \left(\sum_{g \in G} \mathbf{I}_i^g \cdot \int_0^\infty e^{\mathbf{Q}_i \, x} \cdot \bar{F}_g(x) \, dx \right) \cdot \bar{\mathbf{Q}}_i$$

$$\mathbf{c}_i(\mathbf{u}) = \mathbf{u} \cdot \left(\mathbf{I}_i^E - \mathrm{diag}^{-1}(\mathbf{Q}_i^E) \mathbf{Q}_i^E \right)$$

where $\mathbf{Q}_i^E$ is an appropriate projection of matrix $\mathbf{Q}$ and $f_g(x)$ and $\bar{F}_g(x)$ are the PDF and the complementary CDF of the firing distribution time of the general event g enabled in $\mathscr{S}_i$, if any. These terms describe how the process evolves between two regeneration points. Vector $\mathbf{a}_i(\mathbf{u})$ and $\mathbf{b}_i(\mathbf{u})$ are the probability distribution of the next regeneration state reached with the firing of the general event ($\mathbf{a}$), or with the preemption of it ($\mathbf{b}$). Vector $\mathbf{c}_i(\mathbf{u})$ is the probability distribution of the next regeneration state when there are no general events enabled in the starting state.

The advantage of working at the component level is not only the trivial one of solving many small models instead of a single much bigger one, but that the cheapest available solution for each component can be used, as will be indeed the case of MRgP generated from X-PPN. Indeed the work in [6] identifies three component classes, their characterization and the associated solution technique:

[Class $\mathbf{C}_E$:] No general enabled. The component is a CTMC, solved in steady state.

[Class $\mathbf{C}_g$:] A single general transition g is enabled with no internal preemption ($\bar{\mathbf{Q}}_{i,i} = 0 \wedge \mathbf{\Delta}_{i,i} = 0$). The component is a CTMC, solved at time δ^g (duration of g).

[Class $\mathbf{C}_M$:] At least one g enabled, or internal preemption ($\bar{\mathbf{Q}}_{i,i} \neq 0 \vee \mathbf{\Delta}_{i,i} \neq 0$). The component is an MRgP, solved in steady state.

The three techniques for MRgP solution: Explicit, matrix-free and Component Method, have been implemented as part of the GreatSPN [11] solver for MRSPN. The first two can be applied to ergodic and non-ergodic systems, while the last one can be applied only to non-ergodic systems.

GreatSPN Tool GreatSPN is a tool for Petri net definition and analysis developed mainly at the University of Torino during the last 30 years. GreatSPN has been recently renovated to include a new Java-based interface with colored and plain token game simulation, model checking for branching and stochastic logics, a new solver for MRSPN, and additional facilities for model composition and for performing multiple experiments. GreatSPN, including the solution techniques used in this paper, is available upon request visiting its web page [27].

Note that the Component Method is a general technique for non-ergodic MRSPN, so we could avoid the definition of X-PPN and ask the users to directly model using MRSPN, but this is an approach that may not be very useful for a designer that wants to use Petri nets to study a phased system, for which a more structured approach in the description of the system can be useful.

4.4 An Example of MPS Modelling and Evaluation with X-PPN

In this section we consider an example of a *Scheduled Maintenance System* (SMS) model, inspired by [12], and represented using the X-PPN formalism. Figure 4.1 shows the model of the SMS drawn with the GreatSPN GUI. It represents an alternation between a factory work phase and a maintenance phase. The PhN and the SN are drawn into separate boxes. The PhN models the alternation between work and maintenance, for NP consecutive cycles. During the production, raw pieces are loaded from a warehouse, are transformed following a sequence of steps with one of the M available machines, and are then moved back in a storage as finished products. Machines may break, and are therefore subject to a continuous maintenance that is scheduled at fixed intervals during the maintenance phase. Each maintenance phase is scheduled after 4 h of work, and requires an inspection cycle: every step of the pipeline is checked, plus and additional work is needed for every machine that has to be repaired. Repairing may require a variable time, estimated between 1 and 2 h, and it is therefore modelled with a Uniform duration. Numerical solution of general distributions is based on the methods of *alpha-factors* [9]. The subnet *StartProd—InProd—EndProd* that models the production activity is expanded into a subnet, shown on the right of Fig. 4.1, where the production is divided into two separate stages followed by an assembly step. Note the presence of numerous bidirectional arcs: in GreatSPN the dependence of a transition to a specific phase is modeled using test arcs, for example a transition connected with a test arc to the *WorkPhase* place will be enabled only during a work phase.

The SMS is tested in four different configurations. In configuration (A) the system stops when a fixed number of phases is concluded, while varying the number of raw pieces K. Configuration (B) is similar, but runs are performed for an increasing number of both pieces and phases. Objective of the analysis of case (A) and (B) is to monitor the number of products completed before the mission ends. Configuration (C) allows for an unbounded number of phases: place *NumPhases* has been removed and the net is modified so that the system stopswhen *max* products are completed. This allows to investigate how many maintenance phases are required before finishing *max* products. Finally, configuration (D) has both unbounded

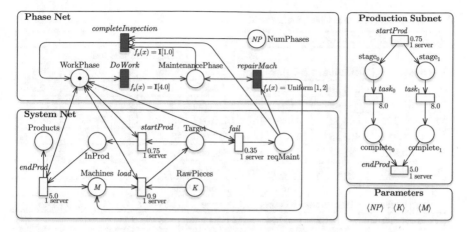

Fig. 4.1 Scheduled Maintenance System MRSPN model drawn in GreatSPN

phases and pieces, and the system stops when *nreps* machine repairs have been performed. This allows to compute the duration of the *repair free operating period* (MFOP) of the system [15, 19] (distribution of the time before the first repair takes place).

Note that, if we assume that all general transitions are deterministic (so no variability in the phases), configurations (A) and (B) produce a PPN, but configurations (C) and (D) require X-PPN, since they have an unbounded number of phases, as the PhN is not a DAG. When solved with the Component Method the (C) and (D) generate components of class C_M, as can be observed from Table 4.1. As we shall see in the next section, C_M component never arise in PPN.

Table 4.1 shows the solver performance of the four configurations of the SMS for varying parameters. Each table reports the test parameters, the number of MRgP states and transitions, and the overall time needed to build the state space (RG time). The table then reports the performance results of the matrix-free solution (matrix-free method) on the MRgP. Data for the explicit case (explicit construction of the EMC) are not reported since they can hardly solve the smallest cases. Finally, the last group of columns evaluates the Component Method, by listing the global number of components, the number of transient components split per class, and the time to compute the steady state distribution of absorbing states. Note that the total number of components contains both the transient and the recurrent components, i.e. it is not just the sum of the C_E, C_g, and C_M transient components.

Case (A) is a non-ergodic MRgP that can be decomposed into a fixed number of C_E and C_g components (half of the phases have an exponential duration). Case (B) is similar, but the number of components grows with the parameters, since they depend on the number of phases NP. In these two cases the Component Method behaves significantly better than the implicit matrix-free one. Note that the aggregation of SCC into components has a significant impact: these systems have hundreds of thousands of SCCs: treating them one at a time would result in

Table 4.1 Matrix-free and Component Method on the SMS model

(A) Finite number of phases $NP = 8$, 3 machines

				Matrix-free		Component Method				
K	States	Transitions	RG time	Iter.	Time	N. comp	C_E	C_g	C_M	Time
10	97,022	215,260	0.971	28	27.667	48	4	44	0	2.978
20	377,419	841,497	4.031	53	211.986	48	4	44	0	12.649
30	739,359	1,644,537	8.088	54	430.079	48	4	44	0	20.673
40	1,109,259	2,463,357	12.079	54	620.940	48	4	44	0	89.832
50	1,479,159	3,282,177	15.706	54	841.496	48	4	44	0	254.051

(B) 3 machines, increasing number of phases NP and pieces K

				Matrix-free		Component Method				
K	NP	States	Transitions	RG time	Iter.	Time	N. comp	C_E	C_g	C_M	Time
10	5	60,344	131,896	0.576	22	11.396	33	4	29	0	1.757
20	10	488,271	1,095,553	5.118	99	482.193	58	4	54	0	27.301
30	15	1,648,076	3,731,570	18.381	185	3031.499	83	4	79	0	252.395
40	20	3,903,594	8,879,596	45.217	186	7360.301	108	4	104	0	1332.887
50	25	7,618,921	17,379,853	94.823	–	–	133	4	129	0	4414.574

(C) Unlimited number of phases, stops after max products, 3 machines

				Matrix-free		Component Method				
max	States	Transitions	RG time	Iter.	Time	N. comp	C_E	C_g	C_M	Time
10	12,225	27,786	0.114	23	1.675	32	1	10	11	0.949
20	55,425	127,026	0.540	51	27.748	62	1	20	21	5.483
30	129,825	298,266	1.258	128	164.952	92	1	30	31	14.763
40	235,425	541,506	2.294	184	468.923	122	1	40	41	29.064
50	372,225	856,746	3.865	184	734.587	152	1	42	43	49.712

(D) Unlimited number of phases, stops after $nreps$ machine repairs

				Matrix-free		Component Method				
nreps	States	Transitions	RG time	Iter.	Time	N. comp	C_E	C_g	C_M	Time
10	14,783	40,822	0.205	38	4.560	31	1	10	10	1.389
20	37,223	103,462	0.550	383	151.630	61	1	20	20	3.620
30	59,663	166,102	0.875	381	287.686	91	1	30	30	5.806
40	82,103	228,742	1.229	488	537.222	121	1	40	40	8.120
50	104,543	291,382	1.551	479	638.227	151	1	50	50	10.218

prohibitive matrix management costs. Both (A) and (B) are totally irreversible nets (no loop among states), which is particularly suitable for the Component Method.

Case (C) is an MRgP that has components of the three classes. The number of C_M components, therefore components that require an MRgP solution, is rather high. Nevertheless, breaking the MRgP solution into smaller blocks eases its solution, since iterations need to be done on smaller matrices. Case (D) has a structure made again with C_M components, whose number strictly follows that of the monitored repairs. The method performances are similar to that of case (C).

Overall, the table shows that the Component Method can be applied to PPN and X-PPN models of medium size, in the number of millions of states, and still being capable of computing the solution in an acceptable amount of time. All computations have been performed on a 2.4 GHz Intel Core Duo processor, with 8Gb of memory.

4.5 GreatSPN vs DEEM on PPN

The work in [13] observes that in a PPN, the sequence of deterministic transition firings in the PhN allows to define a decomposed solution approach. Entering into a marking in the marking graph of the PhN identifies a regeneration point in the MRgP process. Together with the fact that the marking graph is required to be a DAG allows a decomposition of the solution, as explained in the following. This will allow a comparison with the Component Method over PPN. The first step consists in generating the state space of the PhN. By definition each phase i is characterized by a single marking m_i of the PhN marking graph and by a deterministic transition of delay δ_i. Therefore it is possible to solve the whole system as follows (at the beginning $i = 0$):

- Compute the state space $\mathscr{S}_i$ of the SN when the PhN is in state m_i and build the associated rate matrix $\mathbf{Q}_i$.
- Compute the transient solution at time δ_i of the CTMC specified by $\mathbf{Q}_i$ (which gives the distribution of the markings of the SN at the time phase i ends).
- Compute the $\mathbf{\Delta}_{i,j}$ matrices, the branching probability matrix of size $|\mathscr{S}_i| \times |\mathscr{S}_j|$ of going to the states of phase j at the end of phase i. This requires to build the state spaces of the SN for two successive phases.
- The solution at time δ_i of $\mathscr{S}_i$ is mapped, through $\mathbf{\Delta}_{ij}$ to the initial probability of the successive phases.

The above sequence is repeated for successive phases until the target time t of the analysis has been reached. If t is greater than the sum of the duration of all the phases (complete duration of the mission), this corresponds to the distribution of the absorbing states of the system at the end of the mission. If t is smaller, let's say between the end of phase $i - 1$ and i then $\mathbf{Q}_i$ is solved at time $t - t'$, where t' is the sum of all the duration of the previous phases. The above sequence is also reported in [20] as a special case of the very large class of MPS for which the equations of the associated stochastic process are provided.

The theory summarized above has been implemented inside the prototype tool DEEM [13]. Although not recent, DEEM can still be considered the state-of-the-art tool for MPS definition and solution for PPN. DEEM is not only a solver but also has a GUI for the definition of the PhN and of the SN, including facilities for the definition of marking dependent rates, transitions' guards and reward structures.

For what concerns the time complexity, if there are N phases DEEM generates and computes the transient solution of N CTMCs. The space complexity is determined by the need of building N state spaces for the SN (one per phase) and the corresponding N CTMCs, and $N - 1$ state spaces for pairs of successive phases (of size $|\mathscr{S}_i| + |\mathscr{S}_j|$) to build the $\mathbf{\Delta}_{i,j}$ matrices, assuming the phases being in a sequence. If the marking graph is a DAG obviously more than one successive phase should be considered. Since PPNs are a subclass of X-PPNs, we can compare the Component Method of GreatSPN with the solution approach of DEEM outlined above how the Component Method works, when applied to a PPN. If we consider the embedded Markov chain transition matrix $\mathbf{P}$ of a PPN in RNF form, what are the block matrices involved? Since the state space of the PhN forms a DAG, we can certainly find a RNF form for $\mathbf{P}$, in which the states of phase i corresponds to the subset i in the matrix of Eq. (4.1), as it was shown in [20] that the end of a phase corresponds to a regeneration point. For ease of notation each off-diagonal matrix $\mathbf{F}_i$ is split into multiple $\mathbf{F}_{i,j}$ submatrices, with $i < j \le m$. We can then observe that:

$$\mathbf{T}_i = \mathbf{0}, \qquad \mathbf{F}_{i,j} = e^{\mathbf{Q}_i \delta_i} \cdot \mathbf{\Delta}_{i,j}, \qquad \mathbf{R}_i = \mathbf{I}_i - \mathrm{diag}^{-1}(\mathbf{Q}_i) \cdot \mathbf{Q}_i \qquad (4.6)$$

each $\mathbf{T}_i$ will be zero, since $\mathbf{T}_i$ is the submatrix of the probability of remaining in $\mathscr{S}_i$ at the next regeneration point (which is the end of the phase for a PPN), but this is not allowed since the PhN is acyclic. The $\mathbf{F}_{i,j}$ matrices describe the probability of moving from the start of phase i to the start of phase j, and this involves the solution at time δ_i (term $e^{\mathbf{Q}_i \delta_i}$) and the jump probability between the two phases (term $\mathbf{\Delta}_{i,j}$). The $\mathbf{R}_i$ terms corresponds to the recurrent terminal phases (phases with no successors).

The comparison is organized as a set of questions. Given a PPN:

(*Q1*) do the two tools consider the same components?
(*Q2*) do they solve the components with techniques of comparable complexity?
(*Q3*) do they use comparable amounts of memory?
(*Q4*) do they compute the same performance indices?

To answer question *Q1* we should consider that DEEM generates one component per marking of the phase net, and that each marking enables exactly one deterministic transition g. In the Component Method the criteria behind the definition of an optimal partition of the MRP [6, 8] is that, starting from the initial definition of components as SCCs, two components are aggregated together only if they belong to the same complexity class and if the resulting complexity class is the same, given that no cycles among components are introduced by the aggregation. The component construction will indeed put together all states that enable g, whether they correspond to one or more SCCs. This happens because those SCCs are of class $\mathbf{C}_g$ and their union is still of class $\mathbf{C}_g$, since no preemption of the deterministic transitions can take place in a PPN and the firing of the deterministic takes the system outside of all the states that enable g (g is not enabled any longer). Therefore, both DEEM and GreatSPN construct the same components.

The answer to question *Q2* (time complexity) is straightforward given that there is a component per phase, and the component is of class $\mathbf{C}_g$. The $\boldsymbol{\mu}_g$ vector of Eq. (4.2) is therefore computed with a transient solution at time δ_g, that is to say at the time the phase ends, as does the solver of DEEM. Therefore, the two tools do the same numerical computations on the same sub-matrices.

The answer to question $Q3$ (memory) shows a difference in the maximum amount of storage used. Both techniques compute the same $\mathbf{Q}_i$ and $\mathbf{\Delta}_{ij}$ matrices, but the Component Method builds all of them beforehand, to be able to compute the components. DEEM instead builds, for each phase i, the CTMC of the SN at phase i and, for each phase j that is a direct successor of i, it builds its state space to derive the $\mathbf{\Delta}_{ij}$ matrix. So the amount of memory for the Component Method is the size of the whole system, while for DEEM it is the maximum size of the SN for a phase and for its adjacent phases. Difference is at most linear in the number of phases.

The answer to question $Q3$ is that both methods can compute steady-state probabilities while only DEEM can solve for transient. Indeed the transient probability at time t can be easily computed only if the phase net is composed only of deterministic transitions.

In summary we can say that the Component Method "scales well" in time complexity: when the net is a PPN it reaches the same time complexity as the specialized solver of DEEM. Nevertheless the Component Method uses more memory than DEEM, up to at most a factor N, the number of phases, depending on the structure of the marking graph of the PhN. An empirical comparison of DEEM with GreatSPN is provided in [10], an earlier paper that presents an extension of PPN that does not include general distributions. The numerical results confirm better performances of DEEM for what concerns memory, but the DEEM solution is very slow (possibly because of some non-optimal implementation choice) so that, in practice, GreatSPN can solve much bigger PPN than DEEM.

4.6   Conclusions and Future Work

This paper presents X-PPN, a class of MRSPN that are designed to model and solve multiple-phased systems. The definition and solution of X-PPN is supported by the GreatSPN GUI and by three different solvers that use an explicit, matrix-free, and component-based approach. The efficacy of X-PPN as a modelling language for multiple-phased system has been shown on four variations of a scheduled maintenance system. Performances in time and space of the matrix-free and of the Component Method on the four models have been reported and discussed. Results for the explicit method have not been shown since this method, by building explicitly the EMC, suffers from a severe memory bottleneck.

X-PPN are an extension of PPN, the formalism supported by the tool DEEM. A X-PPN allows for an unbounded number of phases, for the presence also of exponential delays in the PhN, for the use of general distributions and for the possibility that an event in the SN to disable a phase. We have discussed, and shown in the example, that these are relevant modelling features. We should remark that certain features of DEEM not available in GreatSPN also have a practical relevance, in particular the rich language for transition guards (to avoid too many crossing arcs for testing and inhibition), the reward structure and the computation of the probability at time t. This last feature is strictly related to the choice of including only deterministic timings in the PhN, as the general case is significantly more complicated. The ideal X-PPN tool should include features of both tools, and in particular it would be very useful to have a tool in which the user can choose his/her own trade-off among modelling power and computable performance indices. It is nevertheless already remarkable that the Component Method is able to scale in complexity to that of the DEEM solvers when the analysed system is expressed by a PPN, and that this simplification is automatic, not requiring any user intervention or net analysis.

While PPN and X-PPN are interesting formalisms for MPS, they both suffer from the limitation that at most one non-exponential transition can be enabled in a state. This restriction is not present in [20] that presents the theoretical framework to partially lift this limitation. This is an interesting and challenging way to pursue: to turn the equations in [20] into a real and scalable solver.

Finally, we have seen that DEEM saves memory by building the state space a few phases at a time: current (unpublished) work on on-the fly generation of MRgP components in the context of the MC4CSLTA model-checker [4], which is also part of GreatSPN, can be a line to follow for the on-the-fly generation of the X-PPN components.

References

1. M. Ajmone Marsan, G. Chiola, On Petri nets with deterministic and exponentially distributed firing times, in *Advances in Petri Nets*. Lecture Notes in Computer Science, vol. 266 (Springer, Berlin, 1987), pp. 32–145
2. M. Ajmone Marsan, G. Conte, G. Balbo, A class of generalized stochastic Petri nets for the performance evaluation of multiprocessor systems. ACM Trans. Comput. Syst. **2**, 93–122 (1984)
3. E.G. Amparore, S. Donatelli, A component-based solution method for non-ergodic Markov Regenerative Processes, in *Computer Performance Engineering*. Lecture Notes in Computer Science, vol. 6342 (Springer, Berlin, 2010), pp. 236–251
4. E.G. Amparore, S. Donatelli, MC4CSL^{TA}: an efficient model checking tool for CSL^{TA}, in *International Conference on Quantitative Evaluation of Systems* (IEEE Computer Society, Los Alamitos, 2010), pp. 153–154
5. E.G. Amparore, S. Donatelli, Revisiting the matrix-free solution of Markov regenerative processes, in *Numerical Linear Algebra with Applications*. Special Issue on Numerical Solutions of Markov Chains, vol. 18 (Wiley, New York, 2011), pp. 1067–1083
6. E.G. Amparore, S. Donatelli, A component-based solution for reducible Markov regenerative processes. Perform. Eval. **70**(6), 400–422 (2013)
7. E.G. Amparore, S. Donatelli, Efficient solution of extended Multiple-Phased Systems, in *10th Valuetools Conference* (EAI, 2016), pp. 125–132
8. E.G. Amparore, S. Donatelli, Optimal aggregation of components for the solution of Markov regenerative processes, in *Quantitative Evaluation of Systems: 13th International Conference, QEST 2016, Quebec City, August 23–25, Proceedings* (Springer International Publishing, Cham, 2016), pp. 19–34
9. E.G. Amparore, S. Donatelli, alphaFactory: a tool for generating the alpha factors of general distributions, in *Proceedings of Quantitative Evaluation of Systems (QEST) 2017* (Springer, Berlin, 2017), pp. 36–51

10. E. Amparore, S. Donatelli, Efficient solution of extended multiple-phased systems, in *Proceedings of 10th VALUETOOLS Conference* (ICST, Brussels, 2017), pp. 125–132
11. E.G. Amparore, G. Balbo, M. Beccuti, S. Donatelli, G. Franceschinis, 30 Years of GreatSPN, in *Principles of Performance and Reliability Modeling and Evaluation: Essays in Honor of Kishor Trivedi* (Springer, Cham, 2016), pp. 227–254
12. A. Bondavalli, I. Mura, K.S. Trivedi, *Dependability Modelling and Sensitivity Analysis of Scheduled Maintenance Systems* (Springer, Berlin, 1999), pp. 7–23
13. A. Bondavalli, S. Chiaradonna, F. Di Giandomenico, I. Mura, Dependability modeling and evaluation of multiple-phased systems using DEEM. IEEE Trans. Reliab. **53**(4), 509–522 (2004)
14. C.G. Cassandras, S. Lafortune, *Introduction to Discrete Event Systems* (Springer, New York, 2006)
15. S. Chew, S. Dunnett, J. Andrews, Phased mission modelling of systems with maintenance-free operating periods using simulated Petri nets. Reliab. Eng. Syst. Saf. **93**(7), 980–994 (2008)
16. H. Choi, V. Mainkar, K.S. Trivedi, Sensitivity analysis of deterministic and stochastic petri nets, in *International Workshop on Modeling, Analysis, and Simulation on Computer and Telecommunication Systems (MASCOTS)*, San Diego (1993), pp. 271–276
17. H. Choi, V.G. Kulkarni, K.S. Trivedi, Markov regenerative stochastic Petri nets. Perform. Eval. **20**(1–3), 337–357 (1994)
18. R. German, *Performance Analysis of Communication Systems with Non-Markovian Stochastic Petri Nets* (Wiley, New York, 2000)
19. C. Hockley, Design for success. J. Aerosp. Eng. **212**(6), 371–378 (1998)
20. I. Mura, A. Bondavalli, Markov regenerative stochastic petri nets to model and evaluate the dependability of Phased Mission Systems dependability. IEEE Trans. Comput. **50**(12), 1337–1351 (2001)
21. I. Mura, A. Bondavalli, X. Zang, K.S. Trivedi, Dependability modeling and evaluation of Phased Mission Systems: a DSPN approach, in *International Conference on Dependable Computing for Critical Applications* (IEEE Press, New York, 1999), pp. 299–318
22. Z. Peng, Y. Lu, A. Miller, Uncertainty analysis of phased mission systems with probabilistic timed automata, in *7th IEEE International Conference on Prognostics and Health Management (PHM'16)* (2016)
23. R. Pyke, *Markov Renewal Processes with Finitely Many States* (Columbia University, New York, 1959)
24. A.K. Somani, J.A. Ritcey, S.H. Au, Computationally-efficient phased-mission reliability analysis for systems with variable configurations. IEEE Trans. Reliab. **41**(4), 504–511 (1992)
25. W.J. Stewart, *Probability, Markov Chains, Queues, and Simulation : The Mathematical Basis of Performance Modeling* (Princeton University Press, Princeton, 2009)
26. K.S. Trivedi, *Probability and Statistics with Reliability, Queuing, and Computer Science Applications* (Wiley, New York, 2002)
27. University of Torino, The GreatSPN tool homepage. http://www.di.unito.it/~greatspn/index.html
28. L. Xing, S. Amari, Reliability of phased-mission systems, in *Handbook of Performability Engineering*, ed. by K. Misra (Springer, London, 2008), pp. 349–368

Chapter 5
Deterministic Network Calculus Analysis of Multicast Flows

Steffen Bondorf and Fabien Geyer

5.1 Introduction

Distributed embedded electronic applications communicating via packet networks have become the norm in various industries such as automotive, avionic, or automation. In such industrial applications, real-time constraints on packet delay and jitter are usually required in order to ensure the specified processes behavior. Due to certification of systems as well as reliability demands, formal methods are applied to validate these timing constraints. They allow for hard guarantees via upper bounds. While different analytical methods have been proposed in the literature, Deterministic Network Calculus (DNC) established itself as common method to analyze asynchronous communications in packet networks. A concrete example of this is Avionic Full-Duplex Ethernet (AFDX), a communication technology based on Ethernet and already deployed in avionic systems. Network calculus has proven

This chapter is an extended version of [4]: Generalizing Network Calculus Analysis to Derive Performance Guarantees for Multicast Flows, in *Proc. of EAI ValueTools* (2016).

Part of this work has been conducted at the Distributed Computer Systems Lab, TU Kaiserslautern (TUK), D-67663 Kaiserslautern, Germany, with support of the Carl Zeiss Foundation.

Part of this work has been conducted at Airbus Group Innovations.

S. Bondorf (✉)
National University of Singapore (NUS), Singapore, Singapore
e-mail: bondorf@comp.nus.edu.sg; bondorf@cs.uni-kl.de

F. Geyer
Technical University of Munich (TUM), München, Germany
e-mail: fgeyer@net.in.tum.de

to be a key method for the certification of deterministic properties of networks used for fly-by-wire [14].

An important property of those industrial networks is that communications are usually based on the multicast paradigm, where packets being sent by one sender are duplicated by switching elements in the network and received by multiple receivers. Using DNC on such multicast protocols requires some adaptations, since this method is restricted to the analysis of unicast communications. As detailed later, in Sect. 5.3, previous attempts for using DNC to analyze multicast communications only circumvented its current restriction. They do not provide a solution to overcome this limitation and cannot benefit from all DNC capabilities to provide accurate end-to-end guarantees.

We address the open issue of multicast flow analysis with DNC. We contribute an approach generalizing the known unicast feed-forward analysis (unicastFFA)— the DNC multicast feed-forward analysis (mcastFFA). Compared to existing approaches, more accurate bounds are obtained since advanced DNC principles can be applied in order to reduce, for instance, overly pessimistic assumptions on flow multiplexing. We numerically evaluate our proposed methods on two AFDX networks and show that our DNC results are on par with other analytical methods or outperform them.

This chapter is organized as follows: Sect. 5.2 presents background on DNC modeling and unicast analysis. In Sect. 5.3, we present related work on multicast flow performance analysis. Section 5.4 contributes a generalization of DNC unicastFFA for the study of multicast flow guarantees. We evaluate our approach in Sect. 5.5. Section 5.6 concludes the chapter and provides an outlook.

5.2 Deterministic Network Calculus Background

Deterministic Network Calculus models resources as bounding functions and provides (min,+)-algebraic operations to derive performance bounds from these. We provide the basic theory applied in this chapter. For a comprehensive description, we refer the reader to [13] and [19]. Bounding functions cumulatively model arrivals or service in interval time. These belong to the set $\mathscr{F}_0$ of non-negative, wide-sense increasing functions:

$$\mathscr{F}_0 = \{f : \mathbb{R} \to \mathbb{R}^+ \mid f(0) = 0, \forall 0 \leq s < t : f(s) < f(t)\}$$

DNC makes use of the concept of arrival curves, which is a function bounding the maximal arrivals of a flow:

Definition 1 (Arrival Curve) Given a flow with input A, a function $\alpha \in \mathscr{F}_0$ is an arrival curve for A iff

$$A(t) - A(s) \leq \alpha(t - s), \forall t, s, 0 \leq s \leq t$$

Minimum service is bounded in a similar way. It is based on the relation between data input and output.

Definition 2 (Service Curve) If the service provided by a server s for a given input A results in an output A', then s offers a service curve $\beta \in \mathscr{F}_0$ iff

$$A'(t) \geq \inf_{0 \leq s \leq t} \{A(t - s) + \beta(s)\}, \forall t$$

The DNC analysis relies on two basic (min,+)-algebraic operations:

Definition 3 ((min, +) Operations) The $(\min, +)$ convolution and deconvolution of two functions $f, g \in \mathscr{F}_0$ are defined as:

$$\text{Convolution: } (f \otimes g)(t) = \inf_{0 \leq s \leq t} \{f(t - s) + g(s)\}$$

$$\text{Deconvolution: } (f \oslash g)(t) = \sup_{s \geq 0} \{f(t + s) - g(s)\}$$

Using these operations, the above definitions translate to $A \otimes \alpha \geq A$ and $A' \geq A \otimes \beta$. Moreover, these operations are used to derive performance bounds.

Theorem 1 (Performance Bounds [19]) *Consider a flow f with arrival curve α traversing a server s with a service curve β. The following bounds can be derived:*

$$\text{Backlog: } Q(t) \leq \sup_{u \geq 0} \{\alpha(u) - \beta(u)\} = (\alpha \oslash \beta)(0)$$

$$\text{Delay: } D(t) \leq \inf\{d \geq 0 \mid (\alpha \oslash \beta)(-d) \leq 0\}$$

$$\text{Output: } \alpha'(d) = (\alpha \oslash \beta)(d)$$

with α' being an output arrival curve for A'.

In advanced network analysis, two further operations are relevant:

Theorem 2 (Concatenation of Servers) *Consider a single flow f crossing a tandem of servers $s_1, \ldots, s_n$ where each server s_i offers a service curve β_i. The overall service curve for f is their concatenation by convolution:*

$$\beta_i \otimes \cdots \otimes \beta_n = \bigotimes_{i=1}^{n} \beta_i$$

Given a strict service curve that guarantees a minimum output of β if data is present at a server, we lower bound the service left-over for a specific flow:

Theorem 3 (Left-Over Service Curve) *Consider a server s that offers a strict service curve β. Let s be crossed by flows f_0 and f_1, with arrival curves α_0, respectively α_1. Then the worst-case residual resource share under arbitrary multiplexing of f_1 at s is:*

$$\beta^{\text{l.o.} f_1} = \beta \ominus \alpha_0$$

with $(\beta \ominus \alpha)(d) = \sup\{(\beta - \alpha)(u) \mid 0 \le u \le d\}$ denoting the non-decreasing upper closure of $(\beta - \alpha)(d)$.

5.2.1 Network Analysis

Using the definitions and theorems presented above, the end-to-end performances of flows interacting on a network of servers can be computed. We call the analyzed flow *flow of interest*, abbreviated foi.

5.2.1.1 Tandems of Servers

The foi's path defines the sequence (tandem) of servers that defines its end-to-end delay. The literature proposes different methods to bound this delay.

Total Flow Analysis (TFA) [19]
The TFA first computes per-server delay bounds. Each one holds for the sum of all the traffic arriving to a server, i.e., these bounds are independent of the foi. The flow's end-to-end delay bound is derived by summing up the individual server delay bounds on its path. The TFA's server-isolating approach constitutes a direct application of Theorem 1; it is known to be inferior to the following analyses [19, 23].

Separated Flow Analysis (SFA) [19]
The SFA is a direct application of other theorems: first compute the left-over service of each server on the foi's path using Theorem 3, then concatenate them using Theorem 2 and finally derive the end-to-end delay bound using Theorem 1. Deriving the end-to-end delay bound using only one service curve will consider the burst term of the foi only once, a property called Pay Burst Only Once (PBOO).

Pay Multiplexing Only Once (PMOO) [23]
The PMOO analysis first convolves the tandem of servers before subtracting the cross-traffic. Using this order, the bursts of the cross-traffic appear only a single time compared to the SFA analysis where the bursts are included at each server. Therefore, multiplexing with cross-traffic is only paid for once. However, [22] showed that the PMOO method does not necessarily outperform the SFA.

5.2.1.2   Feed-Forward Networks

For more complex feed-forward networks, a procedure to combine tandem analyses to a network analysis exists, the unicastFFA. In order to integrate the analysis of multicast flows into DNC, we outline here the structured steps taken by any DNC feed-forward analysis. This structure also serves us to judge and compare different approaches that aim for accurate performance bounds on multicast flows. In previous work, two basic steps of the analysis have already been identified [6]:

unicastFFA Step 1: Cross-Traffic Arrival Bounding
The first unicastFFA step abstracts from the feed-forward network to the foi's path—a tandem of servers that can be analyzed with one of the existing procedures. In detail, this step proceeds as follows:

  (i) Starting at the locations of interference with the foi, cross-flows are back-tracked to their sources. This procedure derives the dependencies between the foi, its cross-flows, their cross-flows, etc., in a recursive fashion. A new instance of this sub-step is started for any cross-flow of the current cross-flow under consideration. Due to the network's feed-forward property, the recursion is guaranteed to terminate.
 (ii) Next, the dependencies are converted into equations, i.e., a sequence of algebraic operations for each location of interference with the foi. They capture the worst-case transformation of flow arrivals towards foi.
(iii) Finally, the equations are solved to obtain the bounds on cross-traffic arrivals.

After these substeps, all cross-flows' arrivals are bounded with arrival curves (arrival bounds).

unicastFFA Step 2: foi Performance Bounding
Given the cross-traffic arrival bounds from step 1, step 2 does not need to consider the part of the network traversed by these flows nor the potentially complex interference patterns they are subject to. The foi's end-to-end delay bound in the feed-forward network is derived with a tandem analysis.

Note that this step provides information required in the previous one. It defines the flow of interest and thus its cross-flows as well as their locations of interference used in step 1(i). This step is strongly based on the tandem analysis that, in turn, is derived with the goal to analyze a unicast flow from end to end. It is not directly applicable to the analysis of multicast flows and thus needs generalization.

5.2.2   Multicast Flows

As mentioned above, flow and network analysis in network calculus have been mostly focused on the modeling of unidirectional and unicast communications. Such a model is not directly applicable to multicast network protocols, where packets are duplicated at certain points of the network in order to provide one-to-

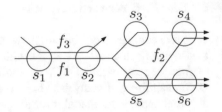

Fig. 5.1 Running example

many communications as illustrated in Fig. 5.1. We define the following terms for describing parts of a multicast flow:

Definition 4 (Trajectory and Fork) A trajectory of a given source-sink pair corresponds to the equivalent unicast flow going from the source to the sink. A fork corresponds to a server where packets are duplicated.

In the following, we will analyze the network of Fig. 5.1 with the given approach. We focus here on the analysis of f_2, which covers all effects relevant to DNC and multicast flows: There is one multicast flow in each step of the unicastFFA, cross-traffic arrival bounding (f_1) as well as flow of interest analysis (f_2). Moreover, a unicast flow is present and this network allows us to observe direct application of the different DNC methods described in Sect. 5.2.1.1, namely TFA, SFA (PBOO effect), and PMOO.

5.3   Related Work

We present three DNC approaches to analyze multicast flows. We focus on how these approaches enable the unicastFFA of the previous section to analyze networks with multicast flows. This work reveals that neither of these approaches constitutes a multicast feed-forward analysis.

5.3.1   unicastFFA Transformation: A Set of Unicast Flows

A first approach to circumvent the issues arises from multicast flows. Each trajectory will become one independent unicast flow, as illustrated in our sample network (Fig. 5.2a) and mentioned in [7].

From a procedural point of view, the unicast transformation does not integrate into the unicastFFA. It only enables for using it by a preceding step that transforms the network. This step is static, i.e., it does not consider the unicastFFA's information like the flow(s) that are under analysis.

The foremost problem of this approach is its overly pessimistic assumption about resource demand of multicast flows. On common sub-paths of a multicast

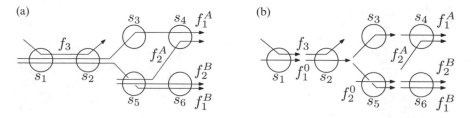

Fig. 5.2 Existing DNC approaches to the multicast analysis applied to the network presented in Fig. 5.1. (**a**) unicastFFA transformation. (**b**) Multicast TFA

flows' trajectories, i.e., the servers before a fork, multiple unicast flows compete for resources. The unicastFFA thus models the worst case with mutual interference between these flows that are not present in the original network model. On the other hand, this approach allows for the PBOO and the PMOO principle in the unicastFFA.

5.3.2 Multicast TFA

Grieu [15] proposes a procedure to apply the TFA presented in Sect. 5.2.1.1 in the analysis of multicast flows. It is tailored to the TFA and shares its inherent isolation of servers. Thus, it does not integrate into the unicastFFA for deriving delay bounds. Figure 5.2b depicts this procedure on the running example network. Flows are cut between all servers, the arrivals are aggregated and a server-local delay bound is computed. In a second step, the server delay bounds on the trajectory of interest are summed up. As this last step is similar to the unicastFFA step 1, it inherits its decisive TFA shortcomings. I.e., neither the PMOO nor the PBOO principle is implemented and the delay bounds are known to be inaccurate.

5.3.3 Explicit Intermediate Bounds (EIB)

An extension of multicast TFA is presented in [4]. The authors propose a different step preceding the unicastFFA analysis. Instead of a per-server delay analysis, it analyzes the tandems of servers between a multicast flow's forks. I.e., a multicast flow is transformed into a set of sub-trajectories. These can then be analyzed individually by computing the left-over service curve on this tandem of servers. Thus, the PBOO as well as the PMOO principle can be applied. In a second step, the analyzed flow's output bounds from all sub-trajectories are derived using their left-over service curves. They are explicitly used as arrival curves after the fork locations at the end of sub-trajectories. Therefore, the approach called Explicit

Fig. 5.3 Application of EIB:
multicast flows are cut into
unicast sub-trajectories

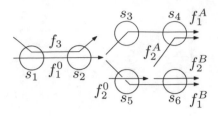

Intermediate Bounds (EIB). Figure 5.3 illustrates the EIB's sub-trajectory approach. Note that the approach cannot implement the PBOO or the PMOO principle on an entire trajectory, even though the foi's left-over service curves will be convolved to attain a valid end-to-end left-over service curve for a trajectory. Moreover note that deriving the left-over service curves required for EIB will itself result in an EIB analysis.

5.3.4 Non-network Calculus Approaches

Current DNC approaches have significant drawbacks such that competing multicast analyses that build on the same modeling as DNC have been proposed.

The *Trajectory Approach* (TA) is an adaptation to the study of network delays of the holistic approach [24]. It was originally developed to give bounds on the scheduling of tasks on a processor. The approach was initially proposed in [21] and later extended to FIFO systems in [20]. Bauer et al. [2] applied TA to the study of avionic networks with multicast flows and showed, via numerical evaluations, that it outperforms the multicast TFA.

The *Forward End-To-End Delay Approach* (FA) has been proposed more recently in [17]. It addresses the shortcomings of the TA. Similarly to the TA, FA is also an adaptation of the holistic approach to the case of FIFO networks. Kemayo et al. [17] applied the FA to the performance evaluation of avionic networks with multicast flows and showed that this approach outperforms the multicast TFA as well.

Although FA sets its focus on the end-to-end analysis—similar to the DNC tandem analyses—neither FA nor TA has been benchmarked against a modern DNC that implements PBOO or PMOO. This can be attributed to the lack of such an analysis for multicast flows. We will provide such benchmarking results in Sect. 5.5.

5.4 A Multicast Feed-Forward Analysis Procedure

In this section, we generalize the unicastFFA presented in Sect. 5.2.1.2 to networks with multicast flows. We call this generalized method *multicast Feed-Forward Analysis*, or mcastFFA. This allows us to make use of the knowledge only available

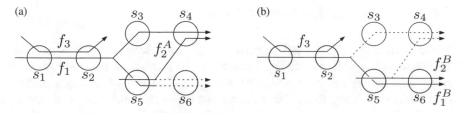

Fig. 5.4 Application of mcastFFA. The dashed lines depict parts of flows that are not considered in the current analysis. (**a**) Running example, trajectory f_2^A. (**b**) Running example, trajectory f_2^B

in the unicastFFA itself. In contrast to the existing DNC approaches and the EIB analysis, no network transformation is amended to the analysis. We do not create a network-wide worst-cast setting for all flows before executing the feed-forward analysis. Instead, our generalization solely constructs a single flow of interest's worst case during analysis—a less pessimistic setting than the static approaches constructing network-wide one for all flows simultaneously. With our approach, the mcastFFA analysis obtains best results by exploiting the PMOO principle end-to-end.

Figure 5.4b illustrates the basic idea behind our solution: If we analyze this multicast flow's trajectory crossing s_5, the other trajectory crossing s_3 becomes irrelevant for the delay bound computation. We neither need to add an entire cross-flow for it nor do we require the output bound from s_1 and s_2. Thus, mcastFFA can treat the multicast trajectories (or unicast flows) of interest in an end-to-end fashion.

The main challenge of this approach is to reduce the network to relevant servers as well as (partial) flows and multicast flow trajectories. This may constitute considerable effort in large networks. Therefore, we present a solution that generalizes the unicastFFA analysis in order to gain from its efficiency [9]. I.e., deriving the sub-network relevant to a specific foi is integral part of the analysis proceedings.

Our mcastFFA solution is mainly based on unicastFFA sub-step 1(i): backtracking of dependencies. Dependencies of flows on others are identified by traversing the network in the opposite direction of links [6]. The entire unicastFFA starts this procedure with the flow of interest. Our mcastFFA will iterate over all n trajectory of interest and execute separate analyses. In case of a unicast flow, we get $n = 1$; for multicast flows n equals the amount of trajectories. Multicast cross-flows are traversed backwards, too, such that their fork locations do not enforce to cut the tandem to analyze; the relevant trajectory of the cross-flow is known and can be treated similar to a unicast cross-flow. The mcastFFA is a generalization of the known unicastFFA. It implicitly restricts the analysis to the trajectory relevant for the analysis. After the backtracking, we know the entire sub-network whose servers and (partial) flows appear in the analysis equation of unicastFFA step 1(ii).

5.4.1   Analysis of the Running Example

We will derive the left-over service curves for f_2's trajectories in order to compare them against the EIB unicastFFA. For brevity, we restrict the depiction to f_2^A's cross-traffic arrival bounding (mcastFFA step 1, Fig. 5.4a) and f_2^B's delay bounding (mcastFFA step 2, Fig. 5.4b). These derivations depict the crucial improvement of mcastFFA's proceedings in both of the analysis steps. They point out the reduction of the network and the increased tandem lengths.

5.4.1.1   mcastFFA Step 1

We consider f_2^A's cross-traffic arrival bounding. Backtracking will be "local" to a single trajectory of a multicast cross-flow. In our example, we finally have established the possibility to apply the PMOO-principle when computing f_1^A's aggregate arrival bound aggrAB at server s_4 [7]. See $\alpha_4^{f_1^A}$ in the following left-over service curve derivation we require to bound cross-traffic arrivals:

$$\beta^{\mathrm{l.o.}f_2^A} = \beta_{\langle 5,4\rangle}^{\mathrm{l.o.}f_2^A} \qquad \text{(only single-hop interference so cutting is fine)}$$

$$= \beta_5^{\mathrm{l.o.}f_2^A} \otimes \beta_4^{\mathrm{l.o.}f_2^A} = \left(\beta_5 \ominus \alpha_5^{f_1^B}\right) \otimes \left(\beta_4 \ominus \alpha_4^{f_1^A}\right)$$

$$= \left(\beta_5 \ominus \left(\alpha^{f_1} \oslash \beta_{\langle 1,2\rangle}^{\mathrm{l.o.}f_1}\right)\right) \otimes \left(\beta_4 \ominus \left(\alpha^{f_1} \oslash \beta_{\langle 1,2,3\rangle}^{\mathrm{l.o.}f_1^A}\right)\right)$$

A cut of $\beta_{\langle 1,2,3\rangle}^{\mathrm{l.o.}f_1^A}$ into $\beta_{\langle 1,2\rangle}^{\mathrm{l.o.}f_1} \otimes \beta_3^{\mathrm{l.o.}f_1^A}$ was needed in the EIB analysis, meaning that PMOO could not be implemented.

This advantage is also depicted in Fig. 5.4a where f_1 retains its multicast shape in the mcastFFA's point of view.

5.4.1.2   mcastFFA Step 2

For the second trajectory of f_2, f_2^B, our mcastFFA derives $\beta^{\mathrm{l.o.}f_2^B} = \beta_{\langle 5,6\rangle}^{\mathrm{l.o.}f_2^B}$. Again, we are not enforced to cut this trajectory's path (see Fig. 5.4b) and in contrast to EIB we can apply alternative tandem analyses:

PBOO: $\qquad \beta^{\mathrm{l.o.}f_2^B} = \beta_{\langle 5,6\rangle}^{\mathrm{l.o.}f_2^B} \qquad \text{(cut enforced by SFA, no single-tandem analysis)}$

$$= \beta_5^{\mathrm{l.o.}f_2^B} \otimes \beta_6^{\mathrm{l.o.}f_2^B} = \left(\beta_5 \ominus \alpha_5^{f_1^B}\right) \otimes \left(\beta_6 \ominus \alpha_6^{f_1^B}\right)$$

$$= \left(\beta_5 \ominus \left(\alpha^{f_1} \oslash \beta_{\langle 1,2\rangle}^{\mathrm{l.o.}f_1}\right)\right) \otimes \left(\beta_6 \ominus \left(\alpha^{f_1} \oslash \beta_{\langle 1,2,5\rangle}^{\mathrm{l.o.}f_1^B}\right)\right)$$

Note that the actual trajectory of the cross-flow, f_1 or f_1^B, was automatically chosen correctly by the backtracking. Moreover, note the contrast to EIB: We can derive f_1^B's arrivals at s_6 with an end-to-end left-over service curve that, in turn, can make use of aggrAB.

PMOO: $\beta^{\mathrm{l.o.}f_2^B} = \beta_{\langle 5,6 \rangle}^{\mathrm{l.o.}f_2^B}$ (there is no enforced cut)

$$= (\beta_5 \otimes \beta_6) \ominus \alpha_5^{f_1} = (\beta_5 \otimes \beta_6) \ominus \left(\alpha^{f_1} \oslash \beta_{\langle 1,2 \rangle}^{\mathrm{l.o.}f_1} \right)$$

where $\beta_{\langle 1,2 \rangle}^{\mathrm{l.o.}f_1}$ can be computed either by applying the left-over service curve derivation of SFA/PBOO or PMOO. This derivation is illustrated in Fig. 5.4b.

5.4.2   Theoretical Evaluation

We conclude this section by a theoretical evaluation of mcastFFA against the related DNC approaches:

- *Relation to unicastFFA (Sect. 5.2.1.2)*: The mcastFFA is a generalization of the unicastFFA. Analysis of unicast flows in either of the two steps remains unaffected (see f_3 in the running example).
- *Relation to unicastFFA transformation (Sect. 5.3)*: Like the unicastFFA transformation, the mcastFFA is able to derive a PMOO end-to-end left-over service curve. However, it does so without the additional cross-traffic assumptions introduced by the unicastFFA transformation. I.e., there are less cross-flows to consider in the analysis, left-over service curves will be larger and delay bounds will be smaller. Thus, mcastFFA outperforms unicastFFA transformation.
- *Relation to EIB unicastFFA*: In comparison to EIB, we gained the ability to operate on end-to-end tandems. This constitutes increased flexibility to cut this tandem during the analysis: Our mcastFFA is compatible with SFA/PBOO, PMOO, aggrAB, or [9] for best attainable left-over service curves. This best solution to cut a tandem and combine sub-tandem results might coincide with EIB's enforced alternative, i.e., mcastFFA is indeed a generalization of EIB unicastFFA.

Before evaluating our contributions, let us briefly clarify their impact on the server backlog bound Q presented in Theorem 1. Deriving these bounds requires the arrival bounds of all flows at a server. I.e., in the DNC analysis procedures, (EIB) unicastFFA and mcastFFA, step 1 is crucial for the result accuracy; step 2 is not required. As shown with the running example, we improved the cross-traffic arrival bounding in case there are multicast flows present. Thus, backlog bounds are also improved by our contribution.

5.5 Numerical Evaluation

In our numerical evaluation, we investigate achieved gains in terms of accuracy of end-to-end delay bounds. To that end, we provide two different comparisons. First, we benchmark our multicast feed-forward analysis (mcastFFA) against the related approaches presented in Sect. 5.3. For our second set of results, a larger network evaluation, we implemented EIB and mcastFFA in the DiscoDNC tool [5].

5.5.1 Comparison to (Non-)Network Calculus Approaches

We study the AFDX network presented in [17]. This allows us to benchmark our proposed approach against the TA and FA since their numerical results are given in the literature. We note that we benchmark against the numerical results of TA and FA without the grouping properties extension since established DNC analyses do not yet take this property into account by default. The grouping property accounts for serialization of packetized flows when crossing links. We leave its implementation in the generalized DNC solutions, potentially based on [11, 15], to future work and restrict our comparison to the non-serialized results. Also, we use a fluid model for our evaluations. To achieve the best comparison possible with the related work on TA and FA, we also model store-and-forward behavior. This is achieved by an additional latency at every server that delays packet forwarding by the time required for full reception of a package of maximum size, $\max(pkg\_size)/R$. Using their parameters defining service and arrivals, this enables us to confirm the DNC delay bounds (see footnote "b" in Fig. 5.5b) given in [17].

(a)   (b)

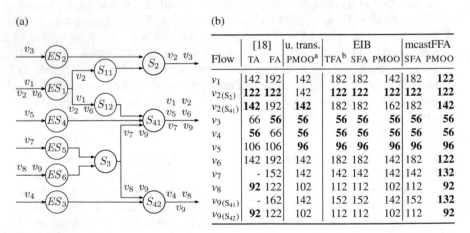

Flow	[18] TA	[18] FA	u. trans. PMOO[a]	EIB TFA[b]	EIB SFA	EIB PMOO	mcastFFA SFA	mcastFFA PMOO
v_1	142	192	142	182	182	142	182	**122**
$v_{2(S_2)}$	**122**	**122**	142	**122**	**122**	**122**	**122**	**122**
$v_{2(S_{41})}$	142	192	**142**	182	182	162	182	**142**
v_3	66	**56**	**56**	**56**	**56**	**56**	**56**	**56**
v_4	**56**	66	**56**	**56**	**56**	**56**	**56**	**56**
v_5	106	106	**96**	**96**	**96**	**96**	**96**	**96**
v_6	142	192	142	182	182	142	182	**122**
v_7	-	152	142	142	142	142	142	**132**
v_8	**92**	122	102	112	112	102	112	**92**
$v_{9(S_{41})}$	-	162	142	152	152	142	152	**132**
$v_{9(S_{42})}$	**92**	122	102	112	112	102	112	**92**

Fig. 5.5 AFDX network evaluation of [17], extended with DNC's EIB and mcastFFA. (**a**) AFDX Network. (**b**) Delay bounds (values given in μs, best in bold). [a]UnicastFFA transformation approach with the stated PMOO end-to-end left-over service curve derivation. [b]Remember, that EIB with TFA corresponds to the multicast TFA analysis presented as related work in Sect. 5.3

Our comparison focuses on mcastFFA with PMOO, TA, and FA delay bounds. We observe (see Fig. 5.5) that mcastFFA at least matches the bounds of the other methods compared here. A maximum gain of 5.86% compared to TA and 18.58% compared to FA is achieved in this small AFDX scenario.

Key observations w.r.t. the performance of DNC analyses confirm our theory:

- mcastFFA with PMOO shows expected gains compared to the multicast TFA (see footnote "b" in Fig. 5.5b).
- EIB with PMOO does not make full use of the PMOO principle end-to-end on trajectories and thus is outperformed by mcastFFA with PMOO in most cases.
- For some trajectories of multicast flows, even TFA results are equal. Then, flow interference is non-existent. These cases do not occur in realistic networks.

5.5.2   An Industry-Scale AFDX Data Network

In order to evaluate our method on a realistic use-case found in industrial applications, we evaluate an AFDX data network. We aim to confirm our hypothesis that the mcastFFA will have a more pronounced advantage over other approaches[2] in larger networks. To that end, we implemented EIB and mcastFFA in the DiscoDNC tool. We also extended the DiscoDNC by an AFDX topology generator following recommendation from [12] and with network parameters according to an Airbus A350 presented in [16]. This also allows us to provide the entire range of DNC analysis configurations pairing EIB or mcastFFA with TFA as well as SFA/PBOO or PMOO $\beta^{l.o.}$ computations. For brevity of presentation, we focus on the most relevant of these combinations, EIB with SFA, EIB with PMOO and mcastFFA with PMOO. All results were computed using aggrAB arrival bounding [7]. Note that this is not a restriction. Using segregated arrival bounding [10] or tandem matching arrival bounding [9] or any combination thereof is possible as well.

The network we generated according to these size parameters resulted in 650 multicast flows with 1112 trajectories in total. In order to compare the gains of mcastFFA PMOO against the other EIB methods, we used the relative difference, namely: $(d^{EIB} - d_{PMOO}^{mcastFFA})/d^{EIB}$. The empirical cumulative distribution (ECDF) over the studied A350-like network is illustrated in Fig. 5.6. Key observations are unaffected: the mcastFFA procedure derives more accurate bounds than EIB. On average, mcastFFA PMOO produces a reduction of 8% of the bound compared to EIB SFA and 6% compared to EIB PMOO. We also observed reduction of up to 25% for some flows. These observations confirm our hypothesis that mcastFFA's potential advantage over other DNC approaches increases with the network size.

We also observe EIB SFA delay bounds that undercut the mcastFFA PMOO (see positive ECDF values for the negative x-axis in Fig. 5.6). The situation stems from

---

[2]Due to a lack of software tools, TA and FA are not included.

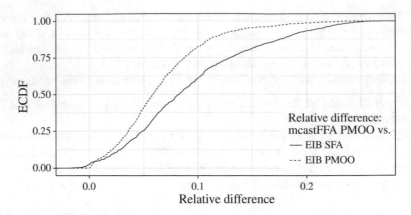

Fig. 5.6 ECDF of the relative difference between mcastFFA and the EIB methods

a well-known phenomenon that allows SFA to theoretically outperform PMOO by an arbitrarily large margin [22]. However, the mcastFFA can be paired with any tandem analysis able to compute output bounds. Doing so with the Tandem Matching Analysis (TMA) proposed in [9] creates a single-best algebraic analysis for arbitrary multiplexing.

5.6 Conclusion and Outlook

In this chapter, we tackled the problem of analyzing multicast flows with deterministic network calculus. DNC was previously tailored to the analysis of unicast flows—a property which was assumed to invariantly hold. Therefore, previous approaches for the DNC analysis of multicast flows tried to adjust to this restriction by, e.g., pessimistic re-modeling of the network. This leads to inaccurate performance bounds and the development of alternative, non-DNC analyses to derive multicast flow guarantees. In contrast, we generalized DNC unicast feed-forward analysis to a multicast one.

In theoretical and numerical evaluations we showed that our contribution results in a single best DNC analysis for multicast flows, the mcastFFA with PMOO. Not only does it outperform any other DNC approach, the evaluation of an AFDX scenario from the literature also shows that DNC achieves at least the results of competitors (Trajectory Approach and Forward Analysis), even outperforming them in a considerable amount of cases.

Existing AFDX networks as deployed in existing Airbus aircraft such as the A380 are larger and more complex than the ones presented in this evaluation [25]. They consist of ~1000 multicast flows (virtual links, VLs) that have an average of ~6.5 trajectories per VL [1]. Therefore, the improvements we achieve with DNC's PMOO in conjunction with mcastFFA is expected to be even larger in practice.

Moreover, the presented mcastFFA has the flexibility to be combined with any DNC tandem analysis and improvement thereof. For instance, [8], [9], FIFO multiplexing service analysis [3] or packetization [11] can tighten guarantees and restriction to finite domains can accelerate the analysis [18].

Acknowledgements The authors would like to thank Bruno Oliveira Cattelan for his work on implementing the explicit intermediate bounds analysis and the multicast feed-forward analysis in the Disco Deterministic Network Calculator.

References

1. H. Bauer, Analyse pire cas de flux hétérogènes dans un réseau embarqué avion. Ph.D. thesis, Université de Toulouse, 2011
2. H. Bauer, J. Scharbarg, C. Fraboul, Applying and optimizing trajectory approach for performance evaluation of AFDX avionics network, in *Proceedings of IEEE ETFA* (2009)
3. L. Bisti, L. Lenzini, E. Mingozzi, G. Stea, Numerical analysis of worst-case end-to-end delay bounds in fifo tandem networks. Springer Real-Time Syst. J. **48**, 527–569 (2012)
4. S. Bondorf, F. Geyer, Generalizing network calculus analysis to derive performance guarantees for multicast flows, in *Proceedings of EAI ValueTools* (2016)
5. S. Bondorf, J.B. Schmitt, The DiscoDNC v2 – a comprehensive tool for deterministic network calculus, in *Proceedings of EAI ValueTools* (2014)
6. S. Bondorf, J.B. Schmitt, Boosting sensor network calculus by thoroughly bounding cross-traffic, in *Proceedings IEEE INFOCOM* (2015)
7. S. Bondorf, J.B. Schmitt, Calculating accurate end-to-end delay bounds – you better know your cross-traffic, in *Proceedings of EAI ValueTools* (2015)
8. S. Bondorf, J.B. Schmitt, Improving cross-traffic bounds in feed-forward networks – there is a job for everyone, in *Proceedings of GI/ITG MMB & DFT* (2016)
9. S. Bondorf, P. Nikolaus, J.B. Schmitt, Quality and cost of deterministic network calculus – design and evaluation of an accurate and fast analysis, in *Proceedings of ACM SIGMETRICS* (2017)
10. A. Bouillard, *Algorithms and Efficiency of Network Calculus*. Habilitation thesis, ENS, 2014
11. M. Boyer, P. Roux, A common framework embedding network calculus and event stream theory, in *Proceedings of IEEE ETFA* (2016)
12. M. Boyer, N. Navet, M. Fumey, Experimental assessment of timing verification techniques for AFDX, in *Proceedings of ERTS* (2012)
13. C.S. Chang, *Performance Guarantees in Communication Networks* (Springer, Berlin, 2000)
14. F. Geyer, G. Carle, Network engineering for real-time networks: comparison of automotive and aeronautic industries approaches. IEEE Commun. Mag. **54**, 106–112 (2016)
15. J. Grieu, Analyse et évaluation de techniques de commutation Ethernet pour l'interconnexion des systèmes avioniques. Ph.D. thesis, Institut National Polytechnique de Toulouse, 2004
16. O. Hotescu, K. Jaffres-Runser, J.L. Scharbarg, C. Fraboul, Towards quality of service provision with avionics full duplex switching, in *Euromicro ECRTS, Work-in-Progress Session* (2017)
17. G. Kemayo, N. Benammar, F. Ridouard, H. Bauer, P. Richard, Improving AFDX End-to-End delays analysis, in *Proceedings of IEEE ETFA* (2015)
18. K. Lampka, S. Bondorf, J.B. Schmitt, N. Guan, W. Yi, Generalized finitary real-time calculus, in *Proceedings IEEE INFOCOM* (2017)
19. J.-Y. Le Boudec, P. Thiran, *Network Calculus: A Theory of Deterministic Queuing Systems for the Internet* (Springer, Berlin, 2001)
20. S. Martin, P. Minet, Schedulability analysis of flows scheduled with FIFO: application to the expedited forwarding class, in *Proceedings of IPDPS* (2006)

21. J. Migge, L'ordonnancement sous contraintes temps-réel un modèle à base de trajectoires. Ph.D. thesis, INRIA Sophia Antipolis, 1999
22. J.B. Schmitt, F.A. Zdarsky, M. Fidler, Delay bounds under arbitrary multiplexing: when network calculus leaves you in the lurch..., in *Proceedings of IEEE INFOCOM* (2008)
23. J.B. Schmitt, F.A. Zdarsky, I. Martinovic, Improving performance bounds in feed-forward networks by paying multiplexing only once, in *Proceedings of GI/ITG MMB* (2008)
24. K. Tindell, J. Clark, Holistic schedulability analysis for distributed hard real-time systems. Microprocess. Microprogramm. **40**, 117–134 (1994)
25. N. Tobeck, Enforcing domain segregation in unified cabin data networks, in *Proceedings of IEEE/AIAA DASC* (2017)

Chapter 6
Modeling Techniques for Pool Depletion Systems

Davide Cerotti, Marco Gribaudo, Riccardo Pinciroli, and Giuseppe Serazzi

6.1 Introduction

In the last years, the development of new programming paradigms was increasingly affected by the evolution of Big Data technologies. Indeed, the enormous amount of data generated by IoT and any other type of input devices requires specific techniques for storage, access, and process. To satisfy the increasing demand of computational capabilities required to process such huge volume of data in a reasonable amount of time, the concepts of parallelism and distributed computation have been extensively adopted. Big Data applications split the data to be analyzed into blocks and generate multiple tasks dedicated to their processing. The tasks are executed in parallel by the resources of a distributed computing environment. At the end of all the executions, newly created tasks concurrently process the intermediate results to produce the final output of the application. This operational structure, proposed originally by Google with Hadoop MapReduce [5, 8], is adopted by many current Big Data applications.

A complete execution of a Big Data application consists of a sequence of parallel computations and their following synchronization. Typically, the computing framework orchestrates the parallel executions of the tasks taking into consideration the configuration of the available architecture. The allocation strategy of the tasks to the distributed systems has a deep influence on the execution time of the applications. Indeed, the load of the systems must be controlled in order to avoid the

D. Cerotti
DiSIT, Università Piemonte Orientale, Alessandria, Italy
e-mail: davide.cerotti@uniupo.it

M. Gribaudo (✉) · R. Pinciroli · G. Serazzi
DEIB, Politecnico di Milano, Milano, Italy
e-mail: marco.gribaudo@polimi.it; riccardo.pinciroli@polimi.it; giuseppe.serazzi@polimi.it

© Springer International Publishing AG, part of Springer Nature 2019   79
A. Puliafito, K. S. Trivedi (eds.), *Systems Modeling: Methodologies and Tools*,
EAI/Springer Innovations in Communication and Computing,
https://doi.org/10.1007/978-3-319-92378-9_6

creation of bottlenecks that severely limit the performance. On the other hand, also the under-load of the resources must be avoided trying to balance their utilization. A policy to allocate the tasks to the distributed systems to minimize the execution time of the entire application was presented in [3].

To this end, we studied *Pool Depletion Systems* (PDS) and in this chapter we describe the techniques to be used for their modeling.

PDS is a framework adopted in [2, 3, 11] to analyze all the systems where a huge amount of tasks, composing a job, must be executed by one or more subsystems with a finite capacity. PDS are studied assuming the tasks may belong to two different classes that are served by subsystem's resources with different service demands. Each PDS is characterized by a *transient behavior* that may be summarized as follows: (1) initially, all the tasks of a job are in the pool and they are admitted gradually into a finite capacity subsystem to be executed; (2) whenever a task is completed, it leaves the subsystem and another task can start being processed; (3) when all the tasks have been executed, the job is completed and a new request can be served. The metric we are interested in is the *depletion time*, i.e., the time required to execute all the tasks. The main goal is to find the allocation policy of all the tasks in the pool that minimizes the depletion time.

Two techniques have been used to model PDS: (1) Markov analysis [2], that provides exact analytic solutions, and (2) discrete events simulation [3, 11]. The former may be affected by the state space explosion when the number of tasks to be considered is large. The latter could take a very long time to complete a simulation run with complex models. In this chapter, we propose a further approach, i.e., the fluid approximation technique, to analyze PDS. It allows the analysis of models with large dimensions (e.g., even million of tasks) in a very short amount of time. Comparisons of time required by the three techniques to solve models of various sizes and the accuracy of their results are provided and analyzed.

This chapter is organized as follows. Sections 6.2 and 6.3 provide a review of the background results and a description of the system architecture, software and hardware, considered. Section 6.4 describes and compares different models to study PDS. Section 6.5 draws conclusions.

6.2   Related Work

In this section we emphasize some results on product-form closed queuing networks [1] with two-class workload and fixed rate single server stations that will be exploited in the analysis of depletion systems developed in the following sections. Two-class workloads are simple enough to analytically deal with, while being representative of several realistic systems' workload. In this case, the workload of models with M stations can be characterized by a matrix of service demands $\mathbf{D} = [D_{rc}]$ where D_{rc} is the time required by station $r \in \{1, \ldots, M\}$ to completely process a class $c \in \{A, B\}$ request. Hereinafter, we will consider a system with two resources and the following service demands:

$$\mathbf{D} = \begin{bmatrix} 0.8 & 0.496 \\ 0.2 & 1.25 \end{bmatrix} \tag{6.1}$$

According to matrix (6.1), class A service demands on stations 1 and 2 are 0.8 and 0.2 time units, respectively, whereas class B demands on the two stations are 0.496 and 1.25 time units, respectively. Note that, since the system considered in this chapter is a separable queuing network, its solution depends only on the product of visits v_r and service time S_r at each resource r (i.e., on the service demand $D_r = v_r \cdot S_r$) and not on the individual factors [9]. For this reason, the topology of the network is arbitrary, thus we assume the two resources to be in series (as shown in Fig. 6.1).

The performance of systems with multi-class workloads depends on the fraction of requests in execution for each class, referred to as *population mix*. Let $\boldsymbol{\beta} = (\beta_A, \beta_B)$ represent the population mix. In the example of matrix $\mathbf{D}$, when $\beta_A = 0$ we have only class B requests and the bottleneck (i.e., the saturated resource) is station 2, whereas with $\beta_A = 1$ (only class A requests) the bottleneck is station 1. Thus, varying the value of $\boldsymbol{\beta}$ a *bottleneck switch* occurs. In these conditions, there exists a value $\boldsymbol{\beta}^*$ such that both stations are equally utilized for any number of requests inside the system; such value is called the *equi-utilization point*. In addition, providing that the service demand matrix values allow the occurrence of a bottleneck switch[1] and when the number of requests is sufficiently large, it is possible to identify an interval of values of the population mix that concurrently saturate the stations. Indeed, for values of $\boldsymbol{\beta}$ belonging to such *common saturation sector* (CSS) the system throughput is maximum and the system response time is minimum.

In [12] it is shown that the equi-utilization point is computed as:

$$\boldsymbol{\beta}^* = \left(\beta_A^* = \frac{\log \dfrac{D_{22}}{D_{12}}}{\log \dfrac{D_{11} D_{22}}{D_{12} D_{21}}}, \beta_B^* = 1 - \beta_A^* \right) \tag{6.2}$$

and it is proved to always belong to the CSS of edges:

$$\begin{cases} \beta_A^L = D_{2A} \dfrac{D_{2B} - D_{1B}}{D_{1A} D_{2B} - D_{2A} D_{1B}} \\[4mm] \beta_A^U = D_{1A} \dfrac{D_{2B} - D_{1B}}{D_{1A} D_{2B} - D_{2A} D_{1B}} \end{cases} \tag{6.3}$$

---

[1] It can be verified by checking if $D_{1A} > D_{2A}$ and $D_{1B} < D_{2B}$.

where β_A^L and β_A^U denote the lower and upper edges, respectively. Moreover, in the same paper the behavior of the per-class throughputs is analytically computed. In particular, we have that for values of β_A inside the CSS the per-class throughputs are:

$$
\begin{aligned}
X_A^{CSS} &= X_A(\mathbf{K}) = \frac{D_{2B} - D_{1B}}{D_{1A} D_{2B} - D_{2A} D_{1B}} \\
X_B^{CSS} &= X_B(\mathbf{K}) = \frac{D_{2A} - D_{1A}}{D_{1B} D_{2A} - D_{1A} D_{2B}}
\end{aligned}
\tag{6.4}
$$

where $\mathbf{K} = (K_A, K_B)$ denotes the number of requests of each class. For values of β_A at the right of the saturation sector (i.e., with $\beta_A > \beta_A^U$) we have:

$$
X_A(\mathbf{K}) = \frac{\beta_A}{D_{1A}} \quad X_B(\mathbf{K}) = \frac{1 - \beta_A}{D_{1B}}
\tag{6.5}
$$

Similar equations can be derived for values of β_A at the left of the saturation sector, thus obtaining the overall trend of the per-class throughputs as function of the population mix.

In [12], Rosti et al. also introduced the equi-load point, referred to as α^*, that is derived as:

$$
\alpha^* = \left(\alpha_A^* = \frac{D_{2B} D_{1B}}{D_{1A} + D_{2B} - D_{1B} - D_{2A}}, \alpha_B^* = 1 - \alpha_A^* \right)
\tag{6.6}
$$

Although two equally loaded stations may be expected to be also equally utilized, it is not the case since $\alpha^* \neq \beta^*$.

6.3 Scenario

We consider a system where each job consists of several independent tasks. Examples of such type of workloads are video transcoding/analysis, applications of business analytics, NoSQL queries and so on [4, 6]. In particular, we focus on jobs composed by two types of tasks, defined as class A and class B. For example, in multimedia stream applications each chunk is processed by a single task and the two classes may represent the computation of audio and video chunks, respectively.

We assume that the tasks of a job are executed sequentially by two resources, denoted as Res_1 and Res_2, e.g., the CPU and the storage of a server. The global time required by resource r for a complete execution of a class c task is referred to as *service demand* D_{rc}. The service demands characterize the workload in terms of total processing requirements to the resources, and their values are considered exponentially distributed. Each resource executes the concurrent tasks according to

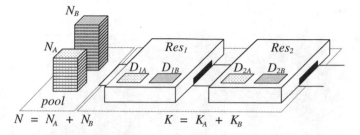

Fig. 6.1 The considered scenario

a *processor sharing* queuing discipline: all tasks are processed by the resources with a service rate proportional to the current number of tasks in service.

We call N_A the number of class A tasks, N_B the number of class B and $N = N_A + N_B$ the total number of tasks of a job. The system can execute at most K tasks concurrently. This limitation may be used to model, for instance, constraints on memory occupancy that prevents all the N tasks to be executed in parallel. In particular, the system is allowed to concurrently execute K_A tasks of class A and K_B of class B, with $K = K_A + K_B$. As soon as one task is completed, another task of the same class is admitted in execution, if available. When all the tasks of one class are completed, the system allows the processing of the remaining tasks of the other class.

Figure 6.1 gives a visual representation of the considered scenario. We are interested in studying the total time T required to complete the execution of all the N tasks of a job that are initially into the pool. With the type of applications considered in this study we have $N_A \geq K_A$ and $N_B \geq K_B$.

Figure 6.2 shows the temporal evolution of the number of tasks in the system. K out of N tasks immediately starts being serviced by the first resource of the system. As long as there are tasks of both classes in the pool waiting to be executed, the number of tasks in the system is constantly kept to $K = K_A + K_B$. We address this phase as Φ_I: after a short initial transient period, $T_{\tau 0}$, required to load K tasks into the first resource, the system behaves as a closed queuing model with K_A and K_B customers, since as soon as one of the task of a class leaves the system, it is replaced by another one of the same class. At time $T_{\Phi 1}$ the tasks in the pool of one class are finished and no new tasks of that class may enter the system (in Fig. 6.2 this happens for class A tasks). At this time the system starts replacing the tasks of the exhausted class with the ones of the other class: phase Φ_{II} begins. After an initial transient (that lasts until time $T_{\tau 12}$) in which all the remaining tasks of the exhausted class are executed, the system behaves as a closed queuing model with K customers (in Fig. 6.2 these customers are of class B). After some time, the tasks in the pool that still need to be executed are exhausted, and the system begins to execute a decreasing number of tasks. We denote this instant of time by $T_{\Phi 2}$, and the period of time in which the server is working with less than K tasks as phase

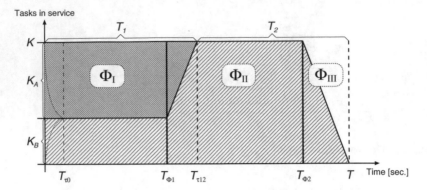

Fig. 6.2 Behavior of the number of tasks in execution with phases characterization

Φ_{III}. The job completes its execution when all its tasks are terminated at time T, referred to as *depletion time*.

6.4 Models Analysis

PDS cannot be analyzed through closed-form formulas when the asymptotic assumptions presented in [3] (i.e., $N \gg K$ and K is large enough to saturate the system) are not satisfied. In this section, PDS are analyzed using analytic, discrete event simulation and fluid models. In particular, the three models are described and compared to each other in order to determine those that may provide the more accurate results in the shorter time.

6.4.1 Markov Analysis

A continuous-time Markov chain (CTMC) model was proposed in [2] to analytically study a PDS. Such model provides the exact results since it analytically describes each phase of a PDS. In the CTMC the state of the system is described by a tuple counting, for each class, the current number of tasks in the pool, in station Res_1 and in station Res_2. The completions of tasks change the system state and are represented in the CTMC as transitions from a value of the tuple to another. Assuming that all such events follow an exponential distribution, the whole state space of the system is built and the resulting CTMC can be analyzed by standard numerical techniques. In particular, the depletion time and the duration of each phase, including the transient ones, are analytically computed through absorption time analysis. Further details can be found in [2] and [10].

Unfortunately, analytic analysis of CTMCs suffers from the well-known state space explosion problem: large values of K and N make the resulting state space grow exponentially, thus making the solution computationally unfeasible. For this reason, also simulation and fluid models are considered for PDS analysis.

6.4.2   Discrete Event Simulation

The single-subsystem PDS may be analyzed recurring to the multi-formalism model in Fig. 6.3.

Such a model consists of a Colored Petri Net (CPN) and a multi-class fork-and-join queuing network. In fact, the adoption of several formalisms allows to use the most appropriate primitives while modeling a system [7]. The workload of the model is composed by tokens and jobs: the former are used by CPN to model the subsystem finite capacity, the latter by queuing network to represent the tasks of a job and their execution. Note that, since a two-class PDS is considered, there are two classes both for tokens and for jobs.

Initially, a single job is in the system starting in the delay station Jobs with zero service time: the purpose of this node is to provide the reference station[2] for computing throughput and response time, and to restart the system after one execution run. The job is immediately split into N_A class A tasks and N_B class B tasks (i.e., $N = N_A + N_B$) by the fork node named Fork. Thus, a task that is waiting to be executed is represented as a token in place Wait: a color class $< C >$ is assigned to each token to identify the class of that task. The capacity $K = K_A + K_B$ of the subsystem is represented by the number of color $< C' >$ tokens in place MaxTasks that are initialized to K_A and K_B for the corresponding task classes. The notation used in Fig. 6.3 is summarized in Table 6.1.

The subsystem starts executing a task when transition Enter fires; in fact, this can occur in one of the four different modes shown in Table 6.2. When a color class A (B) task is waiting in place Wait and at least a token of the same color is

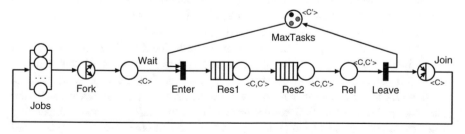

Fig. 6.3 The multi-formalism model of the considered scenario

---

[2]In closed queuing networks throughput and response time can be computed only with respect to a given/specific resource, the so-called reference station [9].

Table 6.1 Color-sets of the discrete event multi-formalism model

Color-set	Description
$< C >$	Task class $C = \{A, B\}$
$< C' >$	Token color $C' = \{A, B\}$
$< C, C' >$	Task class C, token color C'

Table 6.2 Firing modes for transitions `Enter` and `Leave`

Transition	Mode	In$_1$	In$_2$	Out$_1$	Out$_2$	Description
`Enter`		`Wait`	`MaxTasks`	`Res1`		
	1	A	A	(A, A)		Class A task
	2	B	B	(B, B)		Class B task
	3	A	B with `Wait`.$B = 0$	(A, B)		Class A task, depletion
	4	B	A with `Wait`.$A = 0$	(B, A)		Class B task, depletion
`Leave`		`Rel`		`MaxTasks`	`Join`	
	1	(A, A)		A	A	Class A task
	2	(B, B)		B	B	Class B task
	3	(A, B)		B	A	Class A task, depletion
	4	(B, A)		A	B	Class B task, depletion

available in place `MaxTasks`, the system is in phase Φ_I and transition `Enter` fires in mode 1 (2). Instead, if system is operating in phase Φ_{II} or Φ_{III}, one of the two classes of tasks is exhausted and its tokens are used to allow the remaining tasks to be executed. In this case, transition `Enter` fires in mode 3 or 4, according to the color class $< C >$ of tasks still waiting to be executed.

As shown in Table 6.1, the color class of a task is referred to as $< C >$, whereas $< C' >$ is used to represent the color of a token. Instead, notation $< C, C' >$ is used to represent the tasks admitted into the subsystem; indeed, a token of color $< C' >$ is always associated to each of these tasks.

The subsystem is composed by two queuing network primitives, `Res1` and `Res2`. They represent the resources of the subsystem and process the tasks currently into it. The service requirements of each task are determined by its color class $< C >$, while the color $< C' >$ of the token associated to that task is used only to correctly return the token in place `MaxTasks`. When a task has been executed by both the resources, it enters place `Rel` and enables transition `Leave`.

The absence of tokens in place `MaxTasks` means that the subsystem has reached its maximum capacity. In such condition further tasks are not admitted until the completion of at least a task currently in execution. When a task is completed, transition `Leave` fires: a token return to place `MaxTasks` and the task is sent to the join node `Join`. Also transition `Leave` can fire in four different modes depending on the color class of the task that has been completed and the color of its associated token. The four modes are shown in Table 6.2. When all the N tasks generated by `Fork` have been collected by `Join`, the initially job return to the delay station and a new one can start its computation.

Note that the approximations introduced by the multi-formalism model are related to the technique adopted for its simulation.

6.4.3 Fluid Approximation

The fluid approximation is based on the following assumption: except from the transient behavior that occurs whenever the system switches phases, if we focus only on the service components (namely Res1 and Res2), they operate as in a two class closed model with K_A and K_B jobs. This occurs because as soon as a job leaves the server components, a new one of the same class is re-introduced. Let us call $X_A^{K_A, K_B}$ and $X_B^{K_A, K_B}$ the throughput that the two classes would have in a closed model composed by Res1 and Res2, with K_A and K_B jobs. The time $T_{\Phi 1}$ at which the first phase (Φ_I) ends can thus be approximated as:

$$T_{\Phi 1} = \min \left(\frac{N_A - K_A}{X_A^{K_A, K_B}}, \frac{N_B - K_B}{X_B^{K_A, K_B}} \right) \tag{6.7}$$

Without loss of generality, let us suppose that class A ends first: $\frac{N_A - K_A}{X_A^{K_A, K_B}} < \frac{N_B - K_B}{X_B^{K_A, K_B}}$. The number $N_{B\Phi 1}$ of class B jobs that still can enter the system at the end of phase Φ_I, can then be determined as:

$$N_{B\Phi 1} = N_B - K_B - X_B^{K_A, K_B} \cdot T_{\Phi 1} \tag{6.8}$$

For the sake of simplicity, let us suppose that $N_{B\Phi 1}$ is large enough to allow the system to complete the transient part up to time $T_{\tau 12}$ (Fig. 6.2). We approximate the $T_{\tau K_A-1, K_B+1}$ time at which the first of the K_A jobs still in the system ends as:

$$T_{\tau K_A-1, K_B+1} = T_{\Phi 1} + \frac{1}{X_A^{K_A, K_B}} \tag{6.9}$$

and compute the number $N_{B\tau K_A-1, K_B+1}$ of class B jobs that are still in the pool and need to enter the servers as:

$$N_{B\tau K_A-1, K_B+1} = N_{B\Phi 1} - \frac{X_B^{K_A, K_B}}{X_A^{K_A, K_B}} \tag{6.10}$$

At this point the population in the system changes to $K_A' = K_A - 1$ and $K_B' = K_B + 1$ since the class A job is replaced by a class B one. We thus compute $X_A^{K_A-1, K_B+1}$ and $X_B^{K_A-1, K_B+1}$ by solving the corresponding closed model with the new population mix, and compute the time $T_{\tau K_A-2, K_B+2}$ at which the next class A job finishes and the corresponding class B population $N_{B\tau K_A-2, K_B+2}$ that still have to enter the system:

$$T_{\tau K_A-2,K_B+2} = T_{\tau K_A-2} + \frac{1}{X_A^{K_A-1,K_B+1}}$$

$$N_{B\tau K_A-2,K_B+2} = N_{B\tau K_A-1,K_B+1} - \frac{X_B^{K_A-1,K_B+1}}{X_A^{K_A-1,K_B+1}}$$

(6.11)

The process is repeated $K - K_A$ times until time $T_{\tau 12}$ is reached (i.e., all the class A jobs inside the system and in the pool are completed). At that point, the system has only class B jobs. We thus solve the closed model as a single class one (considering only class B) with K jobs, and determine its throughput $X_B^{0,K}$. Let us call $N_{B\tau 12}$ the number of class B jobs still waiting to be executed in the pool (computed with the previous procedure repeating until $N_{B\tau 12} = N_{B\tau 0,K}$). We then approximate the end of the second phase $T_{\Phi 2}$ as:

$$T_{\Phi 2} = T_{\tau 12} + \frac{N_{B\tau 12}}{X_B^{0,K}}$$

(6.12)

Now depletion starts and the number of class B jobs inside the system decreases from K down to zero. Let us call $X_B^{0,k}$ the throughput of the closed model when its population is composed by k class B jobs. The process completion time T is thus approximated as:

$$T = T_{\Phi 2} + \sum_{k=1}^{K} \frac{1}{X_B^{0,k}}$$

(6.13)

The fluid approximation might become slightly more complex when the number of class B jobs waiting in the pool becomes zero while there are still class A jobs in service. In this case we will have again to consider the minimum time between class A jobs and class B jobs, and compute the throughput accordingly.

In practice, depending on the system configuration, all the throughput of a closed model with $k_A + k_B \leq K$ jobs might be required. However, thanks to the properties of the Mean Value Analysis algorithm, all these values can be easily computed with complexity $o(K^2)$, and a minor overhead with respect to the computation of the solution for K_A and K_B alone. Algorithm 1 summarizes the proposed procedure: variables n_A and n_B contain the fluid count of the total population (inside the servers and in the pool) for the two classes, while k_A and k_B the current population inside the server. Lines 3 to 10 consider phases Φ_I and Φ_{II} (see Fig. 6.2), while lines 11 to 19 deal with phase Φ_{III}, since after line 10 either $k_A = 0$ or $k_B = 0$.

Algorithm 1 Fluid approximation of T

1: Compute $X_A^{k_A,k_B}$ and $X_B^{k_A,k_B}$, $\forall k_A, k_B > 0 : k_A + k_B \leq K$ using MVA
2: $k_A = K_A, k_B = K_B, n_A = N_A, n_B = N_B, T = 0$
3: **while** $(k_A > 0$ **and** $k_B > 0)$ **do**
4:    $\Delta T_A = (n_A - k_A + 1)/X_A^{k_A,k_B}$, $\Delta T_B = (n_B - k_B + 1)/X_B^{k_A,k_B}$
5:    **if** $\Delta T_A < \Delta T_B$ **then**
6:        $n_A \mathrel{-}= \Delta T_A \cdot X_A^{k_A,k_B}$, $n_B \mathrel{-}= \Delta T_A \cdot X_B^{k_A,k_B}$, $k_A\mathord{-}\mathord{-}$, $k_B\mathord{+}\mathord{+}$, $T \mathrel{+}= \Delta T_A$
7:    **else**
8:        $n_A \mathrel{-}= \Delta T_B \cdot X_A^{k_A,k_B}$, $n_B \mathrel{-}= \Delta T_B \cdot X_B^{k_A,k_B}$, $k_A\mathord{+}\mathord{+}$, $k_B\mathord{-}\mathord{-}$, $T \mathrel{+}= \Delta T_B$
9:    **end if**
10: **end while**
11: **if** $k_A > 0$ **then**
12:    **while** $(k_A > 0)$ **do**
13:        $\Delta T_A = (n_A - k_A + 1)/X_A^{k_A,0}$, $n_A \mathrel{-}= \Delta T_A \cdot X_A^{k_A,k_B}$, $k_A\mathord{-}\mathord{-}$, $T \mathrel{+}= \Delta T_A$
14:    **end while**
15: **else**
16:    **while** $(k_B > 0)$ **do**
17:        $\Delta T_B = (n_B - k_B + 1)/X_B^{0,k_B}$, $n_B \mathrel{-}= \Delta T_B \cdot X_B^{k_A,k_B}$, $k_B\mathord{-}\mathord{-}$, $T \mathrel{+}= \Delta T_B$
18:    **end while**
19: **end if**

6.4.4   Techniques Comparison

The three techniques previously presented are now compared while analyzing a PDS with a single subsystem characterized by the service demand matrix in Eq. (6.1). In particular, the accuracy of each model in evaluating the PDS performance is analyzed taking into account the time required to compute the results. For this purpose, the number of tasks that must be executed and the capacity of the subsystem increase, while their ratio is constant and set to $K/N = 0.1$.

The shortest depletion time is expected to be observed when the number of tasks into the pool (N_A and N_B) and subsystem (K_A and K_B) are initialized to their optimal population mix, independently of the values of N and K. The pool and subsystem optimal population mixes (i.e., α^* and β^*, respectively) have been shown to coincide with equi-load and equi-utilization points [3] and may be derived through Eqs. (6.6) and (6.2). They only depend on the service demand matrix of the PDS and, considering the one given in Eq. (6.1), are $\alpha^* = (0.556868, 0.443131)$ and $\beta^* = (0.4, 0.6)$.

Results depicted in Fig. 6.4 are obtained studying the PDS with the service demand matrix given in Eq. (6.1) and for pool population mix $\alpha = \alpha^*$. The PDS is studied for $N = 100$ and $K = 10$, while values of K_A and K_B vary. Figure 6.4 depicts the depletion time T of the configuration considered as a function of the subsystem population mix β; as expected, the minimum depletion time is measured by all the models for $\beta = \beta^*$. Moreover, the largest error in estimating the depletion time made by simulation and fluid models with respect to the analytic one are $1\%_0$ and 4%, respectively.

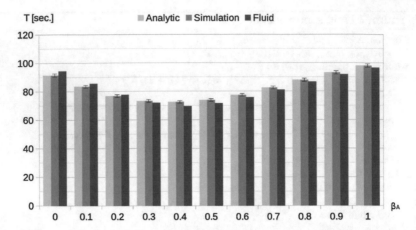

Fig. 6.4 Depletion time of a PDS as a function of the subsystem population mix obtained using three different models

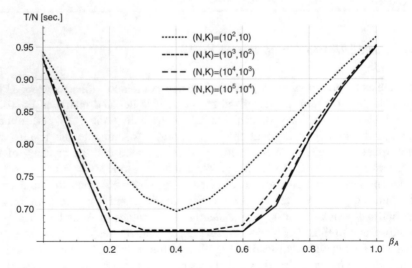

Fig. 6.5 Ratio of depletion time to the number of tasks initially into the pool, for N and K such that $K/N = 0.1$

Figure 6.5 represents the ratio of depletion time to the number of tasks initially into the pool T/N—or normalized depletion time—obtained with fluid model for configurations $(N, K) = \{(100,10), (1000,100), (10000,1000), (100000,10000)\}$. When the pool size and the subsystem capacity are large enough, we can observe the presence of an interval of value of β_a where the normalized depletion time is minimized and constant.

Figure 6.6 shows the normalized depletion time T/N as a function of subsystem population mix, when the number of tasks initially into the pool changes, but the

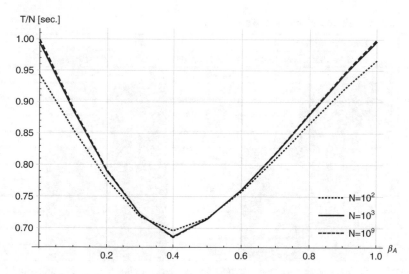

Fig. 6.6 Ratio of depletion time to the number of tasks initially into the pool, for $K = 10$ and different values of tasks initially into the pool

subsystem capacity is always the same (i.e., $K = 10$). In this case, the normalized depletion time behaves at the same way for all the configurations with $N > 1000$. As shown, the depletion time may be shortened by 30% if $N > 1000$ and PDS works with its optimal pool and subsystem population mixes (i.e., α^* and β^*).

In the PDS analysis, the fluid model provides results in the shortest time and is capable to handle very large values of N and K. Furthermore, although analytic model is faster than the simulation one, it cannot manage large values of pool size and subsystem capacity (i.e., $N > 100$ and $K > 10$) due to state space explosion. For these reasons, we used all the models to study configuration $(N, K) = (100, 10)$, simulation and fluid models for considering complex system and only the fluid one while analyzing very large values of N and K. For further details about performance of the three models, the reader is referred to Table 6.3 where the total time and the maximum Mean Absolute Percentage Error (MAPE) in computing the depletion time of each configuration are reported for all the techniques. In particular, the MAPE made by the simulation model with respect to the analytic one (that provides exact solutions) while computing the depletion time of the PDS is derived as:

$$\text{MAPE}_{sim} = \frac{|T_{sim} - T_{anal}|}{T_{anal}} \tag{6.14}$$

where T_{sim} and T_{anal} are the depletion times observed with simulation and analytic models, respectively. Similarly, the MAPE made by the fluid model with respect to the simulation one is:

Table 6.3 Models execution time and accuracy

N	K	Execution time			Errors	
		Markov chain	Simulation	Fluid approx.	$MAPE_{sim}$	$MAPE_{fluid}$
10^2	10	5 min	55 min	38 ms	1‰	4%
10^3	10		9 h	34 ms		2.6%
10^3	10^2		9 h	38 ms		5.5%
10^4	10		3 days	22 ms		5‰
10^4	10^3		7 days	590 ms		6%
10^5	10			67 ms		
10^5	10^4			53 s		
10^6	10			231 ms		
10^7	10			2 s		
10^8	10			17 s		
10^9	10			3 min		
10^9	10^5			4 min		

A cell is gray if it has not been possible to derive these measures due to long execution time

$$\mathrm{MAPE}_{fluid} = \frac{|T_{fluid} - T_{sim}|}{T_{sim}} \tag{6.15}$$

where T_{fluid} is the depletion time estimated by the fluid model.

In Table 6.3, columns 1 and 2 report, respectively, the values of N and K for which the PDS is studied; the time required by each model to provide the depletion time of the system for different population mixes of the subsystem (i.e., $\beta_A = i/10$, with $i = (0, \ldots, 10)$), is shown in columns 3, 4, and 5; finally, the maximum MAPE made by each simulation and fluid models with respect to the analytic and the simulation ones, respectively, are shown in columns 6 and 7. Since the results obtained using discrete event simulation are very similar to those derived through Markov analysis and, differently from analytic model, the simulation one may be used to study configurations with large values of N and K, the fluid model results are compared to those of the simulation model. Gray cells in Table 6.3 mean that results are not available due to the long time required for their computation.

While analytic and simulation models performance is affected by values of N and K, the time required by fluid model to provide results mainly depends on values of K. For this reason, adopting fluid model to analyze PDS allows the users to largely increase pool size N. Moreover, even if K affects the performance of fluid model more than N, it is still possible to consider also large values of K and get results in a short time. Finally note that, although maximum $MAPE_{fluid}$ is between 4% and 6% when $K/N = 0.1$, it decreases (i.e., the fluid model's accuracy improves) when K/N is small enough, such as for $N = 10^4$ and $K = 10$.

6.5 Conclusions

In this chapter the performance of three different techniques used to study PDS was analyzed and compared. The CTMC model provides analytic results, but it cannot deal with large and complex systems due to the well-known state space explosion. Thus, a discrete event multi-formalism model has been adopted to study more complex systems. In fact, it can analyze PDS with large pool size and subsystem capacity, while deriving results with high accuracy. Unfortunately, it may require a long computation time to provide results. Finally, a fluid model has been proposed in order to analyze complex systems in a short time. Although its MAPE with respect to simulation model has been observed between 4% and 6% for larger values of K/N, the results are provided in few minutes also for very large values of N and K, and its accuracy increases when K/N is small.

Although all the models presented in this chapter provide very similar results in different amounts of time, they must still be validated against a real scenario. In fact, the next step of our research will be the validation of the analytic, simulation and fluid models against a PDS deployed on a real cloud environment.

Acknowledgements This research was supported in part by the European Commission under the grant ANTAREX H2020 FET-HPC-671623.

References

1. F. Baskett, K.M. Chandy, R.R. Muntz, F.G. Palacios, Open, closed, and mixed networks of queues with different classes of customers. J. ACM **22**(2), 248–260 (1975)
2. D. Cerotti, M. Gribaudo, R. Pinciroli, G. Serazzi, Stochastic analysis of energy consumption in pool depletion systems, in *Measurement, Modelling and Evaluation of Dependable Computer and Communication Systems - 18th International GI/ITG Conference, MMB& DFT 2016, Münster, April 4–6, 2016, Proceedings* (2016), pp. 25–39
3. D. Cerotti, M. Gribaudo, R. Pinciroli, G. Serazzi, Optimal population mix in pool depletion systems with two-class workload, in *Proceedings of the 10th EAI International Conference on Performance Evaluation Methodologies and Tools on 10th EAI International Conference on Performance Evaluation Methodologies and Tools* (Institute for Computer Sciences, Social-Informatics and Telecommunications Engineering, Brussels, 2017), pp. 11–18
4. V.T. Chakaravarthy, A.R. Choudhury, S. Roy, Y. Sabharwal, Scheduling jobs with multiple non-uniform tasks, in *European Conference on Parallel Processing* (Springer, Berlin, 2013), pp. 90–101
5. J. Dean, S. Ghemawat, Mapreduce: simplified data processing on large clusters. Commun. ACM **51**(1), 107–113 (2008)
6. D. Díaz-Sánchez, A. Marín-López, F. Almenarez, R. Sánchez-Guerrero, P. Arias, A distributed transcoding system for mobile video delivery, in *Wireless and Mobile Networking Conference (WMNC), 2012 5th Joint IFIP* (IEEE, New York, 2012), pp. 10–16
7. M. Gribaudo, M. Iacono, *Theory and Application of Multi-Formalism Modeling* (IGI Global, Hershey, 2013)
8. L. Huang, X.-w. Wang, Y.-d. Zhai, B. Yang, Extraction of user profile based on the hadoop framework, in *5th International Conference on Wireless Communications, Networking and Mobile Computing, 2009. WiCom'09* (IEEE, New York, 2009), pp. 1–6

9. E.D. Lazowska, J. Zahorjan, G.S. Graham, K.C. Sevcik, *Quantitative System Performance: Computer System Analysis Using Queueing Network Models* (Prentice-Hall, Upper Saddle River, 1984)
10. J. Muppala, M. Malhotra, K. Trivedi, Markov dependability models of complex systems: analysis techniques, in *Reliability and Maintenance of Complex Systems*, vol. 154, ed. by S. Ozekici (Springer, Berlin, 1996), pp. 442–486
11. R. Pinciroli, M. Gribaudo, G. Serazzi, Modeling multiclass task-based applications on heterogeneous distributed environments, in *International Conference on Analytical and Stochastic Modeling Techniques and Applications* (Springer, Berlin, 2017), pp. 166–180
12. E. Rosti, F. Schiavoni, G. Serazzi, Queueing network models with two classes of customers, in *Proceedings Fifth International Symposium on Modeling, Analysis, and Simulation of Computer and Telecommunication Systems, 1997. MASCOTS'97* (IEEE, New York, 1997), pp. 229–234

Chapter 7
Performance of a Single Server Queue Supported by an Intermittent Server

Raymond A. Marie

7.1 Introduction

Let us consider a single server queue where the server can be supported by a second one who (1) leaves his current work to join the first server when the number of customers reaches a threshold K, (2) leaves the queuing system when he has no more customers to serve. A typical example of such a situation comes from the banking sector where the unique server from the front office is supported by a second server regularly assigned to the back office who joins the front office as soon as the number of customers reaches a given threshold (denoted here by the integer K). But such a situation could come from a more industrial area. The introduction of an intermittent server allows to decrease the expected waiting times of customers at a lower cost than affecting an extra permanent server. And the aim of this study is to determine the efficiency of such a policy.

A closed situation is one of the supermarket check-out counters where a counter can be activated/deactivated based on the states of the different queues. This larger model is a good example to be used in a course on discrete event simulation as a practical exercise because the queuing model is easy to elaborate and has no (known) analytical solution in its general configuration. This helps students to realize all the advantages of a simulation approach. In addition, such a model is easily adaptable to other fields such as those of telecommunication or of data centers. Nevertheless, when possible, an analytical solution must be looked for since its cost is generally lower than the one of the simulation approaches.

Although most of the research work in the domain of the M/M/r queue with intermittent servers has been done through the use of simulation, we noted some

R. A. Marie (✉)
University Rennes 1, IRISA-INRIA, Rennes Cedex, France
e-mail: raymond.marie@irisa.fr

© Springer International Publishing AG, part of Springer Nature 2019   95
A. Puliafito, K. S. Trivedi (eds.), *Systems Modeling: Methodologies and Tools*,
EAI/Springer Innovations in Communication and Computing,
https://doi.org/10.1007/978-3-319-92378-9_7

developments connected to the subject. In 1971, J. Blackburn published a report [1] relative to an M/G/1 queue in which the server is an intermittent one who starts working when the number of customers crosses some threshold. This threshold is the value realizing the optimum of an objective function. A more recent analytical study investigated the case of an airline check-in counters set in an airport [5]. In this study, Parlar et. al elaborated a Markovian model and its transient solution. A major difference with the supermarket check-out system is that the number of customers to be served is known in advance (number of customers who have a reserved seat for a given flight). The problem is to control the number of open check-in counters such that all the customers that will show up before a deadline T will be served on time (such that the plane can take off on time). But most of the literature involving intermittent servers concerns studies where the activations of the servers depend on reliability/availability of the set of servers rather than on the states of the systems.

Another related class of models is the "coupled processor model" where each processor can help the other when it is idle. The two queues have their own arrival processes and service time distributions. Such a class has been the object of intensive analytical works in the past. Close to that is the case where the behaviors of the servers are no more symmetrical and only one processor can, when it becomes idle, give time to the other processor until its own queue reaches a given threshold (see the intensive study of Osogami et al. [4]). Note also the different model known as "the slow server problem" (see [6]) where, depending on the values of the parameters, the use of the slow server may increase the response time.

The present study is different in the sense that the server who gives some part of his time is not idle but works on tasks which are not directly impacting customers (the notion of response time is in some sense meaningless). This study is less general than the one cited above [4] but produces a closed form solution for the steady state probability distribution and for different metrics such as expected waiting times for customers or expectation of busy periods for the intermittent server. Our objective is to promote a better understanding of the benefits of such a strategy. In particular, we have to consider the trade-off between the help to the customer and the perturbation of the work in the back office. This is achieved thanks to a cost function providing an optimal value of the threshold K as a tool to help a manager in charge of the economical decision.

The chapter is organized as follows: in Sect. 7.2 we present a Markovian model of the investigated system while in the following section we exhibit the steady state probability distribution of the stochastic process and the expression of the mean number of customers (or mean response time) in terms of the different parameters. In Sect. 7.4, we conduct the determination of the expectation of the time spent by the second server in one passage in the back office and those of the expectation of one sojourn time at the front office. In the following section we introduce a cost function allowing us to provide an optimal threshold K^*. Finally, we conclude by summarizing the advantages of using an intermittent server (Sect. 7.6).

7.2 Hypotheses and Model

We consider that the two servers are equivalent and that the service times are independent and identically distributed random variables following an exponential distribution with rate μ. The first server assigned to the front office stays available for serving the arriving customers.

When there are $(K - 1)$ customers, if the server affected to the back office is not already serving in the front office, then this server leaves the back office at the instant of arrival of a new customer and starts serving him in the front office. Once he is in the front office, the second server stays there until he has no more customers to serve and re-integrates the back office.

We assume the customer arrival process is Poisson with rate λ.

Under these hypotheses, the stochastic process modeling the number of customers in the office is a continuous time Markov chain (CTMC) $\{X(t), t \geq 0\}$ [2, 3, 7]. Its transition graph is given in Fig. 7.1.

A couple $(i, 0)$ (respectively $(i, 1)$) denotes a state where i customers are present and where the second server is in the back office (respectively present). State (0) refers to the empty system and, for $i \geq K$, state i denotes the system when i customers and the second server are present. Note that the first server is idle in state $(1, 1)$. In addition, E_0 (respectively E_1) will denote the subset of states where the second server is in the back office (respectively present):

$$E_0 = \{(0), (1, 0), \ldots (K - 1, 0)\}, \quad E_1 = \{(1, 1), \ldots (K - 1, 1), (K), (K + 1), \ldots\} .$$

The steady state probability distribution of this CTMC is determined in the following section.

The case $K = 2$ corresponds to an $M/M/2$ queue with a little specificity: once the queue is empty, the first server deals with the new arrival, the second server arriving only when a new arrival finds the first server busy, and going back as soon as there is no more customer to serve in the front office. But from the customer point of view, this specificity does not affect the performance of the queue.

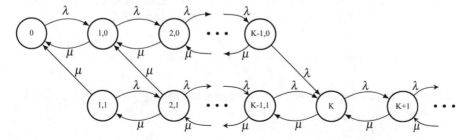

Fig. 7.1 Transition graph of the CTMC

7.3 Steady State Probability Distribution, Mean Number of Customers

7.3.1 Steady State Probability Distribution

For any state e, π_e will denote the steady state probability of state e. Defining $\rho = \lambda/2\mu$, we note that the steady state probability will exist only if $\rho < 1$. Using the Chapmann-Kolmogorov (C-K) equations of states $(i, 0)$, $i = 2, \ldots, K - 1$, it is not difficult to prove by induction the relation:

$$\pi_{K-i,0} = \left(\sum_{j=0}^{i-1} \phi^j\right) \pi_{K-1,0}, \qquad i = 2, \ldots, K - 1, \tag{7.1}$$

where $\phi = \mu/\lambda$. Use of the cut theorem on the partition $\{E_0, E_1\}$ and of the steady state C-K equation of state $(1, 1)$ gives us

$$\pi_{1,1} = \frac{1}{(1 + 2\phi)} \pi_{K-1,0}. \tag{7.2}$$

Then, using Eqs. (7.2) and (7.1) and the C-K equation for state (0), we can express probability $\pi_{K-1,0}$ in terms of π_0 as:

$$\pi_{K-1,0} = \pi_0 \left[\frac{\phi}{(1 + 2\phi)} + \phi\left(\sum_{j=0}^{K-2} \phi^j\right)\right]^{-1}, \tag{7.3}$$

or, for the case $\phi \neq 1$, as:

$$\pi_{K-1,0} = \pi_0 \frac{(1 + 2\phi)(1 - \phi)}{D_0}, \tag{7.4}$$

where $D_0 = \phi[(1 - \phi) + (1 + 2\phi)(1 - \phi^{K-1})]$. Considering now the C-K equations of states $(i, 1)$, $i = 2, \ldots, K - 1$, we can prove by induction that:

$$\pi_{i,1} = \pi_{1,1} \frac{(1 + \rho) - 2\rho^i}{(1 - \rho)} \qquad i = 2, \ldots, K, \tag{7.5}$$

Since $\rho = 1$ is a root of the numerator, let us note that this probability can also be expressed as:

$$\pi_{i,1} = \pi_{1,1} \left(1 + 2\sum_{j=1}^{i-1} \rho^j\right) \qquad i = 2, \ldots, K, \tag{7.6}$$

When $i = K$, we get in particular the probability $\pi_{K,1}$ that we can rename π_K without any ambiguity:

$$\pi_K = \pi_{1,1} \frac{(1+\rho) - 2\rho^K}{(1-\rho)} = \pi_{1,1} \left(1 + 2 \sum_{j=1}^{K-1} \rho^j \right) . \tag{7.7}$$

Then, using Eqs. (7.2) and (7.4), we express the probability π_K as a function of probability π_0 (again for the case $\phi \neq 1$):

$$\pi_K = \pi_0 \frac{(1+\rho) - 2\rho^K}{(1-\rho)} \frac{(1-\phi)}{D_0} . \tag{7.8}$$

Considering the probabilities $\pi_i, i > K$, their expressions are easily obtained thanks to the use of the cut theorem:

$$\pi_i = \rho^{i-K} \pi_K, \quad i > K . \tag{7.9}$$

Let us now consider the normalizing equation that we can write as:

$$S_0 + S_1 = 1 , \tag{7.10}$$

where $S_0 = \pi_0 + \sum_{i=1}^{K-1} \pi_{i,0}$ and $S_1 = \sum_{i=1}^{K-1} \pi_{i,1} + \sum_{i=K}^{\infty} \pi_i$.

Sum S_0 is the steady state probability that the intermittent server is working in the back office and that S_1 is the steady state probability that the intermittent server is working in the front office. This last sum S_1 will be also used later when looking for the optimal threshold.

Using Eqs. (7.1), (7.2), (7.4), (7.5), (7.8), and (7.9), we show in Appendix 1 that the probability π_0 can be written as:

$$\pi_0 = \frac{(1-\rho)(1-\phi)D_0}{D_1} , \tag{7.11}$$

where

$$D_1 = (1-\rho)\{\phi(1-\phi)^2 + (1+2\phi)[(K-1)(1-\phi) - \phi^2(1-\phi^{K-1})]\}$$

$$(1-\phi)^2[K + \rho(K-1)] . \tag{7.12}$$

For the special case where $\phi = 1$, Eqs. (7.1), (7.2), (7.4), and (7.5) reduce to:

$$\pi_{K-1,0} = \frac{3}{3K-2}\pi_0 , \quad \text{and } \pi_{K-i,0} = i\pi_{K-1,0} , \quad i = 2, \ldots, K-1 , \tag{7.13}$$

$$\pi_{1,1} = \frac{1}{3}\pi_{K-1,0} , \qquad \text{and } \pi_{i,1} = (3-2^{-(i-2)})\pi_{1,1} , \qquad i = 2, \ldots, K , \qquad (7.14)$$

while it is shown in Appendix 1 that probability π_0 satisfies:

$$\pi_0 = \frac{2(3K - 2)}{3(K(K + 3) - 2)} . \qquad (7.15)$$

For the case where $K = 2$ some of the equations given for the general case become simpler (in particular because the expression D_0 equals $2\phi(1 - \phi^2)$ when $K = 2$) and it is not difficult to find again the well-known result of the $M/M/2$ queue:

$$\pi_0 = \frac{(1 - \rho)}{(1 + \rho)} . \qquad (7.16)$$

Let us remark that for $\rho = 1/2$, we obtain $\pi_0 = 1/3$. In that case $\phi = 1$, and this result agrees with the one obtained thanks to relation (7.15) when $K = 2$.

7.3.2   Mean Number of Customers, Mean Waiting Time

The determination of the mean number of customers $\mathbb{E}[N]$ is purely technique. For $\phi \neq 1$, it is shown in Appendix 2 that this expectation satisfies the following relation:

$$\mathbb{E}[N] = \frac{(1 - \phi)}{D_1} \left\{ (1 - \rho)(1 + 2\phi) \left(\frac{K(K + 1)}{2} - \frac{K}{(1 - \phi)} + \frac{\phi(1 - \phi^K)}{(1 - \phi)^2} \right) \right.$$

$$\left. + (1 - \phi) \left((1 + \rho)\frac{K(K - 1)}{2} + \frac{K + \rho(K - 1)}{(1 - \rho)} \right) \right\}$$

When $K = 2$, it is not difficult to find again the well-known result of the $M/M/2$ queue:

$$\mathbb{E}[N] = 2\rho/(1 - \rho^2) . \qquad (7.17)$$

For the special case where $\phi = 1$, it is also shown in Appendix 2 that

$$\mathbb{E}[N] = \frac{K(K(K + 3) + 8) - 4}{3(K(K + 3) - 2)} . \qquad (7.18)$$

For $K = 2$, $\mathbb{E}[N] = 4/3$. This result agrees with the one obtained thanks to relation (7.17) when $\rho = 1/2$, i.e., ($\phi = 1$).

Because the aim of using an intermittent server is to decrease the waiting time of the customer in the front office, it is also interesting to consider the expected waiting time $\mathbb{E}[W]$. For that we first obtain the expected response time by use of the Little's formula and then subtract the mean service time:

$$\mathbb{E}[W] = \frac{1}{\lambda}\mathbb{E}[N] - \frac{1}{\mu} . \qquad (7.19)$$

We may prefer to consider what we will call a "normalized" expected waiting time $\mathbb{E}[W_N]$ by taking the mean service time (i.e., $1/\mu$) as the time unit. This gives us:

$$\mathbb{E}[W_N] = \mu\mathbb{E}[W] = \frac{\mu}{\lambda}\mathbb{E}[N] - 1 = \phi\mathbb{E}[N] - 1 .$$

Let us remark that the "normalized" expected waiting time has no dimension and is therefore independent of the initial time unit.

For a given value of ρ we expect that the expected number of customers is greater than the value given by the $M/M/2$ queue. While, as long as ρ is lower than $1/2$, the expected number of customers is lower than the ratio $\dfrac{2\rho}{1 - 2\rho}$, which corresponds to the value given by the $M/M/1$ queue with 2ρ as the utilization factor.

In Fig. 7.2, we have plotted the expectation of the number of customers as a function of ρ, for different values of the integer K. As we would expect, this

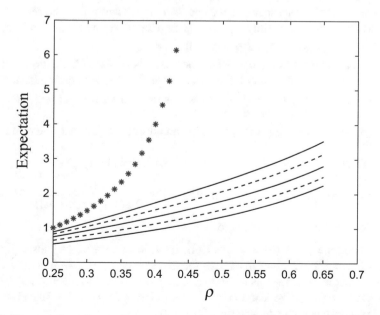

Fig. 7.2 Mean number of customers as a function of ρ. For (bottom-up) $K = 2, 3, 4, 5$, and 6. Curve with stars corresponds to infinite K, the second server being never called

expectation is increasing with ρ and with K. Note that without the second server, the mean number of customers would tend to infinity when ρ tends to $1/2$.

7.4 Pseudo-Idle and Busy Periods of the Intermittent Server

The pseudo-idle period of the second server is defined as the period of time during which this server is working in the back-office. We are interested by the expectation of such a period because we understand that a too short period would have a negative effect on the productivity of the server. Such a period corresponds to a sojourn time of the CTMC in the subset E_0 and therefore we need to obtain the expectation of this sojourn time.

7.4.1 Mean Time of a Passage in the Back Office

First let us determine the probability that a pseudo-idle period starts in state (0) (respectively in state $(1, 0)$). Given that the CTMC is in state $(2, 1)$, if a service completes before a new arrival, the CTMC joins either state $(1, 0)$ if the second server finishes his service first or state $(1, 1)$ in the other case. These two events have equal probabilities (0.5 each). If the CTMC joins state $(1, 1)$ from state $(2, 1)$, this means that the permanent server becomes idle. Then either the second server becomes idle (with probability $\frac{\mu}{\lambda+\mu}$) or the regular server becomes busy again, the CTMC revisiting state $(2, 1)$ (with probability $\frac{\lambda}{\lambda+\mu}$).

So, given a service completes when the CTMC is in state $(2, 1)$, the CTMC goes to state $(1, 0)$ with probability 0.5, goes to state (0) without coming back to state $(2, 1)$ with probability $0.5 \times \dfrac{\mu}{\lambda + \mu}$ or comes back to state $(2, 1)$ with probability $0.5 \times \dfrac{\lambda}{\lambda + \mu}$. Considering these three eventualities, we see that when the CTMC enters subset E_0, it enters it through state (0) with probability $\dfrac{0.5(\mu/(\lambda + \mu))}{0.5(1 + \mu/(\lambda + \mu))}$ or enters it through state $(1, 0)$ with probability $\dfrac{0.5}{0.5(1 + \mu/(\lambda + \mu))}$. These two expressions reducing respectively to $\dfrac{\phi}{1 + 2\phi}$ and $\dfrac{1 + \phi}{1 + 2\phi}$.

Let us assume that $X(0) = 0$. Let T_A be the sojourn time in the subset E_0: $T_A = inf\{t|X(t) = K\}$. In order to express the expectation of T_A, we first consider the random variable T_i defined as the time it takes to the CTMC to reach state $(i + 1, 0)$ given $X(0) = (i, 0)$. We also denote the expectation of T_i by α_i. Introducing the discrete random variable I_i such that, for $i \geq 0$:

$$
I_i = \begin{cases} 1 & \text{if the first transition of the CTMC from state } (i,0) \\ & \text{is a jump to state } (i+1,0); \\ 0 & \text{if the first transition of the CTMC from state} (i, 0) \\ & \text{is a jump to state } (i\text{-}1,0); \end{cases}
$$

we get when conditioning w.r.t. I_i: $\mathbb{E}[T_i|I_i = 1] = \frac{1}{\lambda+\mu}$, and $\mathbb{E}[T_i|I_i = 0] = \frac{1}{\lambda+\mu} + \alpha_{i-1} + \alpha_i$.

For $i = 0$, we have immediately $\mathbb{E}[T_0] = \frac{1}{\lambda}$. Since the departure rate from state $(i, 0)$ equals $(\lambda+\mu)$ while the transition rate from state $(i, 0)$ to state $(i+1, 0)$ equals λ, the probability that the first transition of the CTMC from state $(i, 0)$ is a jump to state $(i + 1, 0)$ is $\mathbb{P}(I_i = 1) = \frac{\lambda}{\lambda + \mu}$. Therefore, deconditioning the expectation $\alpha_i = \mathbb{E}[T_i]$ gives us, for $i > 0$,

$$
\alpha_i = \frac{1}{\lambda + \mu} \frac{\lambda}{\lambda + \mu} + \left(\frac{1}{\lambda + \mu} + \alpha_{i-1} + \alpha_i \right) \frac{\mu}{\lambda + \mu} ,
$$

that reduces to $\alpha_i = \frac{1}{\lambda}(1 + \mu\, \alpha_{i-1})$.

Since $\alpha_0 = \mathbb{E}[T_0] = \frac{1}{\lambda}$, we can compute successfully $\alpha_0, \alpha_1, \alpha_2, \ldots$ It is not difficult to prove that $\alpha_i = \frac{1}{\lambda} \sum_{j=0}^{i} \phi^j$.

In addition, $\mathbb{E}[T_A]$ depends on the way the CTMC enters the subset E_0 since $\mathbb{E}[T_A|X(0) = 0] = \sum_{j=0}^{K-1} \alpha_j$, while $\mathbb{E}[T_A|X(0) = (1,0)] = \sum_{j=1}^{K-1} \alpha_j$.

Therefore, after deconditioning we obtain:

$$
\mathbb{E}[T_A] = \frac{1}{\lambda} \frac{1}{2(1 + \rho)} + \frac{1}{\lambda} \left((K-1) + \sum_{i=1}^{K-1} (K-i)\phi^i \right) . \tag{7.20}
$$

We can scale this result by expressing this time expectation in terms of a number of mean service times:

$$
\mu\mathbb{E}[T_A] = \frac{\phi}{2(1 + \rho)} + \phi \left((K-1) + \sum_{i=1}^{K-1} (K-i)\phi^i \right) , \tag{7.21}
$$

In Fig. 7.3, we have plotted the scaled expectation of the pseudo-idle period of the second server as a function of ρ, for different values of the integer K. We can say that the expectation of the pseudo-idle period of the second server is important when ρ is between 0 and around 0.4. Remember that when $\rho = 0.4$, the utilization factor of the single server of the $M/M/1$ queue equals 0.8. As we would expect, this expectation is decreasing with ρ and increasing with K.

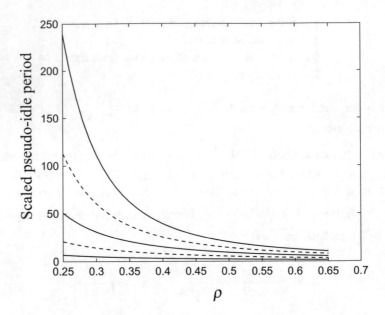

Fig. 7.3 Scaled expectation of the pseudo-idle period of the second server as a function of ρ. For (bottom-up) $K = 2, 3, 4, 5$, and 6

If the manager decides to change the rule by switching from K to $(K + 1)$, then the scaled expectation will be increased of the quantity:

$$\Delta_K(\mu\mathbb{E}[T_A]) = \mu\mathbb{E}[T_A(K + 1)] - \mu\mathbb{E}[T_A(K)] = \phi\left(\sum_{i=0}^{K}\phi^i\right).$$

Even in the case where $\phi = 1$ (i.e., $\rho = 0.5$), this increase can be shown to correspond to $(K + 1)$ mean service times!

7.4.2 Mean Time of a Passage in the Front Office

Now let $\mathbb{E}[T_P]$ be the expectation of a period spent in the front office by the intermittent server. This server starts such a period with the frequency $\lambda\pi_{K-1,0}$. Using the fact that this frequency must be equal to $(\mathbb{E}[T_A] + \mathbb{E}[T_P])^{-1}$, we obtain a first expression for $\lambda\mathbb{E}[T_P]$:

$$\lambda\mathbb{E}[T_P] = [\pi_{K-1,0}]^{-1} - \lambda\mathbb{E}[T_A].$$

Then, starting from Eqs. (7.4) and (7.11) we express the inverse of probability $\pi_{K-1,0}$ as:

$$[\pi_{K-1,0}]^{-1} = \frac{D_1}{(1-\rho)(1+2\phi)(1-\phi)^2},$$

$$= \frac{\phi}{(1+2\phi)} + \frac{(K-1)(1-\phi)-\phi^2(1-\phi^{K-1})}{(1-\phi)^2} + \frac{[1+(K-1)(1+\rho)]}{(1-\rho)(1+2\phi)}.$$

Using Eq. (7.20) we develop the expression of $\lambda\mathbb{E}[T_A]$ as:

$$\lambda\mathbb{E}[T_A] = \frac{1}{2(1+\rho)} + \left((K-1) + \sum_{i=1}^{K-1}(K-i)\phi^i\right)$$

$$= \frac{\phi}{(1+2\phi)} + \left((K-1) + K\sum_{i=1}^{K-1}\phi^i - \sum_{i=1}^{K-1}i\phi^i\right)$$

$$= \frac{\phi}{(1+2\phi)} + \left((K-1) + K\left(\frac{1-\phi^K}{(1-\phi)} - 1\right) - \frac{(K-1)\phi^{K+1}-K\phi^K+\phi}{(1-\phi)^2}\right)$$

$$= \frac{\phi}{(1+2\phi)} + \left((K-1) + \frac{K\phi(1-\phi^{K-1})}{(1-\phi)} - \frac{(K-1)\phi^{K+1}-K\phi^K+\phi}{(1-\phi)^2}\right)$$

$$= \frac{\phi}{(1+2\phi)} + \frac{(K-1)(1-\phi)^2+K\phi(1-\phi^{K-1})(1-\phi)}{(1-\phi)^2} - \frac{(K-1)\phi^{K+1}-K\phi^K+\phi}{(1-\phi)^2}$$

$$= \frac{\phi}{(1+2\phi)} + \frac{(K-1)-(K-1)\phi-\phi^2+\phi^{K+1}}{(1-\phi)^2}$$

$$= \frac{\phi}{(1+2\phi)} + \frac{(K-1)(1-\phi)-\phi^2(1-\phi^{K-1})}{(1-\phi)^2}. \tag{7.22}$$

Subtracting this last expression to the one obtained for $[\pi_{K-1,0}]^{-1}$ we get the expression of $\lambda\mathbb{E}[T_P]$:

$$\lambda\mathbb{E}[T_P] = \frac{[1+(K-1)(1+\rho)]}{(1-\rho)(1+2\phi)} = \frac{\rho}{(1-\rho)}\left((K-1) + \frac{1}{(1+\rho)}\right), \tag{7.23}$$

and then the expression of the expectation scaled in terms of a number of mean service time:

$$\mu\mathbb{E}[T_P] = \frac{1}{2(1-\rho)}\left((K-1) + \frac{1}{(1+\rho)}\right). \tag{7.24}$$

Note that $\mu\mathbb{E}[T_P]$ represents also the expected number of customers served by the intermittent server during a passage in the front office.

In Fig. 7.4, we have plotted the scaled expectation of the pseudo-busy period of the second server as a function of ρ, for different values of the integer K. As we

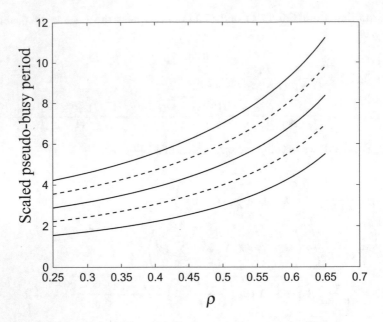

Fig. 7.4 Scaled expectation of the pseudo-busy period of the second server as a function of ρ. For (bottom-up) $K = 2, 3, 4, 5$, and 6

would expect, this expectation is increasing with ρ and with K. Moreover, we can say that the expectation of the pseudo-busy period of the second server is relatively small when ρ is between 0 and around 0.4, when we compare it with the one of the pseudo-idle period (cf. Fig. 7.3). This shows the benefit of the intermittent server since the use of a low percentage of his time significantly decreases the mean waiting time.

7.5   Cost Function

We have to consider two somewhat different situations. The first one is when the second server is not necessary for the system to be stable (i.e., when $\rho < 0.5$). The second situation is when the second server is necessary to the system ($\rho \geq 0.5$).

In the first situation, the second server just helps to decrease the mean waiting time $\mathbb{E}[W]$ seen by the customers. We have to compare this help to the customers with respect to the perturbation of the work done in the back office.

We assume here that there is a fixed penalty c_0 to pay each time the second server has to leave the back office and that the cost per unit of time of this second server is c_1. We also assume that c_2 is the cost per unit of waiting time. During a unit time, the expectation of the cumulative value of the waiting times equals $\lambda \mathbb{E}[W]$; this expectation being nothing else than the expectation of the number of waiting customers in the queue. Let $\mathbb{E}[N_w]$ denote this expectation. The expression of $\mathbb{E}[N_w]$ is deduced from Eq. (7.19):

$$\mathbb{E}[N_w] = \mathbb{E}[N] - 2\rho . \tag{7.25}$$

Then, depending on the value K, the function to minimize corresponds to the expected total variable cost per time unit, and is given by:

$$C(K) = c_0[\mathbb{E}[T_A] + \mathbb{E}[T_P]]^{-1} + c_1 S_1 + c_2 \mathbb{E}[N_w] , \tag{7.26}$$

where here also, S_1 denotes the sum $\sum_{i=1}^{K-1} \pi_{i,1} + \sum_{i=K}^{\infty} \pi_i$. Note that this sum of probabilities S_1 is nothing but the mean time per time unit spent by the second server in the front office.

When the variable K is increased, the first two terms are decreasing while the term $c_2 \mathbb{E}[N_w]$ is increasing. More precisely, considering a cycle of the intermittent server, we start from the relation:

$$S_1 = \frac{\mathbb{E}[T_P]}{(\mathbb{E}[T_A] + \mathbb{E}[T_P])} = \frac{1}{1 + \frac{\mathbb{E}[T_A]}{\mathbb{E}[T_P]}} . \tag{7.27}$$

Considering Eqs. (7.22) and (7.23) we deduce that, when K tends to infinity, the two expectations tend to infinity. Considering now the ratio $\dfrac{\mathbb{E}[T_A]}{\mathbb{E}[T_P]}$ when K tends to infinity, since ϕ satisfies $\phi > 1$, the limit of this ratio is the same as the limit of the following ratio:

$$\lim_{K \longrightarrow +\infty} \frac{\mathbb{E}[T_A]}{\mathbb{E}[T_P]} = \lim_{K \longrightarrow +\infty} \frac{(2\phi - 1)}{(1 - \phi)^2} \frac{\phi^K}{K} = +\infty . \tag{7.28}$$

Therefore, the first two terms of the cost function tend asymptotically to zero when K tends to infinity while the term $c_2 \mathbb{E}[N_w]$ is increasing (from $c_2 2\rho^3/(1-\rho^2)$ when $K = 2$ to the asymptotic value $c_2 4\rho^2/(1 - 2\rho)$ when K tends to infinity). In this situation the optimal K may not be finite if the penalty coefficient c_2 is not large enough.

The second situation is different in the sense that K has to be finite in order to have a stable solution. In this case, the intermittent server has to work in the front office a percentage of time S_1 greater than $(\lambda/\mu - 1)$ in order that the system admits a steady state solution. The maximal feasible value $K_{\max}$ of K is given by $K_{\max} = \max\{K | S_1(K) > \lambda/\mu - 1\}$. Practically, if $K_{\max}$ is large enough (i.e., when $(\lambda/\mu - 1)$ is not close to unity), the cost $c_2 \mathbb{E}[N_w]$ should be large when $K = K_{\max}$ and we may expect the cost function to be convex. However, the convexity of $C(K)$ has not been investigated theoretically. Also, from a practical point of view, the parameter c_2 has again to be not too small with respect to c_0 and c_1 in order to avoid

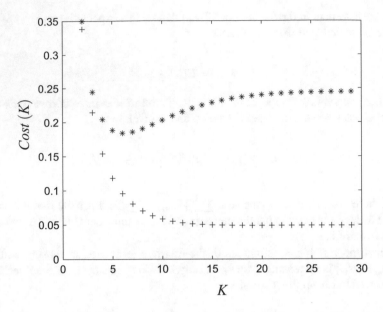

Fig. 7.5 Variable cost function, with $\rho = 0.35$, $c_0 = 0.5$, $c_1 = 1$. Case 1 (stars): $c_2 = 0.15$. Case 2 (sign +): $c_2 = 0.03$

the limit behavior where the second server would come once a year to empty the waiting room.

In Fig. 7.5, we have plotted two sets of values of $C(K)$ when $\rho = 0.35$ (alone the permanent server queue would have a utilization factor of 0.7), for $c_0 = 0.5$, $c_1 = 1$. Case $c_2 = 0.15$ (noted with stars) gives an optimal $K^* = 5$. From Figs. 7.3 and 7.4, we can check that for this optimal solution the mean pseudo-idle period of the second server is around 70 times the mean service time while the mean pseudo-busy period is close to 5 times the mean service time. But case $c_2 = 0.03$ (noted with sign +) gives a decreasing cost for $K \in [2, 30]$.

7.6 Conclusions

We have shown in this paper the importance of intermittent servers in order to reduce the response times without increasing significantly the idle times of servers. For such situations where a single server would satisfy the stability condition ($\lambda < \mu$), a nontrivial result is that the pseudo-idle period of the second server is significantly longer than what would be generally expected by the management and also that the pseudo-busy period stays small; and so the second server can keep his main activity in the back office.

We can think of applications in architectures for quite large telecommunication switches where we have "guard" processors to help the congested input queues on demand. It may also help in the context of network function virtualization (NFV) in which a service might be deployed on demand to face a transient congestion. Not only these results are interesting by themselves if such a situation occurs in a real situation but also, this study can be used to check simulation models used for a more complex situation.

Appendix 1: Determination of Eq. (7.11)

Starting from the normalizing equation: $S_0 + S_1 = 1$, where $S_0 = \pi_0 + \sum_{i=1}^{K-1} \pi_{i,0}$,

and $S_1 = \pi_0 + \sum_{i=1}^{K-1} \pi_{i,1} + \sum_{i=K}^{\infty} \pi_i$, we first consider the partial sum S_0:

$$S_0 = \pi_0 + \sum_{i=1}^{K-1} \pi_{i,0} = \pi_0 + \sum_{i=1}^{K-1} \pi_{K-i,0} ,$$

$$= \pi_0 + \pi_{K-1,0} \sum_{i=1}^{K-1} \left(\sum_{j=0}^{i-1} \phi^j \right) = \pi_0 + \pi_{K-1,0} \sum_{i=1}^{K-1} (K - i) \phi^{i-1} ,$$

$$= \pi_0 + \pi_{K-1,0} \left(K \sum_{i=1}^{K-1} \phi^{i-1} - \sum_{i=1}^{K-1} i \phi^{i-1} \right) ,$$

or (if $\phi \neq 1$):

$$S_0 = \pi_0 + \pi_{K-1,0} \left(K \frac{(1 - \phi^{K-1})}{(1 - \phi)} - \frac{(1 - K\phi^{K-1} + (K - 1)\phi^K)}{(1 - \phi)^2} \right) ,$$

$$= \pi_0 + \frac{\pi_{K-1,0}}{(1 - \phi)} \frac{K(1 - \phi) - (1 - \phi^K)}{(1 - \phi)}$$

$$= \pi_0 + \pi_0 \frac{(1 + 2\phi)}{D_0} \frac{K(1 - \phi) - (1 - \phi^K)}{(1 - \phi)}$$

$$= \pi_0 \left(1 + \frac{(1 + 2\phi)(K(1 - \phi) - (1 - \phi^K))}{(1 - \phi) D_0} \right) ,$$

$$= \frac{\pi_0}{(1 - \phi) D_0} \left(\phi(1 - \phi)^2 + (1 + 2\phi)[(K - 1)(1 - \phi) - \phi^2(1 - \phi^{K-1})] \right) .$$

Considering now the partial sum S_1, i.e., the steady state probability that the back-office server is helping the front-office server, we have:

$$S_1 = \sum_{i=1}^{K-1} \pi_{i,1} + \sum_{i=K}^{\infty} \pi_i = \pi_{1,1} \sum_{i=1}^{K-1} \frac{(1+\rho) - 2\rho^i}{(1-\rho)} + \pi_K \sum_{i=K}^{\infty} \rho^{i-K} ,$$

$$= \pi_{1,1} \frac{(K-1)(1+\rho)}{(1-\rho)} - 2\pi_{1,1} \frac{1}{(1-\rho)} \sum_{i=1}^{K-1} \rho^i + \pi_K \frac{1}{(1-\rho)} ,$$

$$= \pi_{1,1} \frac{(K-1)(1+\rho)}{(1-\rho)} - \pi_{1,1} \frac{2}{(1-\rho)} \frac{(1-\rho^K)}{(1-\rho)} + \pi_{1,1} \frac{(1+\rho) - 2\rho^K}{(1-\rho)} \frac{1}{(1-\rho)}$$

$$= \pi_{1,1} \frac{(K-1)(1+\rho)}{(1-\rho)} - \pi_{1,1} \frac{1}{(1-\rho)} = \pi_{1,1} \frac{1 + (K-1)(1+\rho)}{(1-\rho)} ,$$

$$= \pi_0 \frac{(1-\phi)[K + \rho(K-1)]}{(1-\rho)D_0} .$$

Given that $S_0 + S_1 = 1$, we get the expression of probability π_0 when $\phi \neq 1$:

$$\pi_0 = \frac{(1-\rho)(1-\phi)D_0}{D_1} , \tag{7.29}$$

where D_1 was defined by relation (7.12).

For the special case where $\phi = 1$, it is not difficult, starting from the specific relations between probabilities given at the end of Sect. 7.3.1, to find the following expressions:

$$S_0 = \frac{3K(K+1) - 4}{2(3K-2)} \pi_0 , \quad S_1 = \frac{(3K-1)}{(3K-2)} \pi_0 , \quad \pi_0 = \frac{2(3K-2)}{3(K(K+3) - 2)} .$$

$$\tag{7.30}$$

Appendix 2: Determination of the Mean Number of Customers

In order to obtain the expression, let us start by computing two partial sums (B_0 and B_1), under the condition $\phi \neq 1$:

$$B_0 = \sum_{i=1}^{K-1} i\pi_{i,0} = \sum_{i=1}^{K-1} (K-i)\pi_{K-i,0} = \pi_{K-1,0} \sum_{i=1}^{K-1} (K-i) \left(\sum_{j=0}^{i-1} \phi^j \right) ,$$

$$= \pi_{K-1,0} \sum_{i=1}^{K-1} (K-i) \frac{(1-\phi^i)}{(1-\phi)}$$

$$= \frac{\pi_{K-1,0}}{(1-\phi)} \left(\sum_{i=1}^{K-1} i - K \sum_{i=0}^{K-1} \phi^i + K + \phi \sum_{i=1}^{K-1} i\phi^{i-1} \right) ,$$

$$= \frac{\pi_{K-1,0}}{(1-\phi)} \left(\frac{K(K+1)}{2} - K \sum_{i=0}^{K-1} \phi^i + \phi \sum_{i=1}^{K-1} i\phi^{i-1} \right)$$

$$= \frac{\pi_{K-1,0}}{(1-\phi)} \left(\frac{K(K+1)}{2} - \frac{K}{(1-\phi)} + \frac{\phi(1-\phi^K)}{(1-\phi)^2} \right) ,$$

$$= \frac{(1-\rho)(1-\phi)(1+2\phi)}{D_1} \left(\frac{K(K+1)}{2} - \frac{K}{(1-\phi)} + \frac{\phi(1-\phi^K)}{(1-\phi)^2} \right) ,$$

and secondly:

$$B_1 = \sum_{i=1}^{K-1} i\pi_{i,1} + \sum_{i=K}^{\infty} i\pi_i = \sum_{i=1}^{K-1} i\pi_{i,1} + \pi_K \sum_{i=K}^{\infty} i\rho^{i-K} ,$$

$$= \pi_{1,1} \left(\sum_{i=1}^{K-1} i \frac{(1+\rho)}{(1-\rho)} - \sum_{i=1}^{K-1} \frac{2i\rho^i}{(1-\rho)} \right) + \pi_{1,1} \frac{(1+\rho) - 2\rho^K}{(1-\rho)} \sum_{i=K}^{\infty} i\rho^{i-K} ,$$

$$= \pi_{1,1} \frac{(1+\rho)}{(1-\rho)} \sum_{i=1}^{K-1} i - \frac{2\pi_{1,1}}{(1-\rho)} \sum_{i=1}^{\infty} i\rho^i + \pi_{1,1} \frac{(1+\rho)}{(1-\rho)} \sum_{i=0}^{\infty} (K+i)\rho^i ,$$

$$= \frac{\pi_{1,1}}{(1-\rho)} \left((1+\rho) \frac{K(K-1)}{2} - \frac{\rho}{(1-\rho)} + K \frac{(1+\rho)}{(1-\rho)} \right) , \qquad (7.31)$$

$$= \pi_0 \frac{(1-\phi)}{(1-\rho)D_0} \left((1+\rho) \frac{K(K-1)}{2} + \frac{K + \rho(K-1)}{(1-\rho)} \right)$$

$$= \frac{(1-\phi)^2}{D_1} \left((1+\rho) \frac{K(K-1)}{2} + \frac{K + \rho(K-1)}{(1-\rho)} \right) .$$

From that we get the expression of the expectation of the number of customers:

$$\mathbb{E}[N] = B_0 + B_1 = \sum_{i=1}^{K-1} i\pi_{i,0} + \sum_{i=1}^{K-1} i\pi_{i,1} + \sum_{i=K}^{\infty} i\pi_i ,$$

$$= \frac{(1-\rho)(1-\phi)(1+2\phi)}{D_1} \left(\frac{K(K+1)}{2} - \frac{(K)}{(1-\phi)} + \frac{\phi(1-\phi^K)}{(1-\phi)^2} \right)$$

$$+\frac{(1-\phi)^2}{D_1}\left((1+\rho)\frac{K(K-1)}{2} + \frac{K+\rho(K-1)}{(1-\rho)}\right),$$

$$=\frac{(1-\phi)}{D_1}\left\{(1-\rho)(1+2\phi)\left(\frac{K(K+1)}{2} - \frac{K}{(1-\phi)} + \frac{\phi(1-\phi^K)}{(1-\phi)^2}\right)\right.$$

$$\left.+(1-\phi)\left((1+\rho)\frac{K(K-1)}{2} + \frac{K+\rho(K-1)}{(1-\rho)}\right)\right\}.$$

This last result corresponds to the expression presented in Sect. 7.3.2. In the special situation where $\phi = 1$, let us first consider the sum B_0. Starting from the equality $B_0 = \pi_{K-1,0}\sum_{i=1}^{K-1}(K-i)(\sum_{j=0}^{i-1}\phi^j)$, obtained above, we get:

$$B_0 = \pi_{K-1,0}\sum_{i=1}^{K-1}(K-i)i,$$

$$= \pi_{K-1,0}\left(K\sum_{i=1}^{K-1}i - \sum_{i=1}^{K-1}i^2\right)$$

$$= \pi_{K-1,0}\left(K\frac{K(K+1)}{2} - \frac{(K-1)K(2K-1)}{6}\right),$$

$$= \pi_{K-1,0}\left(\frac{(K-1)K(K+1)}{6}\right)$$

$$= \pi_0\frac{3}{3K-2}\frac{(K-1)K(K+1)}{6} = \pi_0\frac{(K-1)K(K+1)}{2(3K-2)}.$$

Let us now consider the sum B_1. We may start from the expression (7.31) of B_1 obtained above. Since here $\rho = 1/2$, we get:

$$B_1 = \frac{\pi_{1,1}}{2}(3K(K+3)-4) = \frac{\pi_0}{2(3K-2)}(3K(K+3)-4).$$

After summation of B_0 and B_1 and use of the expression of π_0 given by relation (7.30), we are able to exhibit the following expression:

$$\mathbb{E}[N] = \frac{K(K(K+3)+8)-4}{(K(K+3)-2)}.$$

References

1. J.D. Blackburn, Optimal control of queueing systems with intermittent service. Technical report, DTIC Document (1971)
2. E. Cinlar, *Introduction to Stochastic Processes* (Prentice Hall, Englewood Cliffs, 1975)

3. P.G. Harrison, N.M. Patel, *Performance Modelling of Communication Networks and Computer Architecture* (Addison-Wesley, Reading, 1993)
4. T. Osogami, M. Harchol-Balter, A. Scheller-Wolf, Analysis of cycle stealing with switching times and thresholds. Perform. Eval. **61**(4), 347–369 (2005)
5. M. Parlar, M. Sharafali, Dynamic allocation of airline check-in counters: a queueing optimization approach. Manag. Sci. **54**(8), 1410–1424 (2008)
6. M. Rubinovitch, The slow server problem: a queue with stalling. J. Appl. Probab. **22**(4), 879–892 (1985)
7. R.A. Sahner, K.S. Trivedi, A. Puliafito, *Performance and Reliability Analysis of Computer Systems: An Example-Based Approach Using the SHARPE Software Package* (Kluwer Academic Publishers, Norwell, 1996)

Chapter 8
Simulation from the Tail of the Univariate and Multivariate Normal Distribution

Zdravko Botev and Pierre L'Ecuyer

8.1 Introduction

We consider the problem of simulating a standard normal random variable X, conditional on $a \leq X \leq b$, where $a < b$ are real numbers, and at least one of them is finite. We are particularly interested in the situation where the interval (a, b) is far in one of the tails, that is, we assume that $a \gg 0$ (the case where $b \ll 0$ is covered by symmetry). We do not consider the case where $a \leq 0 \leq b$, as it can be handled easily via standard methods, which do not always work well in the tail case $a \gg 0$. Moreover, if we insist on using inversion, the standard inversion methods break down when we are far in the tail. Inversion is preferable to a rejection method (in general) in various simulation applications, for example to maintain synchronization and monotonicity when comparing systems with common random numbers, for derivative estimation and optimization, when using quasi-Monte Carlo methods, etc. [6, 12–15]. For this reason, a good inversion method is needed, even if rejection is faster. We examine both rejection and inversion methods in this paper.

These problems occur in particular for the estimation of certain Bayesian regression models and for exact simulation from these models; see [4, 7] and the references given there. The simulation from the Bayesian posterior requires repeated draws from a standard normal distribution truncated to different intervals, often far in the tail. Note that to generate X from a more general normal distribution with

Z. Botev (✉)
UNSW Sydney, Sydney, NSW, Australia
e-mail: botev@unsw.edu.au

P. L'Ecuyer
Université de Montréal, Montréal, QC, Canada
Inria - Rennes, Rennes, France
e-mail: lecuyer@iro.umontreal.ca

© Springer International Publishing AG, part of Springer Nature 2019
A. Puliafito, K. S. Trivedi (eds.), *Systems Modeling: Methodologies and Tools*,
EAI/Springer Innovations in Communication and Computing,
https://doi.org/10.1007/978-3-319-92378-9_8

mean μ and variance σ^2 truncated to an interval (a', b'), it suffices to apply a simple linear transformation to recover the standard normal problem studied here.

This paper has three main contributions.

1. *Comparison amongst univariate methods.* The first contribution is to review and compare the speed and efficiency of some of the most popular methods [7, 8, 11, 18, 22, 24] for the tail of the univariate normal distribution. These methods are designed to be efficient when $a \gg 0$ and $b = \infty$ (or $a = -\infty$ and $b \ll 0$ by symmetry), and are not necessarily efficient when the interval $[a, b]$ contains 0. We find that these methods may be adapted in principle to a finite interval $[a, b]$, but they may become inefficient when the interval $[a, b]$ is narrow. We also find that the largely ignored (or forgotten) method of Marsaglia [19] is typically more efficient than the widely used accept–reject methods of Geweke [9] and Robert [22].

2. *Accurate inversion for univariate truncated normal.* All of the methods cited above are rejection methods and we found no reliable inversion method for an interval far in the tail (say, for $a > 38$; see Sect. 8.2). Our second contribution is to propose a new accurate inversion method for arbitrarily large a. Our inversion algorithm is based on a numerically stable implementation of the solution of a nonlinear equation via Newton's method.

3. *Rejection method for multivariate truncated normal.* Our third contribution is to propose a simple rejection method in the multivariate setting, where we wish to simulate a vector X with mean zero and covariance matrix $\Sigma \in \mathbb{R}^{d \times d}$, conditional on $X \geq a$ (the inequality is componentwise). We find that, under some conditions, the proposed method can yield an acceptance probability that approaches unity as we move deeper into the tail region.

Simulation methods for exact simulation from multivariate normal distributions conditional on a general rectangular region, $a \leq X \leq b$, were developed recently in [3, 4, 6]. But for sampling in the tail, the proposed sampler in this paper has two advantages compared to the samplers in these previous works. First, it is much simpler to implement and faster, because it is specifically designed for the tail of the multivariate normal. Second, the theoretical results in [3] do not apply when the target pdf is the most general tail density (see (8.9) in Sect. 8.3), but they do apply for our proposal in this paper. On the downside, the price one pays for these two advantages is that the proposed sampler works only in the extreme tail setting ($[a, \infty]$ with $a \gg 0$), whereas the methods in [3, 4, 6] work in more general non-tail settings ($[a, b]$ which may contain 0).

This chapter is an expanded version of the conference paper [5]. The results of Sect. 8.3 are new while those of Sect. 8.2 were contained in [5].

8.2 Simulation from the Tail of the Univariate Normal

In this section, we use ϕ to denote the density of the standard normal distribution (with mean 0 and variance 1), Φ for its cumulative distribution function (cdf), $\overline{\Phi}$ for the complementary cdf, and Φ^{-1} for the inverse cdf defined as $\Phi^{-1}(u) = \min\{x \in \mathbb{R} \mid \Phi(x) \geq u\}$. Thus, if $X \sim N(0, 1)$, $\Phi(x) = \mathbb{P}[X \leq x] = \int_{-\infty}^{x} \phi(y)dy = 1 - \overline{\Phi}(x)$. Conditional on $a \leq X \leq b$, X has density

$$\frac{\phi(x)}{\Phi(b) - \Phi(a)} \qquad \text{for } a < x < b \tag{8.1}$$

We denote this truncated normal distribution by $\text{TN}_{a,b}(0, 1)$.

It is well known that if $U \sim U(0, 1)$, the uniform distribution over the interval $(0, 1)$, then

$$X = \Phi^{-1}(\Phi(a) + (\Phi(b) - \Phi(a))U) \tag{8.2}$$

has exactly the standard normal distribution conditional on $a \leq X \leq b$. But even though very accurate approximations are available for Φ and Φ^{-1}, (8.2) is sometimes useless for simulating X. One reason for this is that whenever computations are made under the IEEE-754 double precision standard (which is typical), any number of the form $1 - \epsilon$ for $0 \leq \epsilon < 2 \times 10^{-16}$ (approximately) is identified with 1.0, any positive number smaller than about 10^{-324} cannot be represented at all (it is identified with 0), and numbers smaller than 10^{-308} are represented with less than 52 bits of accuracy.

This implies that $\overline{\Phi}(x) = \Phi(-x)$ is identified as 0 whenever $x \geq 39$ and is identified as 1 whenever $-x \geq 8.3$. Thus, (8.2) cannot work when $a \geq 8.3$. In the latter case, or whenever $a > 0$, it is much better to use the equivalent form:

$$X = -\Phi^{-1}(\overline{\Phi}(a) - (\overline{\Phi}(a) - \overline{\Phi}(b))U), \tag{8.3}$$

which is accurate for a up to about 37, assuming that we use accurate approximations of $\overline{\Phi}(x)$ for $x > 0$ and of $\Phi^{-1}(u)$ for $u < 1/2$. Such accurate approximations are available, for example, in [2] for $\Phi^{-1}(u)$ and via the error function erf on most computer systems for $\overline{\Phi}(x)$. For larger values of a (and x), a different inversion approach must be developed, as shown next.

8.2.1 Inversion Far in the Right Tail

When $\overline{\Phi}(x)$ is too small to be represented as a floating-point double, we will work instead with the Mills' [21] ratio, defined as $q(x) \overset{\text{def}}{=} \overline{\Phi}(x)/\phi(x)$, which is the inverse of the hazard rate (or failure rate) evaluated at x. When x is large, this ratio can be approximated by the truncated series (see [1]):

$$q(x) \approx \frac{1}{x} + \sum_{n=1}^{r} \frac{1 \times 3 \times 5 \times \cdots \times (2n-1)}{(-1)^n x^{2n+1}}. \tag{8.4}$$

In our experiments with $x \geq 10$, we compared $r = 5, 6, 7, 8$, and we found no significant difference (up to machine precision) in the approximation of X in (8.3) by the method we now describe. In view of (8.3), we want to find x such that $\overline{\Phi}(x) = \Phi(-x) = \overline{\Phi}(a) - (\overline{\Phi}(a) - \overline{\Phi}(b))u$, for $0 \leq u \leq 1$, when a is large. This equation can be rewritten as $h(x) = 0$, where

$$h(x) \overset{\text{def}}{=} \overline{\Phi}(a) - \overline{\Phi}(x) + (\overline{\Phi}(b) - \overline{\Phi}(a))u \tag{8.5}$$

To solve $h(x) = 0$, we start by finding an approximate solution and then refine this approximation via Newton iterations. We detail how this is achieved. To find an approximate solution, we replace the normal cdf Φ in (8.3) by the standard Rayleigh distribution, whose complementary cdf and density are given by $\overline{F}(x) = \exp(-x^2/2)$ and $f(x) = x \exp(-x^2/2)$ for $x > 0$. Its inverse cdf can be written explicitly as $F^{-1}(u) = (-2\ln(1-u))^{1/2}$. This choice of approximation of Φ^{-1} in the tail has been used before (see, for example, [2] and Sect. 8.4). It is motivated by the facts that $F^{-1}(u)$ is easy to compute and that $\bar{\Phi}(x)/\bar{F}(x) \to 1$ rapidly when $x \to \infty$. By plugging $\overline{F}$ and F^{-1} in place of $\overline{\Phi}$ and Φ^{-1} in (8.3), and solving for x, we find the approximate root

$$x \approx \sqrt{a^2 - 2\ln\left(1 - u + u\exp\left((a^2 - b^2)/2\right)\right)}, \tag{8.6}$$

which is simply the u-th quantile of the standard Rayleigh distribution truncated over (a, b), with density

$$f(x) = \frac{x\exp(-(x^2 - a^2)/2)}{1 - \exp(-(b^2 - a^2)/2)} \qquad \text{for } a < x < b. \tag{8.7}$$

The next step is to improve the approximation (8.6) by applying Newton's method to (8.5). For this, it is convenient to make the change of variable $x = \xi(z)$, where $\xi(z) \overset{\text{def}}{=} \sqrt{a^2 - 2\ln(z)}$ and $z = \xi^{-1}(x) = \exp((a^2 - x^2)/2)$, and apply Newton's method to $g(z) \overset{\text{def}}{=} h(\xi(z))$. Newton's iteration for solving $g(z) = 0$ has the form $z_{\text{new}} = z - g(z)/g'(z)$, where

$$\frac{g(z)}{g'(z)} = \frac{h(\xi(z))}{h'(\xi(z))} \cdot \frac{1}{\xi'(z)}, \qquad \text{(by the chain rule)}$$

$$= z\xi(z)\frac{\overline{\Phi}(\xi(z)) - \overline{\Phi}(a) + u(\overline{\Phi}(a) - \overline{\Phi}(b))}{\phi(\xi(z))}$$

$$= x\left(zq(x) - q(a)(1-u) - q(b)u\exp\left(\tfrac{a^2 - b^2}{2}\right)\right),$$

and the identity $x = \xi(z)$ was used for the last equality. A key observation here is that, thanks to the replacement of $\overline{\Phi}$ by q, the computation of $g(z)/g'(z)$ does not involve extremely small quantities that can cause numerical underflow, even for extremely large a.

The complete procedure is summarized in Algorithm 8.1, which we have implemented in Java, MATLAB®, and **R**. According to our experiments, the larger the a, the faster the convergence. For example, for $a = 50$ one requires at most 13 iterations to ensure $\delta_x \leq \delta^* = 10^{-10}$, where δ_x represents the relative change in x in the last Newton iteration.

Algorithm 8.1 : Computation of the u-quantile of $\mathrm{TN}_{a,b}(0, 1)$

Require: Input $u \in (0, 1)$, δ^*
  $q_a \leftarrow q(a)$
  $q_b \leftarrow q(b)$
  $c \leftarrow q_a(1 - u) + q_b u \exp(\frac{a^2 - b^2}{2})$
  $\delta_x \leftarrow \infty$
  $z \leftarrow 1 - u + u \exp(\frac{a^2 - b^2}{2})$
  $x \leftarrow \sqrt{a^2 - 2\ln(z)}$
  repeat
    $z \leftarrow z - x(zq(x) - c)$
    $x_{\mathrm{new}} \leftarrow \sqrt{a^2 - 2\ln(z)}$
    $\delta_x \leftarrow |x_{\mathrm{new}} - x|/x$
    $x \leftarrow x_{\mathrm{new}}$
  until $\delta_x \leq \delta^*$
  return Quantile x

We note that for an interval $[a, b] = [a, a + w]$ of fixed length w, when a increases the conditional density concentrates closer to a. In fact, there is practically no difference between generating X conditional on $a \leq X \leq a + 1$ and conditional on $X \geq a$ when $a \geq 30$, but there can be a significant difference for small a.

8.2.2   Rejection Methods

We now examine *rejection* (or *acceptance-rejection*) methods, which can be faster than inversion. A large collection of rejection-based generation methods for the normal distribution have been proposed over the years; see [7, 8, 11, 24] for surveys, discussions, comparisons, and tests. Most of them (the fastest ones) use a change of variable and/or precomputed tables to speed up the computations. In its most elementary form, a rejection method to generate from some density f uses a hat function $h \geq f$ and rescales h vertically to a probability density $g = h/\int_{-\infty}^{\infty} h(y)\mathrm{d}y$, often called the proposal density. A random variate X is generated from g, is accepted with probability $f(X)/h(X)$, is rejected otherwise, and the procedure is repeated until X is accepted as the retained realization. In practice, more elaborate versions are used that incorporate transformations and partitions of the area under h.

Any of these proposed rejection methods can be applied easily if $\Phi(b) - \Phi(a)$ is large enough, just by adding a rejection step to reject any value that falls outside $[a, b]$. The acceptance probability for this step is $\Phi(b) - \Phi(a)$. When this probability is too small, this becomes too inefficient and something else must be done. One way is to define a proposal g whose support is exactly $[a, b]$, but this could be inefficient (too much overhead) when a and b change very often. Chopin [7] developed a rejection method specially designed for this situation. The method works by juxtaposing a large number of rectangles of different heights (with equal surface) over some finite interval $[a_{\min}, a_{\max}]$, similar to the trapezoidal approximation in numerical quadrature. However, even Chopin's method achieves efficiency only when it uses an exponential proposal with rate $a = a_{\max}$, when $a_{\max}$ is large enough. Generally, Chopin's trapezoidal method is very fast, and possibly the best method, when $[a_{\min}, a_{\max}]$ contains or is close to zero, but it requires the storage of large precomputed tables. This can be slow on hardware for which memory is limited, like GPUs.

It uses an exponential proposal with rate $a = a_{\max}$ (the RejectTail variant of Algorithms 8.2 below) for the tail above $a_{\max}$ or when $a > a'_{\max}$. The fastest implementation uses 4000 rectangles, $a_{\max} \approx 3.486$, $a'_{\max} \approx 2.605$. This method is fast, although it requires the storage of very large precomputed tables, which could actually slow down computations on certain type of hardware for which memory is limited, like GPUs.

Simple rejection methods for the standard normal truncated to $[a, \infty)$, for $a \geq 0$, have been proposed long ago. Marsaglia [19] proposed a method that uses for g the standard Rayleigh distribution truncated over $[a, \infty)$. An efficient implementation is given in [8, page 381]. Devroye [8, page 382] also gives an algorithm that uses for g an exponential density of rate a shifted by a. These two methods have exactly the same acceptance probability,

$$\alpha(a) = a\sqrt{2\pi} \exp(a^2/2)\overline{\Phi}(a), \tag{8.8}$$

which converges to 1 when $a \to \infty$. Geweke [9] and Robert [22] optimized the acceptance probability to

$$\beta(a) = \lambda\sqrt{2\pi} \exp\left(a\lambda - \lambda^2/2\right) \overline{\Phi}(a)$$

by taking the rate $\lambda = (a + \sqrt{a^2 + 4})/2 > a$ for the shifted exponential proposal. However, the gain with respect to Devroye's method is small and can be wiped out easily by a larger computing time per step. For large a, both are very close to 1 and there is not much difference between them.

We will compare two ways of adapting these methods to a truncation over a finite interval $[a, b]$. The first one is to keep the same proposal g which is positive over the interval $[a, \infty)$ and reject any value generated above b. The second one truncates and rescales the proposal to $[a, b]$ and applies rejection with the truncated proposal. We label them by *RejectTail* and *TruncTail*, respectively. TruncTail has a smaller rejection probability, by the factor $1 - \overline{\Phi}(a)/\overline{\Phi}(b)$, but also entails additional overhead to properly truncate the proposal. Typically, it is worthwhile only if this additional overhead is small and/or the interval $[a, b]$ is very narrow, so it improves the rejection probability significantly. Our experiments will confirm this.

Algorithms 8.2, 8.3, 8.4 state the rejection methods for the TruncTail case with the exponential proposal with rate a [8], with the rate λ proposed in [22], and with the standard Rayleigh distribution, respectively, extended to the case of a finite interval $[a, b]$. For the RejectTail variant, one would remove the computation of q, replace $\ln(1 - qU)$ by $\ln U$, and add $X \le b$ to the acceptance condition. Algorithm 8.5 gives this variant for the Rayleigh proposal.

Algorithm 8.2 : $X \sim \mathsf{TN}_{a,b}(0, 1)$ with exponential proposal with rate a, truncated

$K_a \leftarrow 2a^2$
$q \leftarrow 1 - \exp(-(b-a)a)$
repeat
    Generate $U, V \sim \mathsf{U}(0, 1)$, independent
    $X \leftarrow -\ln(1 - qU)$
    $E \leftarrow -\ln(V)$
until $X^2 \le K_a V$
return $a + X/a$

Algorithm 8.3 : $X \sim \mathsf{TN}_{a,b}(0, 1)$ with exponential proposal with rate λ, truncated

$\lambda \leftarrow (a + \sqrt{a^2 + 4})/2$
$q \leftarrow 1 - \exp(-(b-a)\lambda)$
repeat
    Generate $U, V \sim \mathsf{U}(0, 1)$, independent
    $X \leftarrow a - \ln(1 - qU)/\lambda$
until $V \le \exp((X - \lambda)^2/2)$
return $a + X/a$

Algorithm 8.4 : $X \sim \mathsf{TN}_{a,b}(0, 1)$ with Rayleigh proposal, truncated

$c \leftarrow a^2/2$
$q \leftarrow 1 - \exp(c - b^2/2)$
repeat
    Simulate $U, V \sim \mathsf{U}(0, 1)$, independently.
    $X \leftarrow c - \ln(1 - qU)$
until $V^2 X \le a$
return $X \leftarrow \sqrt{2X}$

Algorithm 8.5 : $X \sim \mathsf{TN}_{a,\infty}(0, 1)$ with Rayleigh proposal and RejectTail

---

$c \leftarrow a^2/2$
repeat
    Simulate $U, V \sim \mathsf{U}(0, 1)$, independently.
    $X \leftarrow c - \ln(U)$
until $V^2 X \le a$ and $2X \le b * b$
return $\sqrt{2X}$

When the interval $[a, b]$ is very narrow, it makes sense to just use the uniform distribution over this interval for the proposal g. This is suggested in [22] and shown in Algorithm 8.6. Generating from the proposal is then very fast. On the other hand, the acceptance probability may become very small if the interval is far in the tail and $b - a$ is not extremely small. Indeed, the acceptance probability of Algorithm 8.6 is:

$$\frac{\sqrt{2\pi} \exp(a^2/2)(\overline{\Phi}(a) - \overline{\Phi}(b))}{b-a} = \frac{q(a) - q(b) \exp((a^2 - b^2)/2)}{b-a},$$

which decays at a rate of $1/a$ when $a \to \infty$ while $(b - a) = \mathcal{O}(1)$ remains asymptotically constant ($f(x) = \mathcal{O}(g(x))$ stands for $\lim_{x \uparrow \infty} |f(x)/g(x)| < \infty$). However, when the length of the interval $(b - a) = \mathcal{O}(1/a)$, then the acceptance probability is easily shown to be asymptotically $\mathcal{O}(1)$, rendering Algorithm 8.6 very useful in the tail. In fact, later in Table 8.2 we report that over the interval $[a, b] = [100.0, 100.0001]$ Algorithm 8.6 is decidedly the fastest method.

Algorithm 8.6 : $X \sim \mathsf{TN}_{a,b}(0, 1)$ with uniform proposal, truncated

---

repeat
    Simulate $U, V \sim \mathsf{U}(0, 1)$, independently.
    $X \leftarrow a + (b - a)U$
until $2 \ln V \le a^2 - X^2$
return X

Another choice that the user can have with those generators (and for any variate generator that depends on some distribution parameters) is to either *precompute* various constants that depend on the parameters and store them in some "distribution" object with fixed parameter values, or to *recompute* these parameter-dependent constants each time a new variate is generated. This type of alternative is common in modern variate generation software [16, 17]. The first approach is worthwhile if the time to compute the relevant constants is significant and several random variates are to be generated with exactly the same distribution parameters. For the applications in Bayesian statistics mentioned earlier, it is typical that the parameters a and b change each time a new variate is generated [7]. But there can be applications in which a large number of variates are generated with the same a and b.

For one-sided intervals $[a, \infty)$, the algorithms can be simplified. One can use the RejectTail framework and since $b = \infty$, there is no need to check if $X \leq b$. When reporting our test results, we label this the *OneSide* case.

Note that computing an exponential is typically more costly than computing a log (by a factor of 2 or 3 for negative exponents and 10 for large exponents, in our experiments) and the latter is more costly than computing a square root (also by a factor of 10). This means significant speedups could be obtained by avoiding the recomputing of the exponential each time at the beginning of Algorithms 8.2, 8.3, and 8.4. This is possible if the same parameter b is used several times, or if $b = \infty$, or if we use RejectTail instead of TruncTail.

8.2.3   Speed Comparisons

We report a representative subset of results of speed tests made with the different methods, for some pairs (a, b). In each case, we generated 10^8 (100 million) truncated normal variates, added them up, printed the CPU time required to do that, and printed the sum for verification. The experiments were made in Java using the SSJ library [16], under Eclipse and Windows 10, on a Lenovo X1 Carbon Thinkpad with an Intel Core(TM) i7-5600U (single) processor running at 2.60 GHz. All programs were executed in a single thread and the CPU times were measured using the stopwatch facilities in class `Chrono` of SSJ, which relies on the `getThreadCpuTime` method from the Java class `ThreadMXBean` to obtain the CPU time consumed so far by a single thread, and subtracts to obtain the CPU time consumed between any two instructions.

The measurements were repeated a few times to verify consistency and varied by about 1–2% at most. The compile times are negligible relative to the reported times. Of course, these timings depend on CPU and memory usage by other processes on the computer, and they are likely to change if we move to a different platform, but on standard processors the relative timings should remain roughly the same. They provide a good idea of what is most efficient to do.

Tables 8.1 and 8.2 report the timings, in seconds. The two columns "recompute" and "precompute" are for the cases where the constants that depend on a and b are recomputed each time a random variate is generated or are precomputed once and for all, respectively, as discussed earlier.

ExponD, ExponR, and Rayleigh refer to the TruncTail versions of Algorithms 8.2, 8.3, and 8.4, respectively. We add "RejectTail" to the name for the RejectTail versions. For ExponRRejectTailLog, we took the log on both sides of the inequality to remove the exponential in the "until" condition. Uniform refers to Algorithm 8.6.

InversionSSJ refers to the default inversion method implemented in SSJ, which uses [2] and gives at least 15 decimal digits of relative precision, combined with a generic (two-sided) "truncated distribution" class also offered in SSJ. InverseQuickSSJ is a faster but much less accurate version based on a cruder

Table 8.1 Time to generate $n = 10^8$ variates for $[a, b] = [3.0, 3.1]$ (left pane) and $[a, b] = [7.0, 8.0]$ (right pane)

Method	CPU time (s)		Method	CPU time	
	recom.	precom.		recom.	precom.
Generation in $[a, b)$			*Generation in $[a, b)$*		
ExponD	6.46	6.22	ExponD	11.70	6.16
ExponDRejectTail	23.04	23.20	ExponDRejectTail	6.04	6.08
ExponR	16.63	9.92	ExponR	15.96	8.98
ExponRRejectTail	32.40	32.40	ExponRRejectTail	9.20	9.09
ExponRRejectTailLog	25.10	25.30	ExponRRejectTailLog	7.03	7.02
Rayleigh	10.29	4.60	Rayleigh	9.86	4.27
RayleighRejectTail	15.23	15.33	RayleighRejectTail	3.91	3.99
Uniform	4.26	4.34	Uniform	25.40	25.68
InverseSSJ	15.14	8.14	InverseSSJ	30.67	8.14
InverseQuickSSJ	18.80	3.31	InverseQuickSSJ	n/a	n/a
InverseRightTail	31.12	7.66	InverseRightTail	31.12	7.70
Generation in $[a, \infty)$			*Generation in $[a, \infty)$*		
ExponDOneSide	6.43	6.46	ExponDOneSide	5.90	5.96
ExponROneSideLog	7.05	6.99	ExponROneSideLog	6.80	6.71
RayleighOneSide	4.07	4.41	RayleighOneSide	3.74	4.05
InverseSSJOneSide	18.81	8.20	InverseSSJOneSide	19.00	8.19
InverseRightTailOneSide	18.72	7.64	InverseRightTailOneSide	18.76	7.59

Table 8.2 Time to generate $n = 10^8$ variates for $[a, b] = [100.0, 102.0]$ (left pane) and $[a, b] = [100.0, 100.0001]$ (right pane)

Method	CPU time (s)		Method	CPU time	
	recom.	precom.		recom.	precom.
Generation in $[a, b)$			*Generation in $[a, b)$*		
ExponD	11.68	6.01	ExponD	12.31	6.83
ExponDRejectTail	5.88	5.91	ExponDRejectTail	543.80	546.58
ExponR	15.79	8.86	ExponR	16.47	10.65
ExponRRejectTail	9.13	9.02	ExponRRejectTail	865.24	865.34
ExponRRejectTailLog	6.93	6.96	ExponRRejectTailLog	651.19	648.99
Rayleigh	9.97	4.16	Rayleigh	10.59	5.07
RayleighRejectTail	3.84	3.90	RayleighRejectTail	323.08	322.41
Uniform	650.62	656.42	Uniform	3.59	3.62
InverseMillsRatio	22.31	15.97	InverseMillsRatio	18.03	12.12
Generation in $[a, \infty)$			*Generation in $[a, \infty)$*		
ExponDOneSide	5.77	5.82	ExponDOneSide	5.79	5.83
ExponROneSideLog	6.72	6.63	ExponROneSideLog	6.74	6.63
RayleighOneSide	3.67	3.96	RayleighOneSide	3.66	3.99
InverseMillsRatioOneSide	15.62	15.84	InverseMillsRatioOneSide	15.67	15.84

approximation of $\overline{\Phi}$ from [20] based on table lookups, which returns about six decimal digits of precision. We do not recommend it, due to its low accuracy. Moreover, the implementation we used does not handle well values larger than about 5 in the right tail, so we report results only for small a. InverseRightTail uses the accurate approximation of $\overline{\Phi}$ together with (8.3). InverseMillsRatio is our new inversion method based on Mills ratio, with $\delta^* = 10^{-10}$. This method is designed for the case where a is large, and our implementation is designed to be accurate for $a \geq 10$, so we do not report results for it in Table 8.1. For all the methods, we add "OneSide" for the simplified OneSide versions, for which $b = \infty$.

For the OneSide case, that is, $b = \infty$, the Rayleigh proposal gives the fastest method in all cases, and there is no significant gain in precomputing and storing the constant $c = a^2/2$.

For finite intervals $[a, b]$, when $b - a$ is very small so $\overline{\Phi}(b)/\overline{\Phi}(a)$ is close to 1, the uniform proposal wins and the RejectTail variants are very slow. See right pane of Table 8.2. Precomputing the constants is also not useful for the uniform proposal. For larger intervals in the tail, $\overline{\Phi}(x)$ decreases quickly at the beginning of the interval and this leads to very low acceptance ratios; see right pane of Table 8.1 and left pane of Table 8.2. A Rayleigh proposal with the RejectTail option is usually the fastest method in this case. Precomputing and storing the constants is also not very useful for this option. For intervals closer to the center, as in the left pane of Table 8.1, the uniform proposal performs well for larger (but not too large) intervals, and the RejectTail option becomes slower unless $[a, b]$ is very wide. The reason is that for a fixed $w > 0$, $\overline{\Phi}(a + w)/\overline{\Phi}(a)$ is larger (closer to 1) when $a > 0$ is closer to 0.

8.3 Simulation from the Tail of the Multivariate Normal

Let $\phi_\Sigma(y)$ and

$$\overline{\Phi}_\Sigma(a) = \mathbb{P}[Y \geq a], \quad Y \sim \mathsf{N}(0, \Sigma),$$

denote the density and tail distribution, respectively, of the multivariate $\mathsf{N}(0, \Sigma)$ distribution with (positive-definite) covariance matrix $\Sigma \in \mathbb{R}^{d \times d}$. In the multivariate extension to (8.1), we wish to simulate from the pdf ($\mathbb{I}\{\cdot\}$ is the indicator function):

$$\frac{\phi_\Sigma(y)\mathbb{I}\{y \geq a(\gamma)\}}{\overline{\Phi}_\Sigma(a(\gamma))}, \tag{8.9}$$

where $\max_i a_i > 0$, and γ is a *tail parameter* such that at least one component of $a(\gamma)$ diverges to ∞ when $\gamma \to \infty$ (that is, $\lim_{\gamma \uparrow \infty} \|a(\gamma)\| = \infty$, see [10]). To simulate from this conditional density, we describe a rejection algorithm that uses an optimally designed multivariate exponential proposal. Interestingly, unlike the truncated exponential proposal in the one-dimensional setting (see Algorithms 8.2

and 8.3), our multivariate exponential proposal is not truncated. Before giving the details of the rejection algorithm, we need to introduce some preliminary theory and notation.

8.3.1 Preliminaries and Notation

Define P as a permutation matrix, which maps $(1, \ldots, d)^\top$ into the permutation vector $\boldsymbol{p} = (p_1, \ldots, p_d)^\top$, that is, $P(1, \ldots, d)^\top = \boldsymbol{p}$. Then, $\overline{\Phi}_\Sigma(\boldsymbol{a}(\gamma)) = \mathbb{P}(PY \geq P\boldsymbol{a}(\gamma))$ and $PY \sim N(\boldsymbol{0}, P\Sigma P^\top)$ for any $\boldsymbol{p}$. We will specify $\boldsymbol{p}$ shortly.

First, define the constrained (convex) quadratic optimization:

$$\min_{\boldsymbol{y}} \frac{1}{2} \boldsymbol{y}^\top \left(P\Sigma P^\top \right)^{-1} \boldsymbol{y}$$

$$\text{subject to: } \boldsymbol{y} \geq P\boldsymbol{a}(\gamma).$$

(8.10)

Suppose $\boldsymbol{\lambda} \in \mathbb{R}^d$ is the Lagrange multiplier vector, associated with (8.10). Partition the vector as $\boldsymbol{\lambda} = (\boldsymbol{\lambda}_1^\top, \boldsymbol{\lambda}_2^\top)^\top$ with $\dim(\boldsymbol{\lambda}_1) = d_1$ and $\dim(\boldsymbol{\lambda}_2) = d_2$, where $d_1 + d_2 = d$. In the same way, partition vectors $\boldsymbol{y}, \boldsymbol{a}$, and matrix

$$\Sigma = \begin{pmatrix} \Sigma_{11} & \Sigma_{12} \\ \Sigma_{21} & \Sigma_{22} \end{pmatrix}.$$

(8.11)

We now observe that we can select the permutation vector $\boldsymbol{p}$ and the corresponding matrix P so that all the d_1 active constraints in (8.10) correspond to $\boldsymbol{\lambda}_1 > \boldsymbol{0}$ and all the d_2 inactive constraints correspond to $\boldsymbol{\lambda}_2 = \boldsymbol{0}$. Without loss of generality, we can thus assume that $\boldsymbol{a}$ and Σ are reordered via the permutation matrix P as a preprocessing step. After this preprocessing step, the solution $\boldsymbol{y}^*$ of (8.10) with P = I will satisfy $\boldsymbol{y}_1^* = \boldsymbol{a}_1$ (active constraints: $\boldsymbol{\lambda}_1 > \boldsymbol{0}$) and $\boldsymbol{y}_2^* > \boldsymbol{a}_2$ (inactive constraints: $\boldsymbol{\lambda}_2 = \boldsymbol{0}$).

We also assume that for large enough γ, the active constraint set of (8.10) becomes independent of γ, see [10]. An example is given in Corollary 8.1 below.

8.3.2 The Rejection Algorithm

First, we note that simulating Y from (8.9) is equivalent to simulating $X \sim N(-\boldsymbol{a}(\gamma), \Sigma)$, conditional on $X \geq \boldsymbol{0}$, and then delivering $Y = X + \boldsymbol{a}$. Thus, our initial goal is to simulate from the target:

$$\pi(\boldsymbol{x}) = \frac{\phi_\Sigma(\boldsymbol{x} + \boldsymbol{a}(\gamma)) \mathbb{I}\{\boldsymbol{x} \geq \boldsymbol{0}\}}{\overline{\Phi}_\Sigma(\boldsymbol{a}(\gamma))}.$$

Second, the partitioning into active and inactive constraints of (8.10) suggests the following proposal density: $g(\boldsymbol{x}; \boldsymbol{\eta}) = g_1(\boldsymbol{x}_1; \boldsymbol{\eta})g_2(\boldsymbol{x}_2|\boldsymbol{x}_1)$, $\boldsymbol{\eta} > \boldsymbol{0}$, where

$$g_1(\boldsymbol{x}_1; \boldsymbol{\eta}) = \exp(-\boldsymbol{\eta}^\top \boldsymbol{x}_1)\prod_{k=1}^{d_1}\eta_k, \quad \boldsymbol{x}_1 \geq \boldsymbol{0}$$

is a multivariate exponential proposal, and

$$g_2(\boldsymbol{x}_2|\boldsymbol{x}_1) = \frac{\phi_\Sigma(\boldsymbol{x} + \boldsymbol{a})}{\phi_{\Sigma_{11}}(\boldsymbol{x}_1 + \boldsymbol{a}_1)}$$

is the multivariate normal pdf of $\boldsymbol{x}_2$, conditional on $\boldsymbol{x}_1$ (see [12, Page 146]):

$$\boldsymbol{X}_2|(\boldsymbol{X}_1 = \boldsymbol{x}_1) \sim \mathsf{N}(-\boldsymbol{a}_2 + \Sigma_{12}^\top\Sigma_{11}^{-1}(\boldsymbol{x}_1 + \boldsymbol{a}_1), \ \Sigma_{22} - \Sigma_{12}^\top\Sigma_{11}^{-1}\Sigma_{12}).$$

With this proposal, the likelihood ratio for acceptance–rejection is

$$\frac{\pi(\boldsymbol{x})\overline{\Phi}_\Sigma(\boldsymbol{a}(\gamma))}{g(\boldsymbol{x}; \boldsymbol{\eta})} = \mathbb{I}\{\boldsymbol{x} \geq \boldsymbol{0}\}\frac{\phi_{\Sigma_{11}}(\boldsymbol{x}_1 + \boldsymbol{a}_1)}{g_1(\boldsymbol{x}_1; \boldsymbol{\eta})} = \mathbb{I}\{\boldsymbol{x} \geq \boldsymbol{0}\}\exp\left(\psi(\boldsymbol{x}_1; \boldsymbol{\eta})\right),$$

where ψ is defined as

$$\psi(\boldsymbol{x}_1; \boldsymbol{\eta}) := -\frac{(\boldsymbol{x}_1 + \boldsymbol{a}_1)^\top\Sigma_{11}^{-1}(\boldsymbol{x}_1 + \boldsymbol{a}_1)}{2} + \boldsymbol{\eta}^\top\boldsymbol{x}_1 - \sum_{k=1}^{d_1}\ln(\eta_k)$$

$$-\frac{\ln|\Sigma_{11}|}{2} - \frac{d_1\ln(2\pi)}{2}.$$

Next, our goal is to select the value for $\boldsymbol{\eta}$ that will maximize the acceptance rate of the resulting rejection algorithm (see Algorithm 8.7 below).

It is straightforward to show that, with the given proposal density, the acceptance rate for a fixed $\boldsymbol{\eta} > \boldsymbol{0}$ is given by

$$\overline{\Phi}_\Sigma(\boldsymbol{a}(\gamma))\exp(-\max_{\boldsymbol{x}_1 \geq \boldsymbol{0}}\psi(\boldsymbol{x}_1; \boldsymbol{\eta})).$$

Hence, to maximize the acceptance rate, we minimize $\max_{\boldsymbol{x}_1 \geq \boldsymbol{0}}\psi(\boldsymbol{x}_1; \boldsymbol{\eta})$ with respect to $\boldsymbol{\eta}$. In order to compute the minimizing $\boldsymbol{\eta}$, we exploit a few of the properties of ψ.

The most important property is that ψ is concave in $\boldsymbol{x}_1$ for every $\boldsymbol{\eta}$, and that ψ is convex in $\boldsymbol{\eta}$ for every $\boldsymbol{x}_1$. Moreover, ψ is continuously differentiable in $\boldsymbol{\eta}$, and we have the saddle-point property (see [3]):

$$\min_{\boldsymbol{\eta} > \boldsymbol{0}}\max_{\boldsymbol{x}_1 \geq \boldsymbol{0}}\psi(\boldsymbol{x}_1; \boldsymbol{\eta}) = \max_{\boldsymbol{x}_1 \geq \boldsymbol{0}}\min_{\boldsymbol{\eta} > \boldsymbol{0}}\psi(\boldsymbol{x}_1; \boldsymbol{\eta}). \tag{8.12}$$

Let $\psi^* = \psi(x_1^*; \eta^*)$ denote the optimum of the minimax optimization (8.12) at the solution x_1^* and η^*. The right-hand-side of (8.12) suggests a method for computing η^*, namely, we can first minimize with respect to η (this gives $\eta = 1/x_1$, where the vector division is componentwise), and then maximize over $x_1 \geq 0$. This yields the concave (unconstrained) optimization program for x_1^*:

$$x_1^* = \operatorname{argmax} \left\{ -\frac{(x_1 + a_1)^\top \Sigma_{11}^{-1}(x_1 + a_1)}{2} + \sum_{k=1}^{d_1} \ln x_k \right\}. \tag{8.13}$$

It then follows that $\eta^* = 1/x_1^*$. In summary, we have the following algorithm for simulation from (8.9).

Algorithm 8.7 : $X \sim \mathsf{N}(\mathbf{0}, \Sigma)$ conditional on $X \geq a(\gamma)$, for large γ

Solve (8.10) with P = I and compute the associated Lagrange multiplier λ. Using λ, construct the reordering (permutation) matrix P, if needed.
$a \leftarrow \mathsf{P}a$
$\Sigma \leftarrow \mathsf{P}\Sigma\mathsf{P}^\top$
Let L be the lower triangular Cholesky factor of $\Sigma_{22} - \Sigma_{12}^\top \Sigma_{11}^{-1} \Sigma_{12}$, see (8.11)
Solve the concave optimization problem (8.13) to obtain x_1^*
$\eta_1^* \leftarrow 1/x_1^*$
$\psi^* \leftarrow \psi(x_1^*; \eta^*)$
repeat
  repeat
    Simulate $U_0, U_1, \ldots, U_{d_1} \sim \mathsf{U}(0, 1)$, independently
    $E_k \leftarrow -\ln(U_k)/\eta_k^*$ for $k = 1, \ldots, d_1$
    $X_1 \leftarrow (E_1, \ldots, E_{d_1})^\top$ {simulate $X_1 \sim g_1(x_1; \eta^*)$}
    $E \leftarrow -\ln(U_0)$
  until $E > \psi^* - \psi(X_1; \eta^*)$
  $Z_2 \leftarrow (Z_1, \ldots, Z_{d_2})^\top$, where $Z_1, \ldots, Z_{d_2} \sim \mathsf{N}(0, 1)$, independently.
  $X_2 \leftarrow \mathsf{L}Z_2 - a_2 + \Sigma_{12}^\top \Sigma_{11}^{-1}(X_1 + a_1)$ {simulate $X_2 \sim g_2(x_2|X_1)$}
until $X_2 \geq 0$
$X \leftarrow X + a$ {shift to obtain draw from pdf (8.9)}
$X \leftarrow \mathsf{P}^\top X$ {reverse reordering, if any}
return X

8.3.3 Asymptotic Efficiency

The acceptance rate of Algorithm 8.7 above is

$$\mathbb{P}_g[E > \psi^* - \psi(X_1; \eta); X_2 \geq 0] = \overline{\Phi}_\Sigma(a(\gamma)) \exp(-\psi^*),$$

where $\mathbb{P}_g$ indicates that X was drawn from the proposal $g(x; \eta^*)$. As in the one-dimensional case, see (8.8), it is of interest to find out how this rate depends on the tail parameter γ. In particular, if the acceptance rate decays to zero rapidly as $\gamma \uparrow \infty$, then Algorithm 8.7 will not be a viable algorithm for simulation from the tail of the multivariate Gaussian. Fortunately, the following result asserts that the acceptance rate does not decay to zero as we move further and further into the tail of the Gaussian.

Theorem 8.1 (Asymptotically Bounded Acceptance Rate) *Let y^* be the solution to (8.10) after any necessary reordering via permutation matrix* P. *Define $a_\infty :=$ $\lim_{\gamma \uparrow \infty}(a_2(\gamma) - y_2^*(\gamma))$ with $a_\infty \leq 0$. Then, the acceptance rate of the rejection Algorithm 8.7 is ultimately bounded from below:*

$$\liminf_{\gamma \uparrow \infty} \overline{\Phi}_\Sigma(a(\gamma)) \exp(-\psi^*(\gamma)) \geq \mathbb{P}[Y_2 \geq a_\infty \mid Y_1 = 0],$$

where the probability $\mathbb{P}[Y_2 \geq a_\infty \mid Y_1 = 0]$ is calculated under the original measure (that is, $Y \sim N(0, \Sigma)$) and, importantly, does not depend on γ.

Proof First, note that with the assumptions and notation of Sect. 8.3.1, Hashorva and Hüsler [10] have shown the following:

$$\overline{\Phi}_\Sigma(a(\gamma)) = \frac{\mathbb{P}[Y_2 \geq a_\infty \mid Y_1=0]}{(2\pi)^{d_1/2}|\Sigma_{11}|^{1/2}\prod_{k=1}^{d_1} e_k^\top \Sigma_{11}^{-1} a_1} \exp\left(-\frac{a_1^\top \Sigma_{11}^{-1} a_1}{2}\right)(1+o(1)), \quad \gamma \uparrow \infty,$$

where e_k is the unit vector with a 1 in the k-th position, and $f(x) = o(g(x))$ stands for $\lim_{x \to a} f(x)/g(x) = 0$.

Second, the saddle-point property (8.12) implies the following sequence of inequalities for any arbitrary η: $\psi^* \leq \psi(x_1^*; \eta) \leq \max_{x_1} \psi(x_1; \eta)$. In particular, when $\eta = \Sigma_{11}^{-1} a_1$, then $\max_{x_1} \psi(x_1; \Sigma_{11}^{-1} a_1) = \psi(0; \Sigma_{11}^{-1} a_1)$, and we obtain:

$$\exp(-\psi^*) \geq \exp(-\psi(0; \Sigma_{11}^{-1} a_1)) = \frac{\prod_{k=1}^{d_1} e_k^\top \Sigma_{11}^{-1} a_1}{\phi_{\Sigma_{11}}(a_1)}$$

Therefore, $\overline{\Phi}_\Sigma(a(\gamma)) \exp(-\psi^*) \geq \mathbb{P}[Y_2 \geq a_\infty \mid Y_1 = 0](1+o(1))$ as $\gamma \uparrow \infty$, and the result of the theorem follows. ∎

As a special case, we consider the asymptotic result of Savage [23]:

$$\frac{\overline{\Phi}_\Sigma(\gamma \Sigma c)}{\phi_\Sigma(\gamma \Sigma c)} = \frac{1}{\gamma^d \prod_{k=1}^{d} c_k}(1+o(1)), \quad c > 0, \quad \gamma \uparrow \infty, \tag{8.14}$$

which is the multivariate extension of the one-dimensional Mills' ratio [21]: $\frac{\overline{\Phi}(\gamma)}{\phi(\gamma)} = \frac{1}{\gamma}(1+o(1))$. Interestingly, the following corollary shows that when the tail is of the Savage-Mills type, the acceptance probability not only remains bounded away from zero, but approaches unity.

Table 8.3 Estimates of the acceptance probability, $\overline{\Phi}_\Sigma(a(\gamma))\exp(-\psi^*)$, as a function of γ

γ	10	15	20	25	30	50	100	10^3
Accept. rate	0.009	0.04	0.0815	0.15	0.19	0.34	0.44	0.50

Corollary 8.1 (Acceptance with Probability One) *The acceptance rate of Algorithm 8.7 for simulation from (8.9) with $a = \gamma\,\Sigma c$ for some $c > 0$ satisfies:*

$$\lim_{\gamma\uparrow\infty}\overline{\Phi}_\Sigma(\gamma\,\Sigma c)\exp(-\psi^*(\gamma)) = 1$$

Proof Straightforward computations show that the Lagrange multiplier of (8.10) (with P = I, the identity matrix) is $\lambda = \Sigma^{-1}a = \gamma c > 0$, so that the set of inactive constraints is empty. Then, repeating the argument in Theorem 8.1:

$$\exp(-\psi^*) \geq \exp(-\psi(0, \gamma c)) = \frac{\gamma^d \prod_{k=1}^d c_k}{\phi_\Sigma(\gamma\,\Sigma c)} \overset{(8.14)}{=} \frac{1 + o(1)}{\overline{\Phi}_\Sigma(\gamma\,\Sigma c)},$$

as desired. ∎

As a numerical example, we used Algorithm 8.7 to simulate 10^3 random vectors from (8.9) for $d = 10$, $a = \gamma\mathbf{1}$, and $\Sigma = \frac{9}{10}\mathbf{1}\mathbf{1}^\top + \frac{1}{10}I$ (strong positive correlation) for a range of large values of γ.

Table 8.3 above reports the acceptance rate, estimated by observing the proportion of rejected proposals in line 17 of Algorithm 8.7, for a range of different γ.

The table confirms that as γ gets larger, the acceptance rate improves.

8.4 Conclusion

We have proposed and tested inversion and rejection methods to generate a standard normal, truncated to an interval $[a, b]$, when $a \gg 0$. We also proposed a rejection method for the tail of the multivariate normal distribution.

In the univariate setting, inversion is slower than the fastest rejection method, as expected. However, inversion is still desirable in many situations. Our new inversion method excels in those situations when a is large (say, $a \geq 10$). For a not too large (say, $a \leq 30$), the accurate approximation of [2] implemented in InversionSSJ works well.

When inversion is not needed, the rejection method with the Rayleigh proposal is usually the fastest when a is large enough. especially if a large number of variates must be generated for the same interval $[a, b]$, in which case the cost of precomputing the constants used in the algorithm can be amortized over many calls.

It is interesting to see that in the univariate setting, using the Rayleigh proposal is faster than using the truncated exponential proposal as in [7, 9, 22]. The RejectTail variant is usually the fastest, unless $\bar{\Phi}(b)/\bar{\Phi}(a)$ is far from 0, which happens when the interval $[a, b]$ is very narrow or a is not large (say $a \leq 5$).

However, in the multivariate setting, we show that the *truncated* exponential method of [7, 9, 22] can be extended to help simulate from the multivariate normal tail, provided that we use an *untruncated* multivariate exponential proposal (that is, $X \geq 0$) combined with a shift of the Gaussian mean (that is, $Y = X + a$).

References

1. M. Abramowitz, I.A. Stegun, *Handbook of Mathematical Functions* (Dover, New York, 1970)
2. J.M. Blair, C.A. Edwards, J.H. Johnson, Rational Chebyshev approximations for the inverse of the error function. Math. Comput. **30**, 827–830 (1976)
3. Z.I. Botev, The normal law under linear restrictions: simulation and estimation via minimax tilting. J. R. Stat. Soc. Ser. B (Stat. Methodol.) **79**(1), 125–148 (2017)
4. Z.I. Botev, P. L'Ecuyer, Efficient estimation and simulation of the truncated multivariate Student-t distribution, in *Proceedings of the 2015 Winter Simulation Conference* (IEEE Press, Piscataway, 2015), pp. 380–391
5. Z.I. Botev, P. L'Ecuyer, Simulation from the normal distribution truncated to an interval in the tail, in *10th EAI International Conference on Performance Evaluation Methodologies and Tools*, 25th–28th October 2016 Taormina (ACM, New York, 2017), pp. 23–29
6. Z.I. Botev, M. Mandjes, A. Ridder, Tail distribution of the maximum of correlated Gaussian random variables, in *Proceedings of the 2015 Winter Simulation Conference* (IEEE Press, Piscataway, 2015), pp. 633–642
7. N. Chopin, Fast simulation of truncated Gaussian distributions. Stat. Comput. **21**(2), 275–288 (2011)
8. L. Devroye, *Non-Uniform Random Variate Generation* (Springer, New York, NY, 1986)
9. J. Geweke, Efficient simulation of the multivariate normal and Student-t distributions subject to linear constraints and the evaluation of constraint probabilities, in *Computing Science and Statistics: Proceedings of the 23rd Symposium on the Interface*, Fairfax, VA, 1991, pp. 571–578
10. E. Hashorva, J. Hüsler, On multivariate Gaussian tails. Ann. Inst. Stat. Math. **55**(3), 507–522 (2003)
11. W. Hörmann, J. Leydold, G. Derflinger, *Automatic Nonuniform Random Variate Generation* (Springer, Berlin, 2004)
12. D.P. Kroese, T. Taimre, Z.I. Botev, *Handbook of Monte Carlo Methods* (Wiley, New York, 2011)
13. P. L'Ecuyer, Variance reduction's greatest hits, in *Proceedings of the 2007 European Simulation and Modeling Conference*, Ghent (EUROSIS, Hasselt, 2007), pp. 5–12
14. P. L'Ecuyer, Quasi-Monte Carlo methods with applications in finance. Finance Stochast. **13**(3), 307–349 (2009)
15. P. L'Ecuyer, Random number generation with multiple streams for sequential and parallel computers, in *Proceedings of the 2015 Winter Simulation Conference*, pp. 31–44 (IEEE Press, New York, 2015)
16. P. L'Ecuyer, SSJ: stochastic simulation in Java, software library (2016). http://simul.iro.umontreal.ca/ssj/
17. J. Leydold, UNU.RAN—Universal Non-Uniform RANdom number generators (2009). Available at http://statmath.wu.ac.at/unuran/
18. G. Marsaglia, Generating a variable from the tail of the normal distribution. Technometrics **6**(1), 101–102 (1964)
19. G. Marsaglia, T.A. Bray, A convenient method for generating normal variables. SIAM Rev. **6**, 260–264 (1964)
20. G. Marsaglia, A. Zaman, J.C.W. Marsaglia, Rapid evaluation of the inverse normal distribution function. Stat. Probab. Lett. **19**, 259–266 (1994)

21. J.P. Mills, Table of the ratio: area to bounding ordinate, for any portion of normal curve. Biometrika **18**(3/4), 395–400 (1926)
22. C.P. Robert, Simulation of truncated normal variables. Stat. Comput. **5**(2), 121–125 (1995)
23. R.I. Savage, Mills' ratio for multivariate normal distributions. J. Res. Nat. Bur. Standards Sect. B **66**, 93–96 (1962)
24. D.B. Thomas, W. Luk, P.H. Leong, J.D. Villasenor, Gaussian random number generators. ACM Comput. Surv. **39**(4), Article 11 (2007)

Part II
Applications to Communication Systems and Infrastructures

Chapter 9
A Comparison of Markov Reward Based Resource-Latency Aware Heuristics for the Virtual Network Embedding Problem

Francesco Bianchi and Francesco Lo Presti

9.1 Introduction

The future Internet will embrace the Infrastructure as a Service (IaaS) service model [13]. In this novel scenario Infrastructure Providers (InP) lease network resources to Service Providers (SP) which in turn offer services to end users, i.e., virtual networks (VNs) and/or application services [12, 13]. In this multi-layered architecture, an InP has the challenging task to manage and efficiently allocate network and computational resources of its *substrate* network to SPs in order to maximize its revenue. To this end, for each SP request, which takes the general form of a Virtual Network Request (VNR), each consisting of a set of virtual nodes and links with a required amount of computational capacity, link bandwidth, and a maximum link latency constraint, an InP has to determine, in an on-line fashion, the subset of physical nodes and links to host the VNR request.

In the literature, this problem is known as the Virtual Network Embedding (VNE). Since it is a well-known NP-complete problem (Andersen, Theoretical approaches to node assignment, unpublished manuscript, December (2002)), [21], heuristics are necessary to support on-line operations, for realistic problem instances. Many heuristics have been proposed in the literature (see, for instance, [9, 10, 21–23]).

In this paper, we consider the VNE problem with QoS constraints. Our aim is to select a set of nodes and links resources in a substrate network which guarantee a delay bound on latency between pairs of nodes. This is motivated by a wealth of

F. Bianchi (✉) · F. Lo Presti
Department of Civil Engineering and Computer Science Engineering, University of Rome "Tor Vergata", Rome, Italy
e-mail: f.bianchi@ing.uniroma2.it; lopresti@info.uniroma2.it

© Springer International Publishing AG, part of Springer Nature 2019
A. Puliafito, K. S. Trivedi (eds.), *Systems Modeling: Methodologies and Tools*,
EAI/Springer Innovations in Communication and Computing,
https://doi.org/10.1007/978-3-319-92378-9_9

time critical and/or delay-sensitive applications (according to Cisco, about 90% of Internet traffic is generated by delay-sensitive applications [16]). The contributions of this paper are as follows:

- Focusing on our previous works on QoS-aware VNE embedding [5, 7], we present a comparison of two solutions for the VNE problem with QoS constraints (e.g., latency). The approaches are inspired by [9, 15, 22] which developed a new type of two-stage VNE algorithms referred to as *coordinated* algorithms: in the first stage the VNR nodes are mapped to the physical (also called substrate) nodes; then, in the next stage, the VNR links are mapped into substrate paths between the virtual nodes mapped in the former stage.
- Following the two-stage approach idea, our first approach proposes a novel latency-aware metric for a two-stage coordinated VNE algorithm: MCRR-LA (Markov Chain with Rewards-Latency Aware) [5]. Differently from earlier methods, our ranking metric relies on the accumulated reward of a suitable Markov Chain. It well captures the amount of resources available in a node area and the overall QoS (in term of latency). This powerful metric successfully guides the node mapping of the full VNE algorithm. Building on previous metric, our second QoS-aware VNE approach MCRM (Markov Chain Reward Metrics) [7] adds a proximity metric with regard to other nodes and a node similarity concept which improve the mapping quality. In this paper we also propose two new methodologies for computing the resource-latency aware metric called MCRR-LA2 and MCRR-LA3 based on different ways to penalize resources according to the delay. In addition, we devise a non-iterative version of the MCRM algorithm acting on the node mapping stage (MCRM-B), and we propose a new link mapping strategy (MCRM-I-KSP) to balance the link load using K-Shortest-Paths algorithm [11].
- We have evaluated our algorithms through simulation. Our results show that our solutions are able to greatly reduce the maximum and average path delay of the embedded VNRs almost preserving good performances regarding blocking probability, revenue and revenue to cost ratio.

The rest of the paper is organized as follows. In Sect. 9.3 we formalize the VNE problem. Our first ranking algorithm solution (MCRR-LA) with new computing methods MCRR-LA2 and MCRR-LA3, based on Markov Reward Processes, are presented in Sect. 9.4. Our second mapping algorithm solution (MCRM) along with its new implementations MCRM-B and MCRM-I-KSP are introduced in Sect. 9.5. In Sect. 9.6 we evaluate and compare our algorithms by simulation. Section 9.7 concludes the paper.

9.2   Related Work

The VNE is a notorious problem addressed by many researchers considering various assumptions. The VNE problem has been shown to be NP-complete (Andersen, Theoretical approaches to node assignment, unpublished manuscript, December (2002)), [21], thus most solutions are based on efficient heuristics (e.g., [9, 10, 21–23]).

Many of these solutions consist of two sequential stages, i.e., node and link mapping stages. Lately, a novel category of two-stage VNE algorithms, called coordinated algorithms (e.g., CO-VNE [15]), has been devised. These algorithms (e.g., [9, 15, 22]) take into account the links information needed for the second stage (i.e., link mapping) also during the first one (i.e., node mapping). Specifically, these algorithms consider the topology information, by computing for every node a metric that accounts for the resources of the node itself and the resources of its neighbors. The node mapping follows a greedy strategy: firstly VNR and physical nodes are sorted based on the metric, then the VNR nodes with highest metric are mapped onto the physical nodes with highest metric. It is interesting to note that most of these algorithms were inspired by the PageRank algorithm [8]. In the PageRank algorithm the importance of a web page is determined by the importance of the web pages that have links towards it. This means that a web page is more important if it is pointed by more important web pages. If we model the web surfing as a Random Walk, we can define a Markov Chain generated by the web links in which the web pages are the states and the links are the transitions between states. Then, it follows that the stationary probability of this Markov Chain represents the importance of a web page. Similarly, the CO-VNE algorithms, in order to determine the "importance" of a node with respect to resources, compute a metric which considers the resources of the node itself, the incoming links, and the neighbor nodes. The metric can be calculated as the stationary probability of a suitable Markov Chain, as done by the PageRank algorithm.

Although also our approach is based on Markov Chains, actually, we adopt a completely different method for computing the node metric. We compute the cumulated reward of a suitable Markov Chain, thus we do not use the stationary probability.

Researchers have considered various QoS objectives in solving the VNE problem, e.g., energy efficiency [18], survivability and resiliency [19], etc. [13]. However, the delay-aware VNE has been addressed only in a few works. Chowdhury et al. in [10] formulate a mixed integer programming model which deals indirectly with delay. Specifically, each VNR node, which asks for a specific location constraint, must be mapped within a pre-set maximum distance from the demanded physical location.

Lately, Behrouznia et al. devised a two-stage algorithm [3] which, as done by our approach, addresses the VNE problem under QoS constraints, e.g., latency. Their solution aims to achieve cost minimization and QoS objectives. During the first stage, the substrate nodes are ranked using a metric representing the available resources of each node (residual CPU resources multiplied by the residual bandwidth resources over incoming links). Then, the virtual nodes are mapped sequentially to the physical ones according to a greedy strategy. In the following stage, the VNE algorithm maps each virtual link seeking the K shortest substrate paths available between its previously mapped end nodes. Specifically, the VNE algorithm, for each path which meets the network resource and QoS constraints, determines a quality index of its links, following the approach proposed by Shamsi and Brokmeier in [20]. Intuitively, each VNR link is embedded by the physical path with highest quality index. Differently, we calculate instead a delay and topology aware metric for both the request and substrate network. Our approaches are also supplemented by different mechanisms to shorten the physical paths.

9.3   The VNE Problem

9.3.1   Substrate Network

We model the substrate network, i.e., the provider infrastructure, as an undirected weighted graph $G^s = (N^s, E^s, C^s, B^s, L^s)$. N^s represents the set of infrastructure nodes and E^s is the set of communication links. C^s denotes the set of available computational capacity associated with each node,[1] while B^s and L^s denote the sets of network bandwidth and network delay associated with each link, respectively. Finally, P^s denotes the set of substrate paths (without loops). Figure 9.1 shows an example of substrate network and VNR.

9.3.2   Virtual Network Request

A VNR is also modeled as a weighted undirected graph, $G^v = (N^v, E^v, C^v, B^v, L^v)$. Here N^v and E^v are the sets of virtual nodes and links to be mapped, respectively. The set C^v denotes the amount of CPU capacity required by the VNR nodes while B^v and L^v denote the set of requested link bandwidth and link delay constraints associated to the VNR links. Associated to each VNR, there is also an arrival time t_a and lifetime t_d. We will write $VNR(t_a, t_d)$ for short to denote the VNR arrival time and lifetime. Figure 9.1 shows a VNR consisting of four nodes and four links with the associated resources demands and maximum delay requirements.

---

[1]For the sake of simplicity, in this paper we only consider the computational capacity as node resource metric. Our approach can be extended in case of multiple resources, e.g., I/O bandwidth, memory.

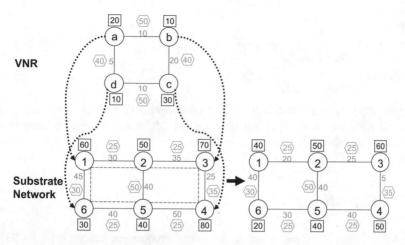

Fig. 9.1 VNE example. A VNR (on the top) is mapped to the substrate network (on the bottom). The resulting remaining substrate resources (on the right). Computational capacity/demand is shown inside square boxes, link capacity/demand in gray. Network delays are shown inside the green hexagons. The dotted and dashed lines show the node and link mapping, respectively

9.3.3 Virtual Network Embedding

The Virtual Network Embedding of a VNR G^v on the substrate network G^s consists in the mapping of the set of virtual nodes N^v onto a subset of substrate nodes N^s and of the set of virtual links E^v onto a subset of substrate links E^s. This procedure consists of two distinct phases: node mapping and link mapping.

In the first phase, virtual nodes are mapped to the substrate ones via the mapping function $M_{node} : N^v \rightarrow N^s$, such that: $\forall n^v, m^v \in N^v$

$$M_{node}(n^v) = n^s \in N^s, \tag{9.1}$$

subject to:

- $M_{node}(n^v) = M_{node}(m^v)$ if and only if $m^v = n^v$;
- $C^v(n^v) \leq C^s(M_{node}(n^v))$.

It is worth observing that in this mapping, different virtual nodes of the same VNR must be mapped to different substrate nodes (but observe that different VNRs nodes can be mapped on the same physical node). The mapping needs also to satisfy the VNR nodes resource demand.

In the second phase, virtual links are mapped to substrate paths, that is, M_{link}: $E^v \rightarrow P^s$, such that: $\forall e^v = (m^v n^v) \in E^v$

$$M_{link}(m^v n^v) = p^s(M_{node}(m^v), M_{node}(n^v)) \in P^s(M_{node}(m^v), M_{node}(n^v)),$$

subject to:

- $B^v(e^v) \leq \min\limits_{e^s \in p^s(M_{link}(e^v))} B^s(e^s);$

- $L^v(e^v) \geq \sum\limits_{e^s \in p^s(M_{link}(e^v))} L^s(e^s).$

Basically, each substrate link in the path should have enough spare capacity to accommodate the virtual link capacity demand. Moreover, the path latency should not exceed the virtual link latency constraint.

9.3.4 Objective

The InP objective is to maximize the long-term time-average revenue [9, 10, 21]. In line with previous works [10, 15, 21–23], we define the InP revenue $R(G^v, t_a, t_d)$ associated to a VNR with lifetime t_d, at time t_a, as:

$$R(G^v, t_a, t_d) = \begin{cases} R_0(G^v, t_a, t_d) \cdot t_d, & \text{if accepted} \\ 0, & \text{otherwise} \end{cases}, \qquad (9.2)$$

where $R_0(G^v, t_a, t_d)$ represents the revenue per time unit for $VNR(t_a, t_d)$ that can be written as:

$$R_0(G^v, t_a, t_d) = \alpha_c \sum_{n^v \in N^v} C^v(n^v) + \alpha_b \sum_{e^v \in E^v} B^v(e^v). \qquad (9.3)$$

In Eq. (9.3), α_c is the unit price for the computational resources and α_b is the unit price for the bandwidth resources.

With the above definition, we have the following expression for the long-term average revenue:

$$R_{\text{tot}} = \lim_{T \to \infty} \frac{\sum_{i=1}^{N_T} R(G^{v,(i)}, t_a^{(i)}, t_d^{(i)})}{T}. \qquad (9.4)$$

Here N_T is the number of requests arrived in the interval $[0, T]$, and $R(G^{v,(i)}, t_a^{(i)}, t_d^{(i)})$ is the revenue associated to the i-th VNR (with arrival time $t_a^{(i)}$ and lifetime $t_d^{(i)}$).

Another important metric, which is relevant to both service provider and users, is the blocking probability or called blocking ratio:

$$BR = \lim_{T \to \infty} \frac{\sum_{j=1}^{N_{REJ_T}} VNR_{rej}^{(j)}(t_a^{(j)}, t_d^{(j)})}{\sum_{i=1}^{N_T} VNR^{(i)}(t_a^{(i)}, t_d^{(i)})}, \qquad (9.5)$$

where $NREJ_T$ is the total number of rejected VNRs within time T and $VNR_{rej}^{(j)}(t_a^{(j)}, t_d^{(j)})$ is the j-th rejected VNR.

We formulate the cost incurred by the InP for accepting a VNR with lifetime t_d, at time t_a, as:

$$CO(G^v, t_a, t_d) = \begin{cases} CO_0(G^v, t_a, t_d) \cdot t_d, & \text{if accepted} \\ 0, & \text{otherwise} \end{cases}, \tag{9.6}$$

where $CO_0(G^v, t_a, t_d)$ represents the cost per time unit for $VNR(t_a, t_d)$ that can be defined as:

$$CO_0(G^v, t_a, t_d) = \sum_{n^v \in N^v} C^v(n^v) + \sum_{e^v \in E^v} |p^s(M_{link}(e^v))| B^v(e^v), \tag{9.7}$$

where $|p^s(M_{link}(e^v))|$ is the physical path length in terms of links. In order to synthesize the efficiency of resource utilization, we adopt the long-term revenue to cost ratio index:

$$RC = \lim_{T \to \infty} \frac{\sum_{i=1}^{N_T} R\left(G^{v,(i)}, t_a^{(i)}, t_d^{(i)}\right)}{\sum_{i=1}^{N_T} CO\left(G^{v,(i)}, t_a^{(i)}, t_d^{(i)}\right)}, \tag{9.8}$$

where $CO(G^{v,(i)}, t_a^{(i)}, t_d^{(i)})$ is the cost incurred for mapping the i-th VNR with arrival time $t_a^{(i)}$ and lifetime $t_d^{(i)}$. If the long-term average revenues of VNE solutions are equal, the lower BR and higher RC are preferred.

Furthermore, as suggested in [1], to capture the performance as perceived by the users, we also consider the average path delay and maximum path delay of accepted VNRs.

9.4   Markov Chain Rewards Based Latency Aware Node Ranking Algorithm

Following previous works in the literature [9, 15, 22], our VNE algorithm consists of two stages: the node mapping stage and the link mapping stage. Upon a VNR arrival, the algorithm first maps virtual nodes on physical nodes with sufficient computational capacity, following a greedy strategy which takes into account both node available computational capacity and the bandwidth and the latency of the attached links; then, in the link mapping stage, virtual links are mapped to substrate network paths with sufficient available link capacity, using a simple shortest-path strategy (using path latency as metric).

In Sect. 9.4.1, we first present our approach to latency aware node ranking, MCRR-LA, which is based on Markov Chains with rewards model [14]. Then in Sect. 9.4.2 we describe two new variants for computing MCRR-LA. Due to space limits we omit to report details of the full mapping algorithm which is possible to find in [5].

9.4.1 MCRR-LA Node Ranking Metric

As noticed in [15], a proper node ranking metric for the VNE problem should take into account the resources of a node neighborhood rather than those of a single node considered in isolation. The intuition is that the higher the aggregate amount of available computational and communication resources in the vicinity of a node, the higher the probability that a VNR can be successfully embedded in that portion of the substrate network. Moreover, since we are also interested in satisfying link delay constraint, we will also account for the links' latency in the definition of the ranking metric.

First of all, we introduce the concept of latency-aware node resources. Without lack of generality, we follow the approach presented in [22][2] and define for each node $n \in N^s$ the available resources $CBL(n)$, as the the amount of available normalized CPU resource multiplied by the sum of the available bandwidth of the attached links, which are weighted by the normalized link latency, as follows:

$$CBL(n) = C^s(n) \cdot \sum_{m \in N(n)} \left[B^s((n, m)) \cdot \left(\frac{max\_lat - L^s((n, m))}{max\_lat - min\_lat} \right) \right]. \qquad (9.9)$$

In Eq. (9.9), for each link (n, m), with $m \in N(n)$, the set of neighbors of node n, the available bandwidth $B^s((n, m))$ is multiplied by a normalized latency term $\frac{max\_lat - L^s((n,m))}{max\_lat - min\_lat} \in [0, 1]$, where $L^s((n, m))$ is the link (n, m) latency and max_lat (min_lat) corresponds to the largest (smallest) substrate link latency. The normalized link latency takes value 1 in case of links with minimum latency and 0 in case of links with maximum latency. These values are then normalized:

$$Res(n) = \frac{CBL(n)}{\sum_{m \in N^s} CBL(m)}. \qquad (9.10)$$

Given the above definition, we define the latency aware ranking metric MCRR-LA of node n, the quantity $V_\gamma(n)$, recursively defined as the weighted sum of the local and the neighboring resources as follows:

---

[2]The ranking metric can be similarly defined using alternative resource metrics, e.g., [15].

$$V_\gamma(n) = (1-\gamma)Res(n) + \gamma \sum_{m \in N(n)} \frac{Res(m)}{\sum_{h \in N(n)} Res(h)} V_\gamma(m), \tag{9.11}$$

where $\gamma \in [0, 1)$ represents the relative weight of the neighbors resources contribution with each neighbor contributing proportionally to its amount of resources (represented by the factor $\frac{Res(m)}{\sum_{h \in N(n)} Res(h)}$). To rewrite (9.11) in a compact form, let $\mathbf{P}$ a $|N| \times |N|$ matrix defined as follows:

$$P(n, m) = \begin{cases} \frac{Res(m)}{\sum_{h \in N(n)} Res(h)} & \text{if } (n, m) \in E^s \\ 0 & \text{otherwise} \end{cases}. \tag{9.12}$$

We can now rewrite (9.11) as:

$$V_\gamma(n) = (1-\gamma)Res(n) + \gamma \cdot \sum_{n' \in N} P(n, n')V_\gamma(n'), \tag{9.13}$$

or in matrix form

$$\mathbf{V}_\gamma = (1-\gamma)\mathbf{Res} + \gamma\,\mathbf{PV}_\gamma, \tag{9.14}$$

where $\mathbf{Res} = (Res(1), Res(2), \ldots, Res(|N|))^T$, and $\mathbf{V}_\gamma = (V_\gamma(1), V_\gamma(2), \ldots, V_\gamma(|N|))^T$.

It is important to observe that (9.14) are the Bellman equations [4] of the discounted cumulative reward of the Markov Chain with state space N and transition probability matrix $\mathbf{P}$ (it is easy to verify that $\mathbf{P}$ is a stochastic matrix) and discount factor γ. The rewards vector is $\mathbf{Rew} = (Rew(1), Rew(2), \ldots, Rew(|N|))$, with $Rew(n) = (1-\gamma)Res(n), n \in N$. This provides an interesting physical interpretation to ranking metric $V_\gamma(n), n \in N$: by construction, $V_\gamma(n)$ is the expected discounted accumulated reward of a Markov Chain with transition probability $\mathbf{P}$, with initial state n, that is:

$$V_\gamma(n) = \lim_{k \to \infty} \mathbb{E}_\mathbf{P}\left[\sum_{i=0}^{k} \gamma^i Rew(n_i)\right], \tag{9.15}$$

where $n_0, n_1, n_2, \ldots$, denotes a sample path with initial state $n_0 = n$.

The computation can be carried out by standard techniques. Indeed, since $\mathbf{P}$ is stochastic, $(\mathbf{I} - \gamma\mathbf{P}), 0 \le \gamma < 1$, is invertible, thus we have

$$\mathbf{V}_\gamma = (\mathbf{I} - \gamma\mathbf{P})^{-1}(1-\gamma)\mathbf{Res}. \tag{9.16}$$

Since the node reward is proportional to the amount of resources, it follows that the higher the node ranking $V_\gamma(n)$, the higher the amount of resources in the neighborhood of node n. Moreover, if we consider nodes with same amount of resources, we will obtain lower values of $V_\gamma(n)$ for the nodes which are located in a high latency neighborhood, since higher latencies negatively impact the (accumulated) reward. In this context, also γ has a physical interpretation as it measures the size of the neighborhood taken into account to determine the node metric $V_\gamma(n)$: when $\gamma = 0$ only the local resources are considered by the metric; as γ increases, larger and larger portions of the graph, in the node neighborhood, are progressively accounted for by the metric.

In the above discussion, we have focused on the substrate network and its resources. The key observation is that the metric $V(n)$ can be applied to the virtual nodes as well. In this case, though, the metric $V_\gamma(n)$ can be regarded as the aggregate amount of resources *required* by a virtual node and its neighbors. As later shown, this equivalence between availability of substrate resources and demand of virtual resources suggests that a simple approach to the embedding problem consists in a greedy strategy whereby we just map virtual nodes with higher resources demand (high values of ranking metric $V(n)$, $n \in N^v$) to substrate nodes with higher resources availability (high values of $V(n)$, $n \in N^s$).

9.4.2   MCRR-LA2 and MCRR-LA3

We developed two new different methodologies for determining the resource latency aware metric called MCRR-LA2 and MCRR-LA3 whose performance will be showed in Sect. 9.6.

In MCRR-LA2 we change the way of computing the amount of available latency-aware resources CBL introduced in Eq. (9.9), involving the discount factor γ in the following way:

$$CBL(n) = C(n) \cdot \sum_{m \in N(n)} \left[B^s((n,m)) \cdot \gamma^{\left(\frac{L^s((n,m))}{min\_lat}\right)} \right].$$

In MCRR-LA3 we compute CBL as in Eq. (9.9), but we introduce the penalization in the transition probabilities computation as follows:

$$P(n,m) = \begin{cases} \dfrac{Res(m)\gamma^{\left(\frac{L^s((n,m))}{min\_lat}\right)}}{\sum_{h \in N(n)} Res(h)\gamma^{\left(\frac{L^s((n,m))}{min\_lat}\right)}} & \text{if}(n,m) \in E^s \\ 0 & \text{otherwise} \end{cases},$$

Everything else remains unchanged. Since the discount factor $\gamma < 1$, the exponential term penalizes links with large delays. In the former case, the larger the link delays, the smaller the CBL (and differently from Eq. (9.9) where we decrease the metric in a linear fashion, here we decrease the metric exponentially fast). In the latter case, the larger the delay, the less likely the path to enter that node and thus less likely to consider its resources in the computation of the metric.

9.5   Markov Chain Rewards Metrics (MCRM) Ranking Algorithm

In a previous work [7] we have introduced an improved latency-aware VNE algorithm, called Markov Chain Reward Metrics (MCRM), which considers also the presence of pinned nodes (i.e., pre-assigned nodes due, for example, to data source nodes physically located in specific areas). Our node mapping procedure is based on a new concept of *similarity* between substrate and VNR nodes. The procedure computes a set of metrics for each substrate and VNR node that accounts for the aggregated available (requested) resources in the vicinity of a substrate (VNR) node and their *proximity* level with respect to the other nodes. Thus, it uses two types of metrics: a resource-latency aware resource metric (MCRR-LA) previously described in Sect. 9.4, and a proximity metric (MCR-P) computed with regards to each pinned node, representing an elegant measure of distance. Then, after computed a *similarity* index between VNR and substrate nodes, the procedure intuitively maps each VNR node to the *most similar* substrate node.

For reasons of space we omit to describe the proximity and similarity metric as well as the full iterative mapping algorithm: for details, please refer to [7]. In this work we refer to it as MCRM-I. In Sect. 9.5.1 we introduce a variant of the iterative algorithm MCRM-I, called MCRM-I-KSP, in which we use K-Shortest-Paths instead of Dijkstra algorithm during the link mapping stage. In Sect. 9.5.2 we introduce a baseline not iterative variant of MCRM-I called MCRM-B which basically maps all the virtual nodes at the same time.

9.5.1   MCRM-I-KSP

In this version of MCRM-I we just modify the link mapping stage in order to use a more advanced strategy. We adopt K-Shortest-Paths algorithm, in terms of latency between each couple of mapped virtual nodes, instead of Dijkstra algorithm. If K-Shortest-Paths is able to find at least a physical path and they meet the virtual delay bound of the virtual link, then the algorithm selects the path that minimizes the number of links and maximizes the minimum bandwidth over links, while maximizing the delay. This method allows to achieve a higher and more uniform use of network resources while meeting the QoS constraint, in order to also avoid the presence of possible unused links as highlighted in [7].

9.5.2 MCRM-B

We also developed a non-iterative version of MCRM-I in order to reduce the
algorithm runtime, while still achieving good performance and QoS indexes results.
As MCRM-I, after mapped the pinned nodes, MCRM-B firstly computes and
normalizes MCRR-LA and MCR-P metrics, secondly it calculates and sorts in
ascending order the similarity metrics (cf. matrix D [7]) between virtual and
substrate nodes. At this time, differently from MCRM-I, MCRM-B maps all the
not yet embedded virtual nodes in one shot. Thus, it sequentially maps each virtual
node onto the physical one with which the similarity metric is lower, provided that
the physical node has enough available resources and has not been already involved
in a mapping.

9.6 Experimental Evaluation

The algorithms have been assessed by the means of simulation. We developed a
VNE simulator to compare the presented MCRR-LA and MCRM based algorithms,
and the algorithm proposed in [2, 3], hereafter referred to as QoS-RS (QoS-Resource
Selection) for brevity. We report the results based on a 32 node substrate network
with 58 links (ANSNET [17], slightly modified as in [22]).

The arrival process of the VNRs is a Poisson process with rate of $\lambda = 5$. The
lifetime of VNRs follows an exponential distribution with an average lifetime
T_{VNR} of 12 time units, representing a medium load scenario. A VNR is modelled
as a random graph whose node number is selected from 4 to 10 according to
a uniform distribution. The virtual link connectivity rate is $\frac{ln(|N^v|)}{|N^v|-1}$ to obtain
$O(ln(|N^v|))$ incident edges for each node. The requested computing resources for
each node and bandwidth resources for each link are randomly selected according
to uniform distribution in the range [4, 8] and [2, 5], respectively. The available
computing and bandwidth resources of the physical network are normalized and
set to nominal value 100. The delays of physical network links are proportional
to the geographical distance between nodes. The latency demanded by each
virtual link is randomly selected in the range [average substrate links latency $\times \delta$,
maximum substrate links latency $\times \delta$], where δ is a multiplicative factor.

We ran each experiment ten times for 5000 time units. For the sake of clarity,
since the intervals of confidence were in general not significant, we omit them.

In the experiments we compare MCRR-LA, MCRR-LA2, MCRR-LA3,
MCRM-I, MCRM-I-KSP, MCRM-B, and QoS-RS. We set the weight assigned
to the resource-latency aware metric (MCRR-LA), w_{V_γ}, to 0 and $\frac{2}{5}$ in the similarity
index computation for algorithms MCRM-I, MCRM-I-KSP, and MCRM-B. For
these two values they achieve on average the worst and best performance results,
respectively. Please remember that we assume that all node proximity MCR-P
weights have the same value, that is $w_{Z_m} = \frac{1-w_{V_\gamma}}{|R^v|}$, where R^v is the set of reference

nodes towards which the proximity metric is computed. VNRs have no pinned nodes for comparison purposes. Following [6], for MCRR-LA we adopt a value of $\gamma = 0.98$ (the discount factor of the Markov Reward Model). In the implementation of the QoS-RS algorithm, we do not consider the node location constraint and the links packet loss constraint which have no counterpart in our proposed solution (we plan to include such constraints in our future work). The K-Shortest-Paths algorithm used by QoS-RS is set with $K = 4$. In order to compute the quality of the physical paths, QoS-RS contemplates the use of weights (summing to 1), as in [20], to balance the importance of the bandwidth (w_1) and delay (w_2) attributes. Since our tests showed that the performance is not considerably affected by the different combination of the weights and since MCRR-LA aims to reduce the latency when it comes to compute the shortest paths, we plot the results of just two combinations: $w_1 = 0.5$, $w_2 = 0.5$ and $w_1 = 0.0$, $w_2 = 1.0$.

In Fig. 9.2a–e we plot the VNR blocking probability, the average revenue, the revenue to cost ratio, the average path delay, and the maximum average path delay of the algorithms, assuming the VNR latency bounds with a multiplicative factor $\delta = 1.5$ (that is, each VNR link latency requirement is randomly drawn in the interval $[1.5 \cdot average\_lat, 1.5 \cdot max\_lat]$, where $average\_lat$ and max_lat are the average and maximum substrate network link latency, respectively). From the figures, we can observe that MCRR-LA and MCRM based solutions considerably outperform QoS-RS in terms of lower blocking probability, higher revenue and lower average path delay. From the simulations, the lower blocking probability (also reflected in terms of higher revenue) is due to a more uniform use of the substrate network resources (both nodes and links).

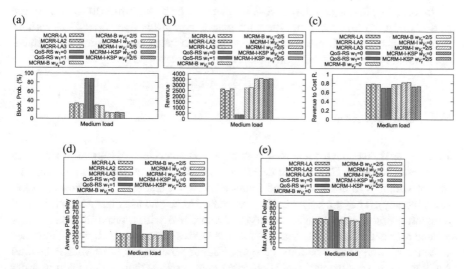

Fig. 9.2 Comparison of algorithms: requested basic latency bounds $\cdot \delta = 1.5$. (**a**) Blocking probability. (**b**) Revenue. (**c**) Revenue to cost ratio. (**d**) Average path delay. (**e**) Max average path delay

The better performance of our MCRR-LA and MCRM based solutions can be ascribed to the use of the latency aware metric also in combination with the proximity metric (in MCRM) in determining the node ranking which gives preference, during the node mapping phase, not only to the node with more resources (computational capacity and link bandwidth) but also with low latency.

Among our solutions, MCRM-I surpasses the other approaches in terms of performance and latency indexes, due to resource-latency aware metric combined with proximity metric and the iterative operation which guide the mapping through the similarity index based on an ever increasing number of reference nodes. MCRM-I-KSP achieves same good revenue and blocking ratio performance of MCRM-I and its load balancing joint with the delay maximization strategy extends on average the path delays and the number of path links, as expected. The last effect is reflected on lower revenue to cost ratio and on slightly higher and more uniform use of links resources, also solving the problem of possible unused substrate links as remarked in [7]. The results of MCRR-LA, MCRR-LA2, and MCRR-LA3 are almost comparable on all the performance and QoS indexes, underlining the same good quality in capturing the resource-latency aware potential of nodes. However, the first and the second solution achieve slightly better performance results. The non-iterative algorithm MCRM-B reaches similar results in terms of delay indexes, but lower revenue and higher blocking probability compared to MCRM-I, as expected, since the similarity index is computed with just one proximity metric value per node. However, its computational cost is lower due to zero iterations, thus experiencing shorter execution time. The overall results of MCRM-B are also slightly better than MCRR-LA based algorithms, underlining the key role of coupling a proximity metric with a resource-latency aware metric.

Indeed, despite the fact that QoS-RS uses a method to determine the quality of the paths during the link mapping, the node mapping phase is first carried out without considering latency. These results stress the importance of including link related QoS metrics in the node mapping phase of two-stage mapping algorithms.

9.7 Conclusions

In this paper, we have presented a comparison overview of our approaches to the Virtual Network Embedding problem based on Markov Reward Processes, the aim of which is to achieve a good trade-off between resource utilization and QoS (e.g., latency). They all rely on a key resource-latency aware metric which is used alone in a first approach (MCRR-LA) and combined with a proximity metric in a second approach (MCRM), with the same objective of ranking and mapping nodes effectively. Along with previously proposed latency-aware solutions MCRR-LA and MCRM, we have introduced and compared new versions that extend them in order to test new performing strategies. Experimental outcomes show that our algorithms achieve good performance objectives and are able to reduce the maximum and average path delay of the embedded VNRs compared to alternative approaches.

References

1. M.T. Beck, C. Linnhoff-Popien, On delay-aware embedding of virtual networks, in *The Sixth International Conference on Advances in future internet, AFIN* (2014)
2. S. Behrouznia, A QoS-based resource selection approach for virtual networks. Master's thesis, Concordia University, April 2015
3. S. Behrouznia, R. Dssouli, M. El Barachi, A QoS-based resource selection approach for virtual networks, in *International Conference on Computer and Information Science and Technology, 2015. CIST'15*, May 2015
4. R. Bellman, *Dynamic Programming*, 1st edn. (Princeton University Press, Princeton, NJ, 1957)
5. F. Bianchi, F. Lo Presti, A latency-aware reward model based greedy heuristic for the virtual network embedding problem, in *Proceedings of InfQ 2016 - New Frontiers in Quantitative Methods in Informatics (in conjunction with VALUETOOLS 2016)*, October 2016
6. F. Bianchi, F. Lo Presti, A Markov reward model based greedy heuristic for the virtual network embedding problem, in *2016 IEEE 24th International Symposium on Modeling, Analysis and Simulation of Computer and Telecommunication Systems (MASCOTS)* (2016)
7. F. Bianchi, F. Lo Presti, A Markov reward based resource-latency aware heuristic for the virtual network embedding problem. SIGMETRICS Perform. Eval. Rev. **44**(4), 57–68 (2017)
8. S. Brin, L. Page, The anatomy of a large-scale hypertextual web search engine. Comput. Netw. ISDN Syst. **30**(1–7), 107–117 (1998)
9. X. Cheng, S. Su, Z. Zhang, H. Wang, F. Yang, Y. Luo, J. Wang, Virtual network embedding through topology-aware node ranking. SIGCOMM Comput. Commun. Rev. **41**(2), 38–47 (2011)
10. N. Chowdhury, M. Rahman, R. Boutaba, Virtual network embedding with coordinated node and link mapping, in *IEEE INFOCOM 2009*, April 2009, pp. 783–791
11. D. Eppstein, Finding the k shortest paths. SIAM J. Comput. **28**(2), 652–673 (1998)
12. N. Feamster, L. Gao, J. Rexford, How to lease the internet in your spare time. SIGCOMM Comput. Commun. Rev. **37**(1), 61–64 (2007)
13. A. Fischer, J. Botero, M. Till Beck, H. de Meer, X. Hesselbach, Virtual network embedding: a survey. IEEE Commun. Surv. Tutorials **15**(4), 1888–1906 (2013)
14. R.G. Gallager, *Stochastic Processes: Theory for Applications* (Cambridge University Press, Cambridge, 2013)
15. L. Gong, Y. Wen, Z. Zhu, T. Lee, Toward profit-seeking virtual network embedding algorithm via global resource capacity, in *2014 Proceedings IEEE INFOCOM*, April 2014, pp. 1–9
16. K. Ivaturi, T. Wolf. Mapping of delay-sensitive virtual networks, in *International Conference on Computing, Networking and Communications (ICNC), 2014*, February 2014, pp. 341–347
17. Z. Li, J.J. Garcia-Luna-Aceves, Finding multi-constrained feasible paths by using depth-first search. Wirel. Netw. **13**(3), 323–334 (2007)
18. L. Nonde, T.E.H. El-Gorashi, J.M.H. Elmirghani, Energy efficient virtual network embedding for cloud networks. J. Lightwave Technol. **33**(9), 1828–1849 (2015)
19. M.R. Rahman, R. Boutaba, SVNE: survivable virtual network embedding algorithms for network virtualization. IEEE Trans. Netw. Serv. Manag. **10**(2), 105–118 (2013)
20. J. Shamsi, M. Brockmeyer, QoSMap: QoS aware mapping of virtual networks for resiliency and efficiency, in *2007 IEEE Globecom Workshops*, November 2007, pp. 1–6
21. M. Yu, Y. Yi, J. Rexford, M. Chiang, Rethinking virtual network embedding: substrate support for path splitting and migration. SIGCOMM Comput. Commun. Rev. **38**(2), 17–29 (2008)
22. S. Zhang, Z. Qian, J. Wu, S. Lu, An opportunistic resource sharing and topology-aware mapping framework for virtual networks, in *2012 Proceedings IEEE INFOCOM*, March 2012, pp. 2408–2416
23. Y. Zhu, M. Ammar, Algorithms for assigning substrate network resources to virtual network components, in *Proceedings INFOCOM 2006. 25th IEEE International Conference on Computer Communications*, April 2006, pp. 1–12

Chapter 10
Delay Efficient Load Balancing Scheme for Component Carrier Selection in Carrier Aggregation in LTE-A

Load Balancing Scheme in LTE-A

Aditi Gupta, Dharmaraja Selvamuthu, and Subrat Kar

10.1 Introduction

The objectives of Long Term Evolution (LTE) started by 3GPP since 2004 are reduced latency, higher user data rate, improved system capacity and coverage, and reduced cost for the operator [1]. Long Term Evolution-Advanced (LTE-A), the evolution of LTE targets for 1 Gbps downlink speed for a user. It is required to increase the bandwidth to achieve these targets as a single carrier of 20 MHz in LTE is not sufficient. Carrier Aggregation(CA) is a feature in LTE-A to aggregate more than one carriers together and provide higher bandwidth. Each channel is of maximum 20 MHz bandwidth, with CA, LTE-A has set the limit to aggregate five carriers providing 100 MHz bandwidth. In future, this limit can be exceeded.

LTE-A carriers can be aggregated in three different ways. The first way is the Intra-band contiguous CA that uses the adjacent CCs. The second way is Intra-band noncontiguous CA in which the CCs aggregated are in the same band but nonadjacent. The third way is inter-band non-contiguous carrier aggregation that uses carriers of different bands. While the first one is the easiest to implement, the latter is the most complicated as it requires the use of multiple transceivers for carriers in different bands. After aggregating, each carrier is known as a Component Carrier (CC). The CC further can be categorized into two types, Primary Component Carrier (PCC) and Secondary Component Carrier (SCC). PCC is the main carrier responsible for exchanging the Radio Resource Control (RRC) signaling messages with the User Equipment (UE). One PCC is always active in the RRC CONNECTED mode while SCCs can be activated or deactivated depending on the usage. After the CCs are assigned, Resource Blocks (RBs) are allocated to

A. Gupta · D. Selvamuthu (✉) · S. Kar
Bharti School of Telecommunication Technology and Management, IIT Delhi, New Delhi, India
e-mail: dharmar@maths.iitd.ac.in; subrat@ee.iitd.ac.in

© Springer International Publishing AG, part of Springer Nature 2019
A. Puliafito, K. S. Trivedi (eds.), *Systems Modeling: Methodologies and Tools*,
EAI/Springer Innovations in Communication and Computing,
https://doi.org/10.1007/978-3-319-92378-9_10

the user. RB is the smallest unit of resources that can be allocated to a user. Each RB consists of 12 sub-carriers, constituting an equivalent bandwidth of 180 kHz in the frequency domain and Transmission Time Interval (TTI) of 1 ms in the time domain.

The goal of CA is to provide enhanced and consistent user experience by maximizing the peak data rate, throughput, and better QoS. CA also allows operators a cost-effective solution to increase their current network throughput and capacity. For better resource utilization and spectrum efficiency the load should be balanced across the carriers [2]. Being an IP based system, for LTE-A based system the bursty and unpredictable nature of packets makes the balancing of load over carriers difficult and thus there is a need for better resource allocation algorithm. While the use of multiple CCs causes load balancing, it also leads to complexity in terms of power requirements and signal processing at the user end. Since the users belong to different classes of traffic, requirements of each user differ. A smart load balancing algorithm for CC selection can thus be utilized to maintain spectral efficiency and QoS requirements. In this research work, we propose an adaptive load balancing algorithm for LTE-A CA based system. The proposed algorithm is capable of balancing the load across different carriers while also considering the service requirements of data.

The rest of this research work is organized as follows. In Sect. 10.2, the research in CC selection methodologies in Radio Resource Management (RRM) framework and the motivation behind this research work are discussed. In Sect. 10.3, the performance model is proposed and the fluid queue analysis is presented. The performance analysis of the proposed model is numerically illustrated in Sect. 10.4. Finally, the conclusions and future work are presented in Sect. 10.5.

10.2 Related Work

CC selection plays an important role in optimizing the system performance with CA. There have been different scheduling algorithms proposed in the literature for CA-based systems. Since this work deals with load balancing and QoS parameters, research in this area is discussed. In [3], it is proposed to assign maximum CCs to the LTE-A user to achieve maximum efficiency. In [4], least load method is introduced in which the data is assigned to the CC that has the least amount of load. For every CC queue length is calculated and the one with the minimum queue length is given the data packet. The users, however, arrive randomly with different sizes of the files for transmission and it is difficult to totally avoid the idle CCs. To overcome this issue CC coupling schemes have been modeled in [5, 6]. In CC coupling, if any of the CC is in busy state, the user can be switched to the other CC. If the idle CC becomes busy the coupling is cut. There are two challenges (1) Handling the CC switch delay. (2) Development of the efficient coupling methods for multiple CC.

Scheduling delay is a more critical factor for real-time data. Thus QoS required is an important factor while implementing the CC allocation algorithm. A scheduler algorithm has been designed to meet the QoS level of real-time traffic in [7]. The data arriving will first be classified into Real-Time (RT) (e.g., live streaming) and Non-Real-Time (NRT) (e.g., emails) traffic by a classifier and divided into RT and NRT queue, respectively. The algorithm proposed aims at optimizing the system overall throughput while maintaining the required QoS of the RT traffic by reserving more RBs for it. A threshold is set for the RT traffic and once that is exceeded the RT packets are dropped. However, to the best of our knowledge, an efficient scheme for load balancing across different carriers has not been proposed. In real-time networks the scenario is quite dynamic and one carrier may be overloaded, the others idle. On switching the data there would be scheduling delay and overhead incurred which would add to the delay and bandwidth consumed. Therefore, the scheduling algorithm should be devised such that there is always an even balance across the carriers to maximize the efficiency. Since LTE-A focuses on reducing the packet delay, it is important to keep the scheduling delay minimum though without complex algorithms. Keeping these requirements in mind, a load balancing CC selection scheme that reduces the delay of the network is proposed in this work.

10.3 Proposed Scheme and Its Performance Model

10.3.1 Radio Resource Management Framework for Carrier Aggregation

The muti-CC operation brings some changes in the RRM framework for LTE-A from LTE [8]. The RRM framework for LTE-A is described in this section (Fig. 10.1). For any user, first CC is selected. One CC, i.e., the PCC is assigned to it and then depending on traffic load and QoS requirements, SCCs are given. It can be noted that the same carrier can be PCC to one UE and SCC to the other. The usage of SCC is configured by the evolved node-b (eNB). After the CCs are allotted, RBs are scheduled to the user. RBs are multiplexed to the users on each CC. The layer-1 transmission containing LA (Link Adaptation) and HARQ (Hybrid ARQ) per CC is carried out independently to provide backward compatibility to co-existing LTE users [9]. In the following subsections, we have described CC selection and RB scheduling in detail.

10.3.1.1 CC Selection and Management

Assigning multiple CCs to the user is the additional feature in RRM framework for LTE-A. There are many factors that can be considered while choosing the carriers for the UE, such as QoS requirements, power capabilities, overall traffic level, load per CC. Other than these, channel quality is also an important factor while allotting the carriers to the user.

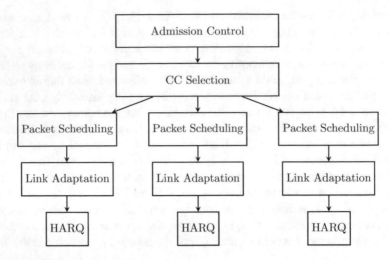

Fig. 10.1 RRM framework of LTE-A

QoS plays an important role in determining the number of CCs, RBs to the UE. The data can be using an RT service or an NRT service and depending on that the optimum number of CCs and their services should be provided to it. The next parameter discussed is the load on the carriers. One of the goals of CA is to balance the load across the carriers to maximize the utilization of spectrum resources. Energy efficiency also plays an important role in the uplink transmissions as the simultaneous transmissions lead to excessive power being spent on the UE terminal [10].

CCs can be dynamically changed, included or removed depending on the signal quality, load or power conditions. Therefore, CC selection and management should be performed to select the appropriate CC and manage them so that resource utilization is maximum and the user is provided the best service.

10.3.1.2 Packet Scheduling

After the CC assignment is completed, the allocation of resource blocks (RB) takes place. This process is similar to that in LTE except that now the user can be connected with RBs belonging to different CCs. Each physical resource block constitutes 12 sub-carriers providing a bandwidth of 180 kHz. Each RB also corresponds to a sub-frame in the time domain, with Transmission Time Interval (TTI) of 1 ms. There have been several algorithms proposed in the literature to select the RB in a CC for the user, the most popular of these being the round robin, and proportional fair (PF) [11]. In PF, the RB is given to that user that has good channel quality. Similar to LTE, dynamic scheduling of a user on an RB is supported in LTE-A. In this case, the user is scheduled on an RB by sending a scheduling grant on PDCCH time multiplexed in each TTI before data channel. LTE-A also introduces cross-carrier scheduling in which eNB can send a scheduling grant on one CC for scheduling the user on another CC [12]. Cross-carrier scheduling makes the control and data channel performance flexible across multiple CCs. Different schemes have been proposed in the literature to reduce scheduling delay by proposing efficient RB allocation schemes. RB grouping method is proposed in [13] where the deduced RBs are grouped into sets and this fixed set is assigned to the user. In [14] scheduling delay is minimized using RB reservation and dispatching frequency to them according to the QoS required.

10.3.2   Performance Model

Whenever CA is performed, data is first assigned to PCC and then given to other SCCs. For the energy efficiency and lesser complexity, it is better to give the load to as minimum of carriers necessary. It is better to balance the load using the least number of carriers needed. Hence, there is a pertinent issue of when to switch the data to other carriers. The first scheduler has the information of the CCs load and decides on the basis of the buffer content, if the buffer content rises above a certain level, the data should be switched to the other carriers.

There is a single threshold involved in these models which indicate for the carrier whether it will receive the data from the scheduler or not. However, there are drawbacks to any single threshold system. If the threshold is crossed often many feedback signals have to be sent to the first scheduler to limit the flow of the data to that carrier. These feedback signals will consume a significant part of the available bandwidth. Also, a single threshold to indicate the heavy loading is not preferred. In this research work, we propose a model with two thresholds for CC selection (Fig. 10.2).

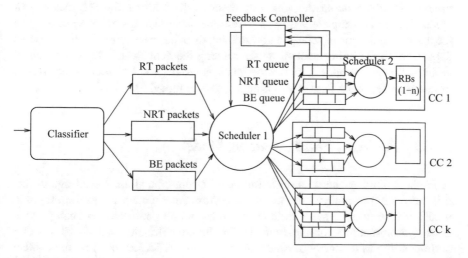

Fig. 10.2 Schematic of the proposed performance model

The two thresholds divide the buffer into three regimes, the relevant rate of the data input to the carrier depending on the regime of the buffer. If the data is above the first threshold denoted by $T^{(1)}$, the rate of the input data should be decreased and routed to another carrier. If the input data rate increases further above the second threshold, denoted by $T^{(2)}$, the input rate turns down to 0. In other words, the data inflow to the carrier is stopped completely.

One of the main requirements in LTE-A is to provide optimum performance with respect to the QoS required for the data. The QoS required for RT data, NRT data or the best effort would be different so the scheduling should be in accordance to that. To provide adaptable service according to the class of traffic, the parameters (input rate in different regimes, the two thresholds value) of the background model should be adjusted.

The proposed scheme is analyzed using the development of performance model through the feedback fluid queue. In fluid queue model, instead of individual customers, a continuous entity called fluid is considered. The fluid flows into the fluid reservoir according to a background Markov process, and flows out dependent on the output rate of the server [15]. Depending on the state of the background process, the input rate to the buffer content changes. For example, if the background process has three states, then fluid will flow into the buffer content in three different rates.

Fluid queues are particularly useful in telecommunication systems because the bursts of data is transmitted in smaller sized cells or data packets. In telecom networks, the variations take place on the burst level rather than the cell level. The fluid queue can thus give good approximations for the actual behavior of network traffic. Also, LTE-A networks are IP based systems with high speed and can be efficiently modelled by fluid queue.

Feedback fluid queue is the type of fluid queue where the input rate to the buffer content depends on the background process and the behavior of the background process also depends on the content present in the buffer [16]. The meaning of feedback here is completely different from the traditional queueing systems. The feedback in fluid queue refers to control signals being sent to originating process depending on the buffer content unlike sending the data back in the conventional systems. Since in proposed model the input rate to the CC depends on the content of its buffer, feedback fluid queue is appropriate to model its behavior.

10.3.3 Feedback Fluid Queueing Model

A feedback fluid queue with an infinite buffer content and a constant output rate c is considered to analyze the proposed performance model. As in the proposed model there cannot be any overflows or losses, infinite buffer system can provide good approximations for the finite buffer. Figure 10.3 shows each CCs buffer corresponding to the users of a particular class of traffic. Let $W(t)$ be the content in the buffer for a class of traffic per CC at time t and the rate at which fluid

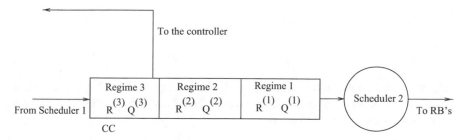

Fig. 10.3 Buffer obtained for corresponding CCs

Table 10.1 The buffer content in different regimes

Buffer content	Regime
$0 \le W(t) < T^{(1)}$	1
$T^{(1)} \le W(t) < T^{(2)}$	2
$T^{(2)} \le W(t) < \infty$	3

enters the queue per unit time is dependent on the current state of a background irreducible continuous-time Markov chain $\{X(t), t \ge 0\}$, defined on a state space $D = \{1, 2, \ldots, d\}$ where d is the total number of states of the background process. In the proposed performance model the background process is the first scheduler that assigns the data to a particular CC. The background process or the first scheduler operates in two states, either it gives data to the CCs buffer (ON state) or it does not (OFF state). The (i, j)th element $(i \ne j)$ of the scheduler generator is given by elements of matrix $Q_{ij}^{(k)}$ ($\tilde{Q}_{ij}^{(k)}$ at the threshold) such that the sum of row elements is 0. Q_{ij} for $j \ne i$ is the transition rate at which the background process jumps from states i to j. $R^{(k)}$ for a regime k is defined as the diagonal matrix with its ith diagonal element be given by $r_i - c$, where r is the input rate to the buffer(r_h and 0 in regime 1, r_l and 0 in regime 2) and c is the constant output rate of the fluid outflow from the buffer(the output rate is constant because some RBs are usually kept reserved for a class of traffic and continuous scheduling for those is carried out). While the output rate is constant, the value of r depends on two parameters. First, on the state of the system, whether it is ON or OFF. If it is ON, then the rate depends on the second parameter, i.e., the regime the buffer. The two thresholds model divides the buffer into three regimes. The input rate is high(in regime 1) as long as the lower threshold $T^{(1)}$ has not crossed. If that happens, the input rate is lowered (in regime 2). Next, if the higher threshold $T^{(2)}$ is reached, the input flow is stopped completely (in regime 3) and the content is let to flow out until it hits the higher threshold (in regime 2) wherein the input is given again.

On the basis of the amount of fluid content in the buffer at times t, the system is divided into three different regimes for the proposed model (Table 10.1).

For all the regimes, the subsets of D consisting of ON state, OFF states, respectively, is defined as:

$$D_+^{(k)} = \{i \in D | r_i^{(k)} > 0\}$$

$$D_-^{(k)} = \{i \in D | r_i^{(k)} < 0\}$$

where $k = 1, 2, 3$ depending on the regime of the buffer content. It is assumed that the input rate is not equal to the output rate for the sake of calculations. The stability condition for the system is given by:-

$$\sum_{i=1}^{D} \pi_i^{(k)} r_i^{(k)} < 0$$

where $\pi_i^{(k)}$ is the stationary distribution of the Markov process with generator $Q^{(k)}$.

Now, we find the expression for distribution of the buffer content, to calculate the performance measures like throughput, delay, etc. Let $F^{(k)}(x)$ be the equilibrium distribution of the buffer where $k = 1, 2, 3$ depending on the regime of the buffer. i.e.,

$$F^{(k)}(x) = \lim_{t \to \infty} \text{Prob}\{W(t) \le x\}, x \ge 0, \quad k = 1, 2, 3.$$

The differential equations satisfied by the distribution can be expressed in matrix form as [17]:

$$\frac{d\mathbf{F}^{(k)}(x)}{dx} R^{(k)} = \mathbf{F}^{(k)}(x) Q^{(k)} + \mathbf{F}^{(k)}(T^{(k-1)})(\tilde{Q}^{(k-1)} - Q^{(k)})$$

$$+ \mathbf{F}^{(k)}(T^{(k-1)}-)(Q^{(k)} - \tilde{Q}^{(k-1)}) + \cdots$$

$$+ \mathbf{F}^{(1)}(T^{(1)}-)(-Q^{(1)} - \tilde{Q}^{(1)})$$

$$+ \mathbf{F}^{(1)}(0)(\tilde{Q}^{(0)} - Q^{(1)}), k = 1, 2, 3. \tag{10.1}$$

The solution of the above differential equations is given by

$$\mathbf{F}_i^{(1)}(x) = a^{(1)} \exp\left[z_{(1)}^{(1)} x\right] v_{(1)}^{(1)} + b^{(1)} v_{(2)}^{(1)} + c^{(1)} \text{ for regime 1}$$

$$\mathbf{F}_i^{(2)}(x) = a^{(2)} \exp\left[z_{(1)}^{(2)} x\right] v_{(1)}^{(2)} + b^{(2)} v_{(2)}^{(1)} + c^{(2)} \text{ for regime 2}$$

$$\mathbf{F}_i^{(3)}(x) = a^{(3)} \exp\left[z_{(1)}^{(3)} x\right] v_{(1)}^{(3)} + c^{(3)} \text{ for regime 3}$$

$(z_i^{(k)}, v_i^{(k)})$ are the eigen value vector pair of $z_i^{(k)} v_i^{(k)} R^{(k)} = v_i^{(k)} Q^{(k)}$ and $a^{(k)}, c^{(k)}$ are the unknown coefficients for $k = 1, 2, 3$, $b^{(k)}$ for $k = 1, 2$ and $i = 1, 2$ depending on the regime of the buffer and on-off state of the system. We obtain in total 11 unknown coefficients in the solution above whose values can be found by the following conditions:-

1. $F_1^{(1)} = 0$.
   We get one equation from this condition.
2. $F_i(T^{(k)}-) = F_i(T^{(k+1)})$ for $i = 1, 2$ and $k = 1, 2$
   This gives four equations from the continuity conditions at the thresholds $T^{(1)}$ and $T^{(2)}$.
3. $0 = \mathbf{c}^{(3)} Q^{(K)} + \mathbf{F}^{(K)}(T^{(K-1)})(\tilde{Q}^{(K-1)} - Q^{(K)}) +$
   $\mathbf{F}^{(K)}(T^{(K-1)}-)(Q^{(K)} - \tilde{Q}^{(K-1)}) +$
   $\ldots + \mathbf{F}^{(1)}(T^{(1)}-)(-Q^{(1)} - \tilde{Q}^{(1)}) + \mathbf{F}^{(1)}(0)(\tilde{Q}^{(0)} - Q^{(1)})$
   will give another equation.
4. $\sum_{j=1}^{3} c^{(j)} = 1$
   is the normalization condition.
5. Substitution of the solution of the balance equations for $k = 1, 2$ in Eq. (10.1) gives four more equations.

The unique solution for the stationary distribution of the buffer content is obtained with which the performance measures have been calculated in the following section.

10.4   Performance Analysis

In the previous section, the density distribution is calculated. In this section, the following issues are considered:

1. Is the two threshold model better than the single threshold model, if so then how?
2. What should be the relation between the thresholds to achieve the optimal performance of the model?
3. How should the threshold and rates be varied according to the QoS required by the user?

To find the answers to above questions, we plot buffer content, throughput, mean delay, and buffer content with respect to the lower threshold $T^{(1)}$ by keeping different values of $T^{(2)}$. For illustration purpose to plot the graphs, the generator matrices and rate vectors are given as follows:

$$Q^{(1)} = Q^{(2)} = Q^{(3)} = \tilde{Q}^{(0)} = \tilde{Q}^{(1)} = \tilde{Q}^{(2)} = \begin{pmatrix} -2 & 2 \\ 1 & -1 \end{pmatrix},$$

$$R^{(1)} = \begin{pmatrix} 15 \\ 0 \end{pmatrix}, R^{(2)} = \begin{pmatrix} 7 \\ 0 \end{pmatrix}$$

With these values and different values of thresholds, first, the corresponding density distribution is to be found and then the performance measures are to be calculated.

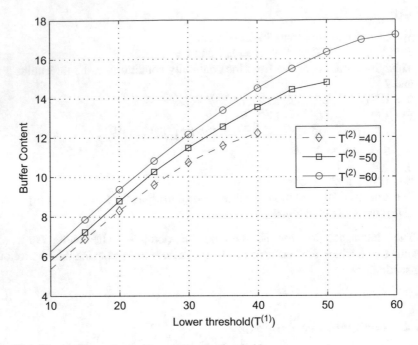

Fig. 10.4 Mean buffer content with respect to the thresholds

10.4.1   Average Buffer Content

Average Buffer content is given by $\int_0^\infty (1 - F(x))dx$. It is plotted in Fig. 10.4. For increasing values of thresholds, the buffer content increases as the accumulations in it increase.

10.4.2   Mean Throughput

Throughput is expressed by the number of data packets (fluid particles in our model) transmitted per unit time. Mean throughput is given by $c \times (1 - F(0))$ where $F(0)$ is the cumulative distribution function when the buffer content is 0. It is observed from Fig. 10.5 that for a given value of higher threshold $T^{(2)}$, the throughput increases as the lower threshold $T^{(1)}$ increases. Also, when $T^{(1)} = T^{(2)}$, which is the maximum limit for $T^{(1)}$, the throughput becomes maximum, this is also the case of a single threshold as both $T^{(1)} = T^{(2)}$. Hence, it is concluded on comparing with a single threshold model (in which the input is given at a single constant rate), proposed model will give lower throughput.

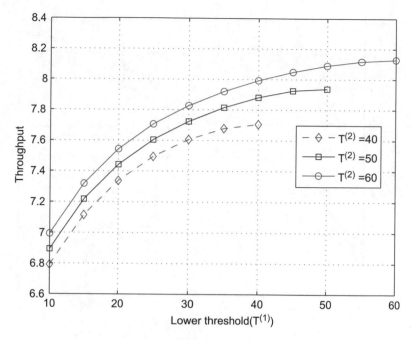

Fig. 10.5 Throughput with respect to the thresholds

10.4.3 Mean Delay

Mean delay is given by Average Buffer content/throughput. In Fig. 10.6, mean delay is plotted against the threshold. It is observed that as the two threshold approach each other, delay increases, delay would be higher than proposed model of two thresholds. Hence, it is concluded that model results in lower delay and lower throughput as compared to model where the CC is assigned based on the single threshold.

Apart from throughput and delay, number of feedback signals sent to the first scheduler also plays role in adjusting the parameters for the incoming traffic. The feedback signals being sent result in causing overheads so it is an important factor while designing the system. It is observed from the above results that throughput increases when the difference between $T^{(1)}$ and $T^{(2)}$ increases though leading to more feedback signals being sent.

Keeping in mind the delay, throughput and amount of feedback involved the model can be designed. The real-time data has the minimum delay requirement, thus for those packets, a lesser value higher threshold should be chosen. In other words, the switching of data to other users should be for lesser value of buffer content. The lower threshold can then be adjusted to provide the delay and throughputrequired.

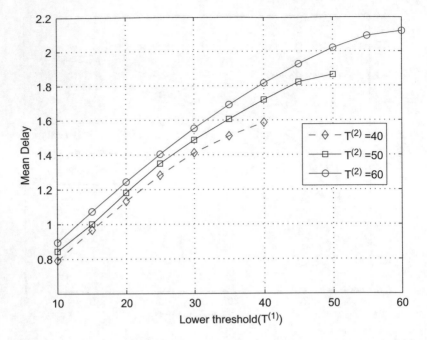

Fig. 10.6 Delay with respect to thresholds

The lower it is, lesser would be the delay as that results in lesser accumulations in the buffer. The bigger the difference between both the thresholds, the lesser will be the feedback signals sent.

The data sent as the best effort on the network such emails, file sharing do not require as least as delay possible, thus the other parameters such as throughput can be maximized for this class of data. Since the data does not require a minimal delay, so higher threshold can be kept more in this case. Thus, a single carrier is sufficient for such data packets.

10.5 Conclusions and Future Work

A delay efficient load balancing scheme for CC selection in CA in LTE-A is proposed. The proposed scheme involving double threshold can be adapted according to different QoS requirement of the user. Performance analysis of the proposed model is presented with fluid queue and the measures such as throughput and mean delay are compared with single threshold system. It can be concluded that the proposed scheme is capable of balancing load across carriers while keeping the delay lesser than the single threshold model. The impact of the this model and implementation on higher layers such as TCP can be studied as future work.

Acknowledgements   One of the authors (SD) would like to thank the National Board for Higher Mathematics, India, for financial support given to them during the preparation of the paper.

References

1. A. Hashimoto, H. Yoshino, H. Atarashi, Roadmap of IMT-advanced development. IEEE Microw. Mag. **9**(4), 80–88 (2008)
2. L. Zhang, K. Zheng, W. Wang, L. Huang, Performance analysis on carrier scheduling schemes in the long-term evolution-advanced system with carrier aggregation. IET Commun. **5**(5), 612–619 (2011)
3. Y. Wang, K.I. Pedersen, P.E. Mogensen, T.B. Sorensen, Carrier load balancing methods with bursty traffic for LTE-Advanced systems, in *20th International Symposium on Personal, Indoor and Mobile Radio Communications* (IEEE, New York, 2008), pp. 22–26
4. L. Chen, W. Chen, X. Zhang, D. Yang, Analysis and simulation for spectrum aggregation in LTE-advanced system, in *70th Vehicular Technology Conference Fall (VTC 2009-Fall)* (IEEE, New York, 2009), pp. 1–6
5. L. Zhang, F. Liu, L. Huang, W. Wang, Traffic load balance methods in the LTE-Advanced system with carrier aggregation, in *International Conference on Communications, Circuits and Systems (ICCCAS)* (IEEE, New York, 2010), pp. 63–67
6. Y. Li, L. Zhange, X. Tan, B. Cao, An advanced spectrum allocation algorithm for the across-cell D2D communication in LTE network with higher throughput. China Commun. **13**(4), 30–37 (2016)
7. Y.L. Chung, L.J. Jang, Z. Tsai, An efficient downlink packet scheduling algorithm in LTE-advanced systems with carrier aggregation. in *Consumer Communications and Networking Conference (CCNC)* (IEEE, New York, 2011), pp. 632–636
8. K.I. Pedersen, F. Frederiksen, C. Rosa, H. Nguyen, L.G.U. Garcia, Y. Wang, Carrier aggregation for LTE-advanced: functionality and performance aspects. IEEE Commun. Mag. **49**(6), 89–95 (2011)
9. 3rd Generation Partnership Project (3GPP), Further advancements of E-UTRA physical layer aspects. TR 36.814 (2009)
10. F. Liu, K. Zheng, W. Xiang, H. Zhao, Design and performance analysis of an energy-efficient uplink carrier aggregation scheme. IEEE J. Sel. Areas Commun. **32**(2), 197–207 (2014)
11. Y. Wang, K.I. Pedersen, P.E. Mogensen, T.B. Sorensen, Resource allocation considerations for multi-carrier LTE-Advanced systems operating in backward compatible mode. *20th International Symposium on in Personal, Indoor and Mobile Radio Communications* (IEEE, New York, 2009), pp. 370–374
12. K.I. Pedersen, F.J. Frederiksen, C. Rosa, H. Nguyen, L.G.U. Garcia, Y. Wang, Carrier aggregation for LTE-advanced: functionality and performance aspects. IEEE Commun. Mag. **49**(6), 89–95 (2011)
13. G. Galaviz, D.H. Covarrubias, A.G. Andrade, On a spectrum resource organization strategy for scheduling time reduction in carrier aggregated systems. IEEE Commun. Lett. **15**(11), 1202–1204 (2011)
14. Y.L. Chung, L.J. Jang, Z. Tsai, An efficient downlink packet scheduling algorithm in LTE-advanced systems with carrier aggregation, in *IEEE Consumer Communications and Networking Conference (CCNC)* (IEEE, New York, 2011), pp. 632–636
15. D. Anick, D. Mitra, M.M. Sondhi, Stochastic theory of a data-handling system with multiple sources. Bell Syst. Tech. J. **61**(8), 1871–1894 (1982)
16. W.R.W. Scheinhardt, Markov-modulated and feedback fluid queues. Universiteit Twente (1998)
17. M. Mandjes, D. Mitra, W. Scheinhardt, Models of network access using feedback fluid queues. Queueing Syst. **44**(4), 365–398 (2003)

Chapter 11
Modeling Security Requirements for VNE Algorithms: A Practical Approach

Ramona Kühn, Andreas Fischer, and Hermann de Meer

11.1 Introduction

Network virtualization is the primary enabling technology to overcome ossification effects in today's networks. It allows network administrators to deploy multiple Virtual Networks (VN) on a single Substrate Network (SN). The respective resource assignment problem is called Virtual Network Embedding (VNE). It describes how a Virtual Network (VN) can be embedded or mapped on the given SN. The networks can be represented as a graph with nodes connected by links, where the virtual nodes or links pose demands for certain resources. Then, they have to be mapped on appropriate hardware components offering these resources.

So far, VNE approaches focus mostly on optimizing the performance of the embedding. Approaches to make the embedding more secure remain mostly abstract and do not easily lend themselves to practical application. Nevertheless, security is a major request nowadays, either to meet legal requirements, to protect own data, or for a network provider to satisfy the needs of the customers. In contrast to other approaches, this work focuses on concrete security mechanisms like firewalls, Network Intrusion Detection Systems (NIDS), and Trusted Hardware (TH), and discusses how they can be included in the embedding process. It considers VNE problems with unsplittable links and focuses on offline evaluation of VNE algorithms.

R. Kühn · H. de Meer (✉)
University of Passau, Passau, Germany
e-mail: ramona.kuehn@uni-passau.de; hermann.demeer@uni-passau.de

A. Fischer
Deggendorf Institute of Technology, Faculty of Electrical Engineering, Media Technology and Computer Science, Deggendorf, Germany
e-mail: andreas.fischer@th-deg.de

© Springer International Publishing AG, part of Springer Nature 2019
A. Puliafito, K. S. Trivedi (eds.), *Systems Modeling: Methodologies and Tools*,
EAI/Springer Innovations in Communication and Computing,
https://doi.org/10.1007/978-3-319-92378-9_11

In this chapter, it is demonstrated how new constraints of security mechanisms can be incorporated into the common algorithm evaluation process with minimal changes to the embedding algorithms themselves. This enables researchers to easily extend their evaluations to include new problems including security requirements, thereby speeding up research in this area. This chapter shows the implementation and a proof-of-concept embedding that uses a common VNE simulator tool called ALEVIN [4, 10]. Furthermore, the concepts of VNE and an overview of ALEVIN are presented, then an extension to the tool to support security requirements of VN. An overview of typical security requirements and a use case are discussed to show the usability of the tool to support security in virtual environments. Furthermore, to show that secure embedding is still practicable, a performance analysis is conducted where different security mechanisms are compared with already existing resources.

The remainder of this chapter is structured as follows: Sect. 11.2 provides background information about the VNE problem. Section 11.3 describes the problem of modeling security mechanisms for VNE with an exemplary use case. The respective security requirement modeling approach is presented in Sect. 11.4. The concrete implementation of the security mechanisms and constraints in a common VNE simulation framework is described in Sect. 11.5. In addition, the necessary changes to the simulation framework and the embedding process are presented. This approach is evaluated in Sect. 11.6, regarding the performance impact on the embedding process. A discussion of related work is presented in Sect. 11.7. Finally, Sect. 11.8 presents a conclusion and a discussion of next steps.

11.2 The Virtual Network Embedding Problem

The VNE problem describes how nodes and links of a VN should be mapped to the nodes and links of an SN. Both nodes and links are considered to provide resources in the case of substrate elements, and pose respective demands in the case of virtual elements.

VN come in the form of Virtual Network Requests (VNR): requests of users for instantiation of a particular network. The VNE algorithm decides whether all requests can be supported by the given SN, and if so, how the individual elements should be mapped. The SN and VNR are both commonly modeled as labeled graphs. Substrate labels indicate the available resources on nodes and links, whereas labels in the VNR indicate the respective resource demand. The widely used interpretation of numerical resources are bandwidth as link resource and CPU time as node resource.

Figure 11.1 shows an example for this problem with two VNR to be mapped onto an SN with four nodes. A feasible embedding is already depicted. Each virtual element poses a demand on its respective substrate element, and the sum of the demands of all virtual elements hosted on a substrate element may not exceed the available resources. If not enough resources exist, only a partial solution can be found. Solutions (partial or complete) are not necessarily unique. Multiple

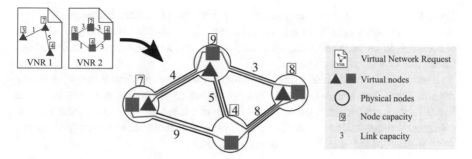

Fig. 11.1 Example for a VNE scenario (adapted from Fischer et al. [11])

simulation frameworks are available for VNE evaluation. Yu et al. propose the VNE Simulator [16]. Chowdhury et al. describe Vineyard [6, 7]. Papagianni et al. propose CVI-Sim [15]. In this work, the ALEVIN simulator [4, 10] is used, due to its flexibility and extensibility. Its flexible resource/demand model can be adapted to model security requirements. ALEVIN also offers functions such as developing new algorithms, flexible creation of embedding scenarios, and defining new metrics for the comparison and evaluation of VNE algorithms. A wide set of algorithms and metrics are already implemented. These features enable easy experimentation even in a non-standard setting such as when taking security requirements into account.

11.3 Problem Description

The implementation of security mechanisms and constraints requires a proper formulation of these requirements for VNE simulation. In this section, an overview is provided, specifying the necessity to formulate concrete security features. A motivational example helps to delineate the problem. The respective requirements are extracted and classified.

11.3.1 Overview

Network security requirements are substantially different from conventional embedding constraints such as bandwidth or CPU time. They typically do not refer to a consumable resource, but rather to a specific set of features that need to be available. For example, a customer might require one of his virtual nodes to be executed in a particularly safe environment, requiring specific protection from the underlying substrate node.

An abstract approach to this problem is to define security levels for substrate and virtual nodes, requiring the embedding algorithm to match these levels appropriately (cf. [9, 13, 14]). However, in practice the concept of strictly hierarchical levels proves to be too abstract. In an environment with multiple involved parties (as it is common in cloud computing), it is difficult to find a comprehensive definition of levels that can satisfy each party.

Instead, it is more likely that customers specify a particular set of security requirements, for example protection by intrusion detection software or by a firewall. A cloud provider, on the other hand, can label its equipment such that the customer's requests can be mapped appropriately. A motivational example is given next, to demonstrate the use case of such a scenario.

11.3.2 Motivational Example

The motivational example in Fig. 11.2 illustrates the application of security requirements in VNE. A cloud provider offers computing resources distributed over three data centers. Two of those data centers are protected by a firewall; one of those offers two separated subnets. In this cloud infrastructure, a client wants to implement a web service which consists of a load balancer, two web servers, a database, and an authentication service.

Each of the components has its specific demands which have to be adhered to by the cloud provider. For example, the web servers have to be protected from the internet by a firewall. Since a firewall cannot prevent all attacks, a NIDS should

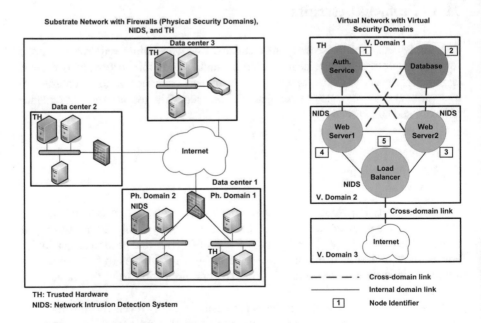

Fig. 11.2 Cloud provider infrastructure and a VN for a web service

provide information about potential malicious actions. The authentication service requires TH, as it is highly security critical. Both the authentication service and the database should be protected from the web servers by a firewall.

VNE algorithms can help to identify how the virtual infrastructure can be mapped while adhering to these requirements. However, a suitable model applicable for VNE algorithms must be found. This chapter discusses how these requirements can be modeled in a public VNE simulator for experimentation with VNE algorithms.

11.3.3   Classification of Requirements

VNE constraints discussed in the literature focus mostly on quantitative resources and demands such as bandwidth for links and CPU capacity for nodes. Security mechanisms, however, are often qualitative in nature: A particular feature or mechanism is required from the SN. This feature or mechanism is not consumed by a virtual entity, but provided for all potential entities.

Security mechanisms, such as demanded by the discussed web service, can be roughly classified into three different types. Similar to conventional constraints, there are requirements that are specific to a single node or link. However, in addition there are also more complex requirements that refer to a part of the topology. In the following, these three types are discussed in detail.

Node Requirements When a virtual node demands a security mechanism, the physical node has to offer this mechanism to be a possible candidate for mapping. Examples for this node-to-node mapping are TH, encrypted data storage, and Virtual Machine Introspection (VMI). Here, TH is used in the motivational example. This means, if a virtual node demands TH, it can only be mapped on a physical node offering TH. If a virtual node does not demand the mechanism, a mapping on a physical TH node is still possible. The virtual node does not have to use the property of the physical node. It only has to be ensured that for all further virtual nodes demanding TH there are still enough physical nodes offering it, so that a mapping is possible.

Link Requirements There are also security mechanisms that affect the links between two nodes. A virtual link might demand that it can only be mapped on a physical link that offers a specific security mechanism, for example a link that provides data encryption.

Topological Requirements A new kind of requirement are security mechanisms that affect not only an individual network entity, but also a part of the topology. Both nodes and links are affected and the topological structure has to be taken into account. Firewalls and NIDS are examples for these security mechanisms. In the motivational example, several components demand the protection by a firewall. They have to be grouped into domains to identify the parts of the affected topology. For example, it is not allowed that one node protected by the firewall is connected to another node in a different domain of the network via a link that does not pass the firewall. If this were the case, the protection of the firewall would be obsolete. This has to be prevented during the embedding.

Therefore, the network has to be separated explicitly into different network domains. On the one hand, there are domains which have to be protected by a firewall or a NIDS. On the other hand, there may be domains where such a security mechanism is not needed. The traffic between those domains has to be exclusively routed through the firewall.

A security-aware VNE takes the requirements described above into account. Therefore, it has to recognize the topology of the network to be able to divide it such that the firewall offers full protection and is not circumvented. The way how the presented requirements have to be translated for VNE simulation frameworks is discussed in the next section.

11.4 Modeling Security Requirements with Resource/Demand Pairs

In VNE, the consumable capabilities of the physical network entities are represented as resources that are attached to physical nodes and links, such as CPU, memory, and bandwidth. The basic model presented in Sect. 11.2 can be extended to also model specific properties of the physical network elements such as security mechanisms available in physical nodes. Likewise, the requirements of the VN can be modeled as demands attached to virtual entities in the VNR.

In this chapter, a demand is formulated to request a certain capacity of a consumable resource or request a certain property in the physical network entity. The resource/demand model depicted in Fig. 11.3 is used to represent the relationship between a virtual demand and a physical resource. The figure shows a substrate node and a virtual node. The substrate node provides certain features (here: a NIDS and a TH), whereas the virtual node requests these features. The mapping can succeed only if all features requested by the virtual node are present on the substrate node. A similar model is adopted for virtual and substrate links.

Fig. 11.3 Resource and demand pairs

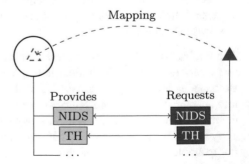

This concept already provides a base for modeling security capabilities and their corresponding demands, allowing the experimenter to model various security requirements of VN. The resource/demand concept previously used to model capacity is adapted to represent the security requirements here. Mapping of these security demands, therefore, does not occupy capacity. The mapping of security requirements of virtual nodes rather depends on available properties in the corresponding physical nodes. Likewise, the mapping of the security requirements of the virtual links depends on the type of the requirements and has to check for certain properties along the physical path that maps a particular virtual link.

Here, the concepts described in the motivational example are discussed and it is shown how the resource/demand model can be adapted to implement them. The concept of resources is re-interpreted to create "pseudo-resources" that are not consumed by their corresponding demand. This allows to model TH, NIDS, and firewalls:

A **TH** provides a trusted computing base to the hosted Virtual Machine (VM) such as a virtualized Trusted Platform Module (TPM) [5], for example. This is a simple node-based requirement that can be modeled by creating a special resource/demand pair "trusted hardware" in which the demand does not consume the resource.

A **NIDS**, as discussed above, represents a topological requirement. However, when the requirement is reformulated from "A NIDS is present" to "The node is protected by a NIDS," the requirement can actually be reformulated as a node-based requirement. It is then modeled similar to a TH node.

The **firewall**, however, is more complex to model. It is not defined explicitly in the VNR. Instead, the VNR has to specify the respective network domains that should be protected and separated from each other. Using this information, Cross-domain Links (CDL) can be identified by the embedding algorithm. As such, the demands in the VNR actually refer to domains.

The SN, on the other hand, provides firewall nodes. This can be simply modeled with a "firewall" resource. Any CDL is required to cross such a firewall node to ensure that nodes are properly separated. Intra-domain links, on the other hand, are preferably mapped to the same subnet.

The resource/demand model can be used to model these requirements. While TH and NIDS are straightforward to implement, the concept of firewalls requires more work. The disparate resource/demand pair has to be combined properly. Appropriate checks for CDL have to be performed. An implementation of these mechanisms in a public VNE simulation framework is described in the following section.

11.5   Implementation of Security Requirements

The evaluation of VNE algorithms under security constraints requires the implementation of security requirements checks in the employed simulation framework. Here, the implementation in the ALEVIN framework and the realization of the use case from Sect. 11.3.2 are demonstrated.

11.5.1 Implementation of Resource/Demand Pairs

The resource/demand model is implemented in ALEVIN using the visitor pattern to represent the occupying relationship between a virtual demand and a physical resource. ALEVIN has a generic structure that facilitates adding new resource/demand pairs, for example security requirements for nodes and links, as discussed in Sect. 11.4. Topological requirements prove to be more involved, though. Here, the simple resource/demand model has to be extended. The simulator has to check the validity of a particular mapping between a virtual link and its respective path in the SN. The mapping is considered valid only if the path can satisfy the security requirements of the virtual link. Firewall demands are implemented through the definition of different domains. These domains are represented as identifiers that are attached to the nodes. Firewall resources are attached to the respective substrate nodes.

However, in addition to this, a check for CDL is necessary. As an example, a check for firewall constraints is presented in Algorithm 1. The check is performed during the link mapping stage and forces all CDL to go through a firewall. First, the algorithm filters the VNR to find CDL by comparing the domain identifiers of the source and destination nodes of the link. Then, for each link, a set of possible physical paths is selected according to the link mapping method. The possible paths are then checked to assert if at least one of the nodes along the path provides a firewall service.

Data: A SN and a VNR to embed
Result: True, if embedding is possible, False otherwise
foreach *Virtual Link VL* ∈ *VNR* **do**
    CandidatePaths CP = findCandidatePaths(VL, SN);
    if *isCrossDomainLink(VL)* **then**
        foreach *Path P* ∈ *CP* **do**
            if *pathContainsFirewall(P)* **then**
                mapLinkToPath(VL, P);
                return True;
            end
        end
    else
        mapLinkToPath(VL, P);
        return True;
    end
    return False;
end

Algorithm 1: Embedding algorithm for CDL

The generic embedding algorithm enforces in particular the following embedding constraints: The virtual domain is not split by a firewall and is mapped in one physical domain. CDL are forced to go through a firewall. Virtual nodes that require a TH are mapped only to substrate nodes that offer it. Virtual nodes that require a NIDS are mapped only in domains in which at least one substrate node offers NIDS. The implementation in the simulator itself allows to stay agnostic of the employed VNE algorithm. It is, thus, possible to evaluate scenarios with security constraints with any VNE algorithm.[1]

11.5.2   Realization of the Motivational Scenario

Figure 11.4 shows the results of the mapping when implementing the motivational scenario for security-aware VNE from Fig. 11.2. The mapping results of the VN on the SN are depicted in Fig. 11.4. To ensure readability, only mapped CDL are represented.

The depicted scenario is realized in ALEVIN to test the functionality of the new security-aware VNE structure and algorithm. For demonstration, the commonly known vnmFlib algorithm by Lischka and Karl [12] is used to perform the actual embedding. When the original topology does not contain firewall resources, the mapping procedure will not succeed since CDL can only be mapped over nodes containing a firewall. However, when a firewall is added to the node that connects the first data center to the internet, the mapping is successful.

Fig. 11.4 Motivational scenario for security-aware VNE

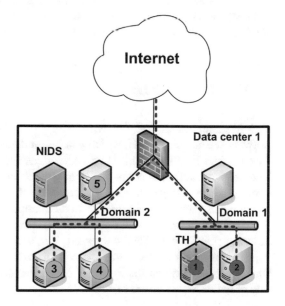

---

[1] It should be noted, though, that algorithms that do not optimize for security constraints will likely produce suboptimal results in many cases.

11.6   Evaluation

This section shows that the embedding does not only work with a specific case. Therefore, the embedding of VN with security demands is compared with the embedding of VN with conventional demands (here CPU), where network size and demands vary.

A fixed SN consisting of ten nodes offering different resources and security mechanisms was created. To test the algorithm, 10,000 VN with randomized topologies containing 4–10 nodes were generated and mapped on the SN. Values for CPU substrate resources were set to random uniform values between 10 and 100, whereas the respective virtual demands were set to random uniform values between 1 and 10. TH was offered by five nodes and demanded by two nodes, independent of the size of the VN. The SN contained two firewalls, corresponding to two distinct domains in the VN that should be protected by a firewall. To guide the analysis, the following hypothesis was developed:

H_1: *The increase of runtime of the embedding of additional security mechanisms will not be higher than 50% of the time of a CPU embedding.* The respective null hypothesis H_0 arises out of H_1. However, this evaluation wants to demonstrate that even if the runtime is higher for the embedding of security mechanism, the mapping can still be efficient and practicable. The runtime is a dependent variable. The independent variables are the distribution of the demands, the amount of available resources, and the size of the networks as well as the underlying computing power. All tests were conducted under the same conditions on the same machine, with 8 GB RAM and an Intel(R) Core-i7-4702MQ 2.2 GHz processor.

11.6.1   CPU vs. TH

The average runtime of VN with only demands for CPU was compared with networks demanding a TH, which can be modelled similar to a CPU requirement. The runtime is shown in Fig. 11.5. All values are depicted with a 95% confidence interval. The runtime is not higher than 50% of the CPU embedding, therefore H_1 is accepted. The average runtime of networks with a TH is higher, because in addition to CPU demands, it is necessary to find a mapping adequate for both CPU and TH. However, all networks with a CPU demand can be successfully mapped whereas several networks with TH are rejected: But, with a size of 8 nodes, only 1.01% are rejected, and with 10 nodes only 5.65% of the networks cannot be mapped. Even if the difference between the runtimes is significant, they are in an acceptable range and only few networks are rejected.

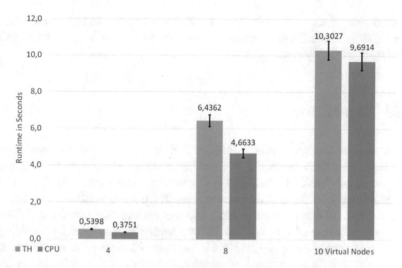

Fig. 11.5 Average runtime of mapping with CPU vs. mapping with TH

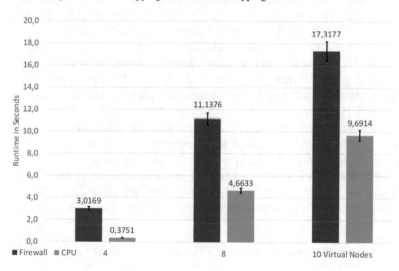

Fig. 11.6 Average runtime of mapping with CPU vs. mapping with Firewall

11.6.2   CPU vs. Firewall

In the second test, the average runtime of VNR demanding only CPU resources was compared with VNR demanding also a firewall, which is a completely new topological requirement. With the growth of the VNR, the runtime increases significantly, which is shown in Fig. 11.6. This is the result of more CDL and the algorithm that tries to find alternative mappings. If the algorithm is not able to find a valid mapping, the VN is rejected. In the case of firewalls, the amount of rejected

VN consisting of four nodes is 70.64%, with eight nodes 76.06%, and with ten nodes 85.78%. The difference in runtime between firewall and CPU exceeds 50%. Accordingly, H_1 is rejected.

11.6.3   Test Results

An overview of the test results is provided in Table 11.1: The first and second columns indicate the size of the VN and the applied resource or mechanism. It should be noted that CPU demands and resources were present in all three test runs, additionally to the security mechanisms for a more realistic scenario. The third and fourth column reports average and median runtime, and the last column shows the percentage of rejected VN. This evaluation shows that the embedding of security mechanisms causes an increase in runtime and more rejected VN but with an acceptable price and without any additional extensive effort. When comparing CPU, TH, and firewall demands directly, it is remarkable that the difference both in runtime and rejected VN between CPU and TH is far less than the difference between CPU and firewalls. Even if TH performs a little bit worse, it is necessary to mention that the amount of available CPU resources was optimal, because every node of the SN offered at least the amount that was demanded by the virtual nodes. When adapting the CPU demand in a way such that the demand exceeds the offered resources, more VN will be rejected, because this imposes a further constraint that cannot be met then.

It is assumed that security mechanisms that are modelled like convenient resources can be taken into account in the mapping process at an affordable price. However, it is also remarkable that the performance of topological requirements lags behind. Nevertheless, it was shown that the problem of the embedding of security mechanisms is still feasible, even if the solution can be improved regarding runtime and accepted networks.This means that there is still work to do. It is necessary to

Table 11.1 Runtime of the embedding of virtual networks with different sizes

VNR size	Requirement	Runtime		Rejected in %
		Avg. (s)	Median (s)	
4 Nodes	CPU	0.3751	0.0	0
	TH	0.5398	1.0	0
	FW	3.0169	3.0	70.64
8 Nodes	CPU	4.6633	5.0	0
	TH	6.4362	6.0	1.01
	FW	11.1376	12.0	76.06
10 Nodes	CPU	9.6914	10.0	0
	TH	10.3027	9.0	5.65
	FW	17.3177	18.0	85.78

foster the development of appropriate algorithms that are able to consider distinctly security mechanisms modelled as a topological requirement during the mapping process.

11.7   Related Work

The VNE problem, first mentioned by Andersen[2], is known to be NP-hard [1]. Efficient approaches, therefore, require the application of heuristics. Many VNE algorithm have been proposed up to now in the literature [11]. Most VNE algorithm are performance-oriented, optimizing for cost or for the number of VNR to be mapped onto an SN. In contrast, security considerations for VNE have been an aspect that has received only small interest, so far. This is despite the fact that the virtualized environment poses specific challenges and threats that need to be considered [3].

Based on a position paper by Fischer and De Meer [9], Liu et al. present an embedding approach using abstract security requirements with different security levels or classes [13, 14]. One application of this is data protection: It is shown that the concept of such security classes can help to define a control flow for different kinds of data (business or personal) and define a location-based resource allocation to fulfill legal requirements [8]. This means that the virtual resources are mapped on hardware resources that comply with an adequate level of protection. However, the definition of such security classes remains abstract. Bays et al. provide a more concrete embedding approach, considering end-to-end cryptography, point-to-point cryptography, and avoidance of co-hosted networks as security requirements [2]. However, the paper does not discuss further concrete security requirements.

11.8   Conclusion and Future Work

Security requirements are highly relevant for VN deployed in public environments. It is necessary to extend VNE to incorporate these requirements and the respective security capabilities of the SN so that the embedding algorithms can satisfy them. Preferably, this is done without having to change the embedding algorithms themselves. This chapter introduced some security requirements of VN and presented a generic methodology for modeling them. Topological constraints were identified as a new type of constraint that requires additional support by the simulation framework. In this chapter, a proof-of-concept implementation of a security-aware VNE model in the ALEVIN simulator was demonstrated, showing that, with some modifications, the existing resource/demand model can be adapted to implement

---

[2]D.G. Andersen, Theoretical Approaches to Node Assignment (unpublished manuscript, 2002).

security requirements. A motivational scenario that represents a web service has been discussed and implemented to demonstrate the applicability of the concept. Furthermore, an evaluation was provided that showed how efficient the mapping with additional security mechanisms can be, especially if they can be modelled similar to convenient resources like CPU resources.

Future work will focus on generalizing the security constraint model to be able to adapt to more types of security requirements. Moreover, now that a generic implementation is available, multiple VNE algorithms can be evaluated in a security-aware environment. Especially, the focus should lie on the implementation of security-aware algorithms that can handle cross-domain links to make the mapping of topological requirements more efficient.

References

1. E. Amaldi, S. Coniglio, A.M. Koster, M. Tieves, On the computational complexity of the virtual network embedding problem. Electron Notes Discrete Math. **52**, 213–220 (2016). {INOC} 2015 7th International Network Optimization Conference
2. L.R. Bays, R.R. Oliveira, L.S. Buriol, M.P. Barcellos, L.P. Gaspary, Security-aware optimal resource allocation for virtual network embedding, in *Proceedings of the 8th International Conference on Network and Service Management, CNSM '12* (International Federation for Information Processing, Laxenburg, 2013), pp. 378–384
3. L.R. Bays, R.R. Oliveira, M.P. Barcellos, L.P. Gaspary, E.R. Mauro Madeira, Virtual network security: threats, countermeasures, and challenges. J. Internet Serv. Appl. **6**(1), 1 (2015)
4. M.T. Beck, A. Fischer, F. Kokot, C. Linnhoff-Popien, H. De Meer, A simulation framework for virtual network embedding algorithms, in *6th International Telecommunications Network Strategy and Planning Symposium (Networks 2014)* (IEEE, New York, 2014), pp. 1–6
5. S. Berger, R. Cáceres, K.A. Goldman, R. Perez, R. Sailer, L. van Doorn, vtpm: kirtualizing the trusted platform module, in *Proceedings of the 15th Conference on USENIX Security Symposium - Volume 15*, Berkeley, 2006
6. N. Chowdhury, M. Rahman, R. Boutaba, Virtual network embedding with coordinated node and link mapping, in *IEEE INFOCOM 2009* (2009), pp. 783–791
7. M. Chowdhury, M. Rahman, R. Boutaba, Vineyard: virtual network embedding algorithms with coordinated node and link mapping. IEEE/ACM Trans. Networking **20**(1), 206–219 (2012)
8. B. Doll, D. Emmerich, R. Herkenhöner, R. Kühn, H. de Meer, *On Location-Determined Cloud Management for Legally Compliant Outsourcing* (Springer Fachmedien Wiesbaden, Wiesbaden, 2015), pp. 61–73
9. A. Fischer, H. De Meer, Position paper: secure virtual network embedding. Praxis der Informationsverarbeitung und Kommunikation **34**(4), 190–193 (2011)
10. A. Fischer, J.F. Botero, M. Duelli, D. Schlosser, X. Hesselbach, H. De Meer, ALEVIN - a framework to develop, compare, and analyze virtual network embedding algorithms. Electron. Commun. EASST **37**, 1–12 (2011)
11. A. Fischer, J.F. Botero, M.T. Beck, H. De Meer, X. Hesselbach, Virtual network embedding: a survey. IEEE Commun. Surv. Tutorials **15**(4), 1888–1906 (2013)
12. J. Lischka, H. Karl, A virtual network mapping algorithm based on subgraph isomorphism detection, in *VISA '09: Proceedings of the 1st ACM Workshop on Virtualized Infrastructure Systems and Architectures* (ACM, New York, 2009), pp. 81–88
13. S. Liu, Z. Cai, H. Xu, M. Xu, Security-aware virtual network embedding, in *2014 IEEE International Conference on Communications (ICC)* (2014), pp. 834–840

14. S. Liu, Z. Cai, H. Xu, M. Xu, Towards security-aware virtual network embedding. Comput. Netw. **91**, 151–163 (2015)
15. C. Papagianni, A. Leivadeas, S. Papavassiliou, V. Maglaris, C. Cervello-Pastor, A. Monje, On the optimal allocation of virtual resources in cloud computing networks. IEEE Trans. Comput. **62**(6), 1060–1071 (2013)
16. M. Yu, Y. Yi, J. Rexford, M. Chiang, Rethinking virtual network embedding: substrate support for path splitting and migration. SIGCOMM Comput. Commun. Rev. **38**(2), 17–29 (2008)

Chapter 12
Performance Analysis of Data Traffic in Small Cells Networks with User Mobility

Philippe Olivier, Florian Simatos, and Alain Simonian

12.1 Introduction

To address the permanent increase of mobile traffic, the capacity of networks can be upgraded by a massive deployment of small cells. This solution is notably envisaged by network operators for the LTE-A heterogeneous networks [9] or Ultra Dense Networks scenarios for future 5G networks [15]. In dense networks, however, the amount of handover generated by users mobility will increase with a notable impact on signaling overhead, and possibly on the throughput of data transfers. In this context, the present paper aims at evaluating the impact of *inter-cell mobility* on the performance of data traffic in dense networks. Specifically, considering small cells enables us to neglect the possible spatial variations of cell capacities and thus to focus on the impact of inter-cell mobility. Furthermore, we decouple the performance evaluation problem from the modeling of user displacement, the latter topic being out of scope of the present paper (see [13, 17] for current displacement models).

Mobility is here supposed to be captured through the distribution of the users residual sojourn time in a cell, that is, the time a mobile user is physically present in the cell once its transmission has started. Given this distribution, we construct a flow-level queuing model that allows us to derive the essential performance metrics in each cell, namely the mean throughput and the handover probability. The generic tool of our model is a multi-class Processor Sharing (PS) queue with "impatient"

P. Olivier (✉) · A. Simonian
Orange Labs, Châtillon, France
e-mail: phil.olivier@orange.com; alain.simonian@orange.com

F. Simatos
ISAE Supaero, Toulouse, France
e-mail: florian.simatos@isae-supaero.fr

© Springer International Publishing AG, part of Springer Nature 2019
A. Puliafito, K. S. Trivedi (eds.), *Systems Modeling: Methodologies and Tools*,
EAI/Springer Innovations in Communication and Computing,
https://doi.org/10.1007/978-3-319-92378-9_12

customers; the impatience here accounts for the mobility of customers from cell to cell. This generic model can then be applied to each individual cell to solve the set of flow equations, which characterize the handover rates between cells, and compute the performance indicators.

To our knowledge, the PS queue with impatience has been mainly addressed in terms of asymptotic regimes for the reneging probability for one customer class [8] or for several classes in overload [10]. The analysis of the stable multi-class PS queue with distinct impatience rates, however, has not received so far a significant contribution. For this multi-class queuing system, we here provide proofs for the stability condition and for regularity properties of the empty-system probability.

Throughput gains induced by mobility in cellular networks have been generally related so far to the spatial variations of capacity inside the cells, which permits an opportunistic use of favorable transmission conditions by mobile users [1, 5–7, 11]. These papers base their evaluation on flow-level traffic modeling and address mobility through a spatial Markov process where users jump between distinct capacity zones in the cells. Due to the complexity of the latter approach, performance indicators can be derived through suitable bounds or approximations only. In the present work, by decoupling the queuing and mobility models, we alternatively formulate the problem in terms of an equilibrium regime for the handover flows, the existence of which is assessed in the case of a homogeneous network.

The paper is organized as follows. A generic one-cell Markovian model is first constructed in Sect. 12.2 and the stability and regularity properties are stated and proved; Sect. 12.3 presents our approach to model networks with mobility; Sect. 12.4 presents numerical results, including simulation, and their discussion; finally, Sect. 12.5 draws conclusions and summarizes our main achievements.

12.2 Generic Queueing Model

As a first step, we consider a single cell model which is used as the generic tool to further analyze the impact of inter-cell mobility in a network.

12.2.1 A PS Queue with Impatience

The considered cell is supposed to be "small," i.e., of limited range so that its transmission capacity C can be assumed spatially constant; this capacity is viewed as an input parameter accounting for radio and interference conditions in the considered cellular network. We suppose that capacity C is equally shared among all active users present in its service area, as implemented by means of a Round-Robin scheduler. Following this fair sharing policy, the system occupancy at the flow level can then be modeled by a Processor-Sharing (PS) queue [4].

We consider K classes of users which generate requests for transmission according to Poisson processes with respective arrival rate λ_k, $k = 1, \ldots, K$. Class-k users have i.i.d. transmission requests of data volume Σ_k with mean σ_k, hence a service rate $\mu_k = C/\sigma_k$. Since customers may actually leave the cell during their communications, we call T_k the *remaining* sojourn time of a mobile user, i.e., the time duration he physically stays in the cell after the transmission has started. We finally denote by $\theta_k = 1/\mathbb{E}(T_k)$ the mean cell departure rate of class-k users, called class-k *mobility rate*; any class k where $\theta_k = 0$ will be called *static*. The cell occupancy can then be described by the K-dimensional process $\mathbf{N}(t) = (N_1(t), \ldots, N_K(t))$, $t \geq 0$, where $N_k(t)$ denotes the number of ongoing class-k data transfers at time t. This process evolves as the occupancy of a PS queue with impatience, the "impatient" customers here corresponding to mobile users that may leave the system before their service completion within the given cell.

We assume that Σ_k and T_k are exponentially distributed with parameters $1/\sigma_k$ and θ_k, respectively. The process $(\mathbf{N}(t))_{t \geq 0}$ is then Markovian in the state space $\mathbb{N}^K$; from state $\mathbf{n} = (n_1, \ldots, n_K)$ and for $\mathbf{e}_k = (0, \ldots, 1, \ldots 0)$ with 1 at the k-th component, it can reach state $\mathbf{n} + \mathbf{e}_k$ with transition rate λ_k, or state $\mathbf{n} - \mathbf{e}_k$ with transition rate $n_k \mu_k / L(\mathbf{n}) + n_k \theta_k$, denoting by $L(\mathbf{n}) = \sum_{1 \leq j \leq K} n_j$ the total number of active users. Let $\rho_k = \lambda_k / \mu_k$ be the offered load of class k and S (resp. M) denote the set of static (resp. mobile) classes.

In stationary regime, the equilibrium equations of process $(\mathbf{N}(t))_{t \geq 0}$ read

$$\sum_{k=1}^{K} \left[\lambda_k + n_k \left(\frac{\mu_k}{L(\mathbf{n})} + \theta_k \right) \right] \mathbb{P}(\mathbf{N} - \mathbf{n}) = \sum_{k=1}^{K} \lambda_k \, \mathbb{P}(\mathbf{N} = \mathbf{n} - \mathbf{e}_k)$$

$$+ \sum_{k=1}^{K} (n_k + 1) \left(\frac{\mu_k}{L(\mathbf{n}) + 1} + \theta_k \right) \cdot$$

$$\mathbb{P}(\mathbf{N} = \mathbf{n} + \mathbf{e}_k) \qquad (12.1)$$

with $\sum_{\mathbf{n} \in \mathbb{N}^K} \mathbb{P}(\mathbf{N} = \mathbf{n}) = 1$. Process $(\mathbf{N}(t))_{t \geq 0}$ is not reversible unless all classes are static; its stationary distribution is thus not amenable to a simple closed form. Nevertheless, a general conservation law between the average arrival and departure rates can be stated as follows: for given k, multiplying each equation of (12.1) by n_k and then summing over all state vectors $\mathbf{n} \in \mathbb{N}^K$ provides

$$\lambda_k = \mu_k \, \mathbb{E} \left(\frac{N_k \mathbf{1}_{N_k > 0}}{L(\mathbf{N})} \right) + \theta_k \, \mathbb{E}(N_k), \quad 1 \leq k \leq K. \qquad (12.2)$$

Proposition 2.1 *The Markov process* $(\mathbf{N}(t))_{t \geq 0}$ *has a stationary regime if and only if*

$$\rho_S = \sum_{k \in S} \rho_k < 1. \qquad (12.3)$$

Proof First assume that process $(\mathbf{N}(t))_{t \geq 0}$ has a stationary distribution; applying conservation law (12.2) to each static class k with $\theta_k = 0$, then summing over all $k \in S$, gives $\rho_S = \sum_{k \in S} \rho_k = \sum_{k \in S} \mathbb{E}(N_k \mathbf{1}_{N_k > 0}/L(\mathbf{N})) < 1$, so that condition (12.3) is necessary.

Conversely, assume that (12.3) holds. For any test function $f : \mathbb{N}^K \to \mathbb{R}^+$, the infinitesimal generator $\mathcal{Q}$ of the Markov process $(\mathbf{N}(t))_{t \geq 0}$ is given by

$$\mathcal{Q}f(\mathbf{n}) = \sum_{1 \leq k \leq K} \lambda_k \left[f(\mathbf{n} + \mathbf{e}_k) - f(\mathbf{n}) \right]$$

$$+ \sum_{1 \leq k \leq K} \left(\frac{\mu_k n_k}{L(\mathbf{n})} + \theta_k n_k \right) \mathbf{1}_{n_k > 0} \left[f(\mathbf{n} - \mathbf{e}_k) - f(\mathbf{n}) \right], \quad \mathbf{n} \in \mathbb{N}^K.$$

Applying [16, Proposition 8.14], the process $(\mathbf{N}(t))_{t \geq 0}$ is ergodic if there exists a so-called Lyapunov function $\Lambda : \mathbb{N}^K \to \mathbb{R}^+$ and positive constants η, δ such that

(a)   the set $\{\mathbf{n} \in \mathbb{N}^K, \Lambda(\mathbf{n}) \leq \eta\}$ is finite,
(b)   random variables $\sup_{0 \leq t \leq 1} \Lambda(\mathbf{N}(t))$ and $\int_{[0,1]} |\mathcal{Q}\Lambda(\mathbf{N}(t))| dt$ are integrable,
(c) $\mathcal{Q}\Lambda(\mathbf{n}) \leq -\delta$ as soon as $\Lambda(\mathbf{n}) > \eta$.

Consider the function $\Lambda : \mathbf{n} \in \mathbb{N}^K \mapsto \Lambda(\mathbf{n})$ defined by $\Lambda(\mathbf{n}) = s^2 + m^2$ with $s = \sum_{i \in S} n_i/\mu_i$, $m = \sum_{j \in M} n_j/\mu_j$. We successively verify conditions (a), (b), and (c):

- (a) is clearly fulfilled by Λ and any finite η;
- if $A_k(t)$ is the number of class-k user arrivals within interval $[0, t]$, we readily have $N_k(t) \leq A_k(t) \leq A_k(1)$ for $0 \leq t \leq 1$, where variable $A_k(1)$ has finite first and second moments. The latter inequalities thus ensure the validity of (b) for Λ;
- denoting by $\rho_M = \sum_{j \in M} \rho_j$ the mobile load, the above definition of $\mathcal{Q}$ yields

$$\mathcal{Q}\Lambda(\mathbf{n}) = \sum_{1 \leq k \leq K} \frac{\rho_k}{\mu_k} + \frac{1}{L(\mathbf{n})} \sum_{1 \leq k \leq K} \frac{n_k}{\mu_k} + 2(\rho_S - 1)s$$

$$+ \frac{2(s - m)}{L(\mathbf{n})} \sum_{j \in M} n_j + 2m \left(\rho_M - \sum_{j \in M} n_j \frac{\theta_j}{\mu_j} \right) + \sum_{j \in M} n_j \frac{\theta_j}{\mu_j^2}$$

for $\mathbf{n} \neq \mathbf{0}$. Setting $\mu_* = \min_{1 \leq k \leq K} \mu_k$, $\mu^{**} = \max_{1 \leq k \leq K} \mu_k$ and $A = \min_{j \in M} \theta_j/\mu_j$, $B = \max_{j \in M} \theta_j/\mu_j^2$, we then derive the upper bound

$$\mathcal{Q}\Lambda(\mathbf{n}) \leq \sum_{1 \leq k \leq K} \frac{\rho_k}{\mu_k} + \frac{1}{\mu_*} + 2(\rho_S - 1)s + m \left[2\rho_M + 2\frac{\mu^{**}}{\mu_*} + B\mu^{**} \right] - 2A\mu_* m^2.$$

$$(12.4)$$

As $\rho_S < 1$ by condition (12.3), we deduce from (12.4) that $\mathcal{Q}\Lambda(\mathbf{n})$ is asymptotically smaller than $-2A\mu_* m^2$ when m tends to infinity. Thus, for any given $\delta > 0$, there exists a constant $m_0 > 0$ such that $\mathscr{D}\Lambda(\mathbf{n}) < -\delta$ as soon as $m > m_0$. Now,

- if $s \leq m$, it is sufficient that $s + m > 2m_0$ to ensure that $m > m_0$;
- if $s > m$ and $m \leq m_0$, all terms depending on m in (12.4) are bounded and $\mathscr{D}\Lambda(\mathbf{n})$ is then asymptotically smaller than $2(\rho_S - 1)s$ when s tends to infinity.

There exists thus a constant $s_0 > 0$ such that $\mathscr{D}\Lambda(\mathbf{n}) < -\delta$ as soon as $s > s_0$. Fixing the constant $\eta = (\max(2s_0, 2m_0))^2$ and using $(s + m)^2 \geq s^2 + m^2$, we conclude that $\mathscr{D}\Lambda(\mathbf{n}) < -\delta$ when $\Lambda(\mathbf{n}) > \eta$, thus fulfilling requirement (c).

Conditions (a), (b), and (c) being verified, Λ is therefore a Lyapunov function for process $(\mathbf{N}(t))_{t \geq 0}$ and condition (12.3) is thus also sufficient.   □

Note that condition (12.3) does not depend on the traffic intensity of mobile users, since the latter leave the cell after a finite time and thus cannot cause overload. Now, given (12.3), we define two performance indicators per user-class, the *average throughput* and the *handover probability*. Considering data (elastic) traffic, the user-perceived QoS can be measured by the average throughput defined as the ratio of the mean volume of transferred data to the mean transfer time [4]. We also define the *handover probability* for class-k users as the proportion of users that exit the cell before the completion of their transmission, i.e., the ratio of the mean handover rate λ_k^{Out} to the mean flow arrival rate λ_k. The latter definitions read

$$\Gamma_k \triangleq \frac{\mathbb{E}(X_k)}{\mathbb{E}(\Delta_k)}, \qquad H_k \triangleq \frac{\lambda_k^{Out}}{\lambda_k}, \qquad 1 \leq k \leq K, \qquad (12.5)$$

where X_k denotes the part of the total data volume Σ_k which is actually transferred by a class-k user during its transmission time Δ_k ($\leq T_k$) in the cell. The following proposition is easily derived, which proof has been given in [14].

Proposition 2.2 *The throughput Γ_k and the handover probability H_k are given by*

$$\Gamma_k = C\left(\frac{\rho_k}{\mathbb{E}(N_k)} - \frac{\theta_k}{\mu_k}\right), \qquad H_k = \frac{\mathbb{E}(N_k)\,\theta_k}{\lambda_k}, \qquad 1 \leq k \leq K \qquad (12.6)$$

which depend on the mean number of class-k users only. They satisfy the remarkable identity

$$H_k = \frac{\theta_k\,\sigma_k}{\Gamma_k + \theta_k\,\sigma_k}, \qquad 1 \leq k \leq K.$$

12.2.2 Regularity Properties of the Empty-System Probability

Monotonicity and continuity of the empty-system probability as a function of any arrival rate λ_k will prove essential in Sect. 12.3 to solve the equilibrium equations of handover flows in a network. We claim that such regularity properties require a specific proof in the present queuing system with infinite state space and no closed form solution for the stationary distribution.

Denote by $\mathscr{A}_K(\lambda_K)$ the PS queuing system with impatience and define

$$Q(\lambda_K) = \mathbb{P}(\mathbf{N} = \mathbf{0}), \quad \lambda_K \geq 0, \tag{12.7}$$

the empty-queue probability as a function of the rate λ_K of class-K users, all other parameters kept constant (distinguishing here class K, be it a static class or not).

Proposition 2.3 *Function $Q(.)$ is strictly decreasing over its definition interval.*

Proof The definition domain of function Q corresponds to those values of λ_K such that $\rho_S < 1$, according to (12.3). To prove the proposition, we proceed in four steps.

A) We compare the empty queue probabilities $Q(\lambda_K)$ and $Q(\lambda'_K)$ of systems $\mathscr{A}_K(\lambda_K)$ and $\mathscr{A}_K(\lambda'_K)$ with $\lambda'_K = \lambda_K + \Delta\lambda_K$, $\Delta\lambda_K > 0$. To do this, it proves convenient to introduce a supplementary user class by defining a new system $\overline{\mathscr{A}}_{K+1}$ with $K+1$ classes, where the first K classes are identical to that in $\mathscr{A}_K(\lambda_K)$ and where the $(K+1)$th class has input rate $\lambda_{K+1} = \Delta\lambda_K$, service rate $\mu_{K+1} = \mu_K$, and impatience rate $\theta_{K+1} = \theta_K$. The occupancy of system $\overline{\mathscr{A}}_{K+1}$ is now defined by the $(K+1)$-dimensional vector $\overline{\mathbf{N}}(t) = (\overline{N}_1(t), \ldots, \overline{N}_K(t), \overline{N}_{K+1}(t)), t \geq 0$.

It proves that the K-dimensional Markov occupancy process, deduced from system $\overline{\mathscr{A}}_{K+1}$ by gathering populations of classes K and $K+1$, has the same transition rates, and thus the same stationary distribution, as the process $(\mathbf{N}'(t))_{t\geq 0}$ of system $\mathscr{A}_K(\lambda'_K)$. This result holds essentially because of the Poisson nature of all the arrival processes and of the "PS + Impatience" form of the departure processes, for which the departure rates do not depend on the fact that some classes are gathered or not.

System $\overline{\mathscr{A}}_{K+1}$ (with occupancy $\overline{\mathbf{N}}$) can therefore be considered in place of system $\mathscr{A}_K(\lambda'_K)$ for the evaluation of the stationary empty-system probability, and we have

$$Q(\lambda'_K) = \mathbb{P}(\overline{\mathbf{N}} = \mathbf{0}). \tag{12.8}$$

B) We now make use of a sample path argument to state that, at any time, each population size $N_k(t)$, $1 \leq k \leq K$, is no greater than the corresponding population size $\overline{N}_k(t)$, assuming that both systems $\mathscr{A}_K(\lambda_K)$ and $\overline{\mathscr{A}}_{K+1}$ are empty at $t = 0$ (for convenience, we set $N_{K+1}(t) = 0$ for all $t \geq 0$).

This may be thoroughly proved by induction on the sequence of all consecutive events (arrivals or departures) occurring in either system $\mathscr{A}_K(\lambda_K)$ or $\overline{\mathscr{A}}_{K+1}$. The result essentially holds because 1) there are supplementary arrivals in system $\overline{\mathscr{A}}_{K+1}$ and, 2) the (PS) per-customer service rate is lower between consecutive events in system $\overline{\mathscr{A}}_{K+1}$, compared to $\mathscr{A}_K(\lambda_K)$.

C) From the above results, we deduce that event $(\overline{\mathbf{N}}(t) = \mathbf{0})$ implies event $(\mathbf{N}(t) = \mathbf{0})$ for all $t \geq 0$. In the stationary regime, we derive that $(\overline{\mathbf{N}} = \mathbf{0}) \subset (\mathbf{N} = \mathbf{0})$ and thus $Q(\lambda'_K) \leq Q(\lambda_K)$. We conclude that $\lambda_K \mapsto Q(\lambda_K)$ is a decreasing function.

D) From the same inclusion argument, we deduce that

$$\mathbb{P}(\mathbf{N} = \mathbf{0}) = \mathbb{P}(\overline{\mathbf{N}} = \mathbf{0}) + \mathbb{P}(\mathbf{N} = \mathbf{0}, \ \overline{\mathbf{N}} \neq \mathbf{0}). \tag{12.9}$$

Noting that $(\overline{N}_{K+1} > 0; \ \forall \ k \in \{1, \dots, K\}, \overline{N}_k = 0) \subset (\overline{\mathbf{N}} \neq \mathbf{0}; \ \forall \ k \in \{1, \dots, K\}, N_k = 0)$ where the inclusion follows by the property derived in **B)**, we deduce that

$$\mathbb{P}(\mathbf{N} = \mathbf{0}, \ \overline{\mathbf{N}} \neq \mathbf{0}) \geq \mathbb{P}(\overline{N}_{K+1} > 0; \ \forall \ k \in \{1, \dots, K\}, \overline{N}_k = 0) > 0, \tag{12.10}$$

since the distribution of $\overline{\mathbf{N}}$ gives positive weight to any subset of its range. After (12.8), (12.9), and (12.10), we derive the strict decreasing behavior of $Q(.)$, as claimed. $\qquad \square$

Proposition 2.4 *Function $Q(.)$ is continuous over its definition interval.*

Proof The derivation proceeds according to the following steps.

A) Keeping the same notation, the right-continuity of $Q(.)$ at a given point λ_K will follow if it is shown that $\mathbb{P}(\mathbf{N} = \mathbf{0}, \ \overline{\mathbf{N}} \neq \mathbf{0})$ tends to 0 when $\Delta\lambda_K$ tends to 0.

Consider the joint Markov process $(\mathbf{N}(t), \overline{\mathbf{N}}(t))_{t\geq 0}$, starting with empty queues, and a cycle of given duration $\overline{\tau}$ starting at $t = 0$, without loss of generality. First note, in view of property **B)** above, that a cycle of this joint process is identical to a cycle of process $(\overline{\mathbf{N}}(t))_{t\geq 0}$. Then the event $(\mathbf{N}(t) = \mathbf{0}, \ \overline{\mathbf{N}}(t) \neq \mathbf{0})$ for given $t \in [0, \overline{\tau}]$ implies that the date τ_0 of the first arrival from class $(K + 1)$ is no greater than t, that is, $\mathbf{1}_{(\mathbf{N}(t)=\mathbf{0}, \ \overline{\mathbf{N}}(t)\neq\mathbf{0})} \leq \mathbf{1}_{t\geq\tau_0}$. We thus derive that

$$\mathbb{E}\left(\int_0^{\overline{\tau}} \mathbf{1}_{(\mathbf{N}(t)=\mathbf{0}, \ \overline{\mathbf{N}}(t)\neq\mathbf{0})} \, dt\right) \leq \mathbb{E}\left(\int_0^{\overline{\tau}} \mathbf{1}_{(t\geq\tau_0)} \, dt\right) \leq \mathbb{E}\left(\overline{\tau} \cdot \mathbf{1}_{(0\leq\tau_0\leq\overline{\tau})}\right).$$

We now apply the cycle formula [3, Chap. IV, Theorem 8.4] to the ergodic process $(\mathbf{N}(t), \overline{\mathbf{N}}(t))_{t\geq 0}$ and use the Cauchy-Schwarz inequality to get

$$\mathbb{P}(\mathbf{N} = \mathbf{0}, \ \overline{\mathbf{N}} \neq \mathbf{0}) = \frac{1}{\mathbb{E}(\overline{\tau})} \cdot \mathbb{E}\left(\int_0^{\overline{\tau}} \mathbf{1}_{(\mathbf{N}(t)=\mathbf{0}, \ \overline{\mathbf{N}}(t)\neq\mathbf{0})} \, dt\right)$$

$$\leq \frac{\mathbb{E}(\overline{\tau}^2)^{1/2}}{\mathbb{E}(\overline{\tau})} \cdot \mathbb{P}(0 \leq \tau_0 \leq \overline{\tau})^{1/2}.$$

B) Decomposing the cycle duration $\overline{\tau}$ as the sum $\overline{\tau} = \overline{\tau}_B + \overline{\tau}_I$ of a busy period duration $\overline{\tau}_B$ and of the following idle period duration $\overline{\tau}_I$, it is shown that the first and second moments of $\overline{\tau}$ are locally (i. e., around $\Delta\lambda_K = 0$) lower- and upper-bounded, respectively. First, $\overline{\tau}_I$ is exponentially distributed with parameter $\Lambda + \Delta\lambda_K$ (where $\Lambda = \lambda_1 + \ldots + \lambda_K$), so that $\mathbb{E}(\overline{\tau}) \geq \mathbb{E}(\overline{\tau}_I) = 1/(\Lambda + \Delta\lambda_K)$. Second, noting that random variables $\overline{\tau}_B$ and $\overline{\tau}_I$ are independent (due to the Poisson arrival processes), we have $\mathbb{E}(\overline{\tau}^2) = \mathbb{E}(\overline{\tau}_B^2) + 2\mathbb{E}(\overline{\tau}_B)\,\mathbb{E}(\overline{\tau}_I) + \mathbb{E}(\overline{\tau}_I^2)$.

To ensure that $\mathbb{E}(\overline{\tau}_B)$ and $\mathbb{E}(\overline{\tau}_B^2)$ are finite and locally bounded functions of $\Delta\lambda_K$, we state that they are upper-bounded by the corresponding moments of τ_B^*, the busy period of the same system without impatience. Then, by gathering all $K + 1$ classes into a single class, τ_B^* is also the busy period of a one-class $M/G/1$ PS queue with a compound distribution for the data volume $\overline{B}$. Since the system without impatience is work-conserving, the distribution of τ_B^* is independent of the actual service discipline; its first and second moments are continuous functions of the load and moments of $\overline{B}$, as shown by formulas given in [12, Vol. I, Chap. 5, Section 5.8, Equ. (5.141) and Equ. (5.142)]. As a consequence, the moments of $\overline{\tau}_B$, and thus $\mathbb{E}(\overline{\tau}^2)$, are locally upper-bounded.

C) It remains to show that $\mathbb{P}(0 \leq \tau_0 \leq \overline{\tau})$ tends to 0. For any $A > 0$, write

$$\mathbb{P}(\tau_0 \leq \overline{\tau}) = \mathbb{P}(\tau_0 \leq \overline{\tau}, \tau_0 \leq A) + \mathbb{P}(A \leq \tau_0 \leq \overline{\tau}) \leq \mathbb{P}(\tau_0 \leq A) + \mathbb{P}(A \leq \overline{\tau}). \tag{12.11}$$

By the Markov inequality, we first have $\mathbb{P}(A \leq \overline{\tau}) \leq \mathbb{E}(\overline{\tau}^2)/A^2$ so that, since $\mathbb{E}(\overline{\tau}^2)$ is locally bounded, we can select A such that $\mathbb{P}(A \leq \overline{\tau})$ is arbitrarily small, say, lower than ε for any given $\varepsilon > 0$. For this value of A, let now $\Delta\lambda_K$ tend to 0. The marginal distribution of τ_0, which is independent from the cycle duration, is a compound of an atom at 0 and an exponential distribution, hence $\mathbb{P}(\tau_0 \leq A) = \alpha + (1 - \alpha)(1 - e^{-A\,\Delta\lambda_K})$ with $\alpha = \Delta\lambda_K/(\Lambda + \Delta\lambda_K)$. This probability can thus be made lower than ε for small enough $\Delta\lambda_K$, thus making $\mathbb{P}(\tau_0 \leq \overline{\tau})$ lower than 2ε after (12.11). This finally justifies the right-continuity of function Q at point λ_K.

D) A similar reasoning shows that function Q is also left-continuous. □

12.3   Network with Inter-Cell Mobility

We now address the description of a whole network of cells where users move from one cell to neighboring ones due to possible handovers.

12.3.1   A Closed Network of Queues

Consider a cellular network of I cells with possibly distinct capacities. Users from K traffic classes may appear and move during their communications. When leaving a cell during transmission, users join one of the neighboring cells according to some routing probabilities. They consequently generate supplementary flows of new arrivals, hereafter called *handover arrivals*, which are to be added to *fresh arrivals* in each cell. Assume that class-k users generate requests for transmission in cell i according to a Poisson process with rate $\lambda_{i,k}^0$, $i = 1, \ldots, I$, $k = 1, \ldots, K$; this corresponds to the fresh traffic offered to cell i. To account for class-k users that became active outside cell i and experienced some handovers, the total flow arrival to cell i is

$$\lambda_{i,k} = \lambda_{i,k}^0 + \lambda_{i,k}^{In} = \lambda_{i,k}^0 + \sum_{j \neq i} p_k(j, i) \cdot \lambda_{j,k}^{Out}, \tag{12.12}$$

where $\lambda_{i,k}^{In}$ denotes the handover arrival rate from neighboring cells, $\lambda_{j,k}^{Out}$ is the handover departure rate from cell j, and $p_k(j, i)$ are the routing probabilities from cells $j \neq i$ to cell i. For all i and k, we will assume that the handover arrival process to cell i from class-k users can be approximated by a Poisson process so that it can be superposed to the fresh arrivals to build up a total Poisson arrival process with rate $\lambda_{i,k}$ given in (12.12). All Poisson processes introduced above are supposed to be mutually independent, an assumption which notably simplifies the global description of the system by reducing it to a network of queues which is *closed* regarding the handover flows. This is, in particular, in contrast to the overall multi-class multi-cell process considered in other papers [5–7, 11].

The rate $\lambda_{j,k}^{Out}$ can in turn be considered as an output of the generic queuing model considered in Sect. 12.2 for cell j, and be calculated by means of some *performance function* $\mathscr{F}_{j,k}(.)$, that is,

$$\lambda_{j,k}^{Out} = \mathscr{F}_{j,k}\left(\lambda_{j,1}^{In}, \ldots, \lambda_{j,K}^{In}\right) = \theta_{j,k}\mathbb{E}(N_{j,k}) \tag{12.13}$$

for $j = 1, \ldots, I$ and $k = 1, \ldots, K$, where $\theta_{j,k}$ (resp. $N_{j,k}$) denotes the mobility rate from cell j emanating from class-k ongoing transfers (resp. the number of class-k ongoing transfers in cell j). In (12.13), only handover arrival rates are considered as variables, all other intrinsic parameters (such as cell capacities, per-class offered traffic and mobility rates) being kept constant. From (12.12) and (12.13), it follows that a stationary network regime can be characterized by a system of $I \times K$ flow equations with the handover arrival rates $\lambda_{i,k}^{In}$ as unknowns, namely

$$\lambda_{i,k}^{In} = \sum_{j \neq i} p_k(j, i) \cdot \mathscr{F}_{j,k}\left(\lambda_{j,1}^{In}, \ldots, \lambda_{j,K}^{In}\right). \tag{12.14}$$

The problem of existence and uniqueness of a solution to the non-linear system (12.14) is out of the scope of the present paper. As the performance functions $\mathscr{F}_{j,k}$ may not be explicit in terms of input parameters, the practical determination of a solution to (12.14) involves a numerical iterative procedure, e.g. a fixed-point algorithm.

12.3.2   The Case of a Homogeneous Network

Now assume that the network is *homogeneous* in the following sense:

1. all intrinsic parameters (capacities, arrival rates, ...) are the same for all cells, so that performance functions do not depend on the cell, that is, $\mathscr{F}_{j,k}(.) = \mathscr{F}_k(.)$;
2. the routing of handover flows is symmetric, i.e., for each class k, cell i receives handover traffic from a set $\mathscr{J}_k(i)$ of neighboring cells with identical probability $p_k(j, i) = 1/J_k$, where J_k is the common cardinal of sets $\mathscr{J}_k(i)$.

   Clearly, any set of rates $\lambda_{i,k}^{In} = \lambda_k^{In}, \forall i, k$, verifying the simpler system

$$\forall k \in \{1, \ldots, K\}, \quad \lambda_k^{In} = \lambda_k^{Out} = \mathscr{F}_k\left(\lambda_1^{In}, \ldots, \lambda_K^{In}\right), \tag{12.15}$$

will provide a particular solution to (12.14), hence *the* solution if uniqueness is ensured. For a homogeneous network, the problem thus reduces to the study of a single *representative cell*, where the ingress and outgoing handover traffics balance exactly.

In this context, we define the total and per-class loads by referring to the fresh traffic, that is, $\rho_k^0 = \lambda_k^0 \sigma_k/C$, $1 \leq k \leq K$, and $\rho^0 = \sum_{k=1}^{K} \rho_k^0$. We now claim that the equilibrium of this system is characterized by $\rho^0 < 1$, a condition which is well understood since mobile users re-enter the system until their transfer is completed.

Proposition 3.1

A) *In the homogeneous network with inter-cell mobility,*

$$\rho^0 < 1 \tag{12.16}$$

   is a necessary condition for the existence and uniqueness of a fixed-point solution to equilibrium equations (12.15), that is, $\lambda_k^{In} = \lambda_k^{Out}$, $1 \leq k \leq K$.

B) *In the specific case of a single mobile class, this condition is also sufficient.*

Proof

A) Assume that there exists a solution to equilibrium equations (12.15). For the system associated with that solution, we apply (12.2) to any class k; recalling that $\lambda_k = \lambda_k^0 + \lambda_k^{In}$ by (12.12) and that $\lambda_k^{In} = \lambda_k^{Out} = \theta_k \mathbb{E}(N_k)$ by (12.13), we then obtain $\rho_k^0 = \mathbb{E}\left(N_k \mathbf{1}_{N_k>0}/L(\mathbf{N})\right)$, $1 \leq k \leq K$. The side-by-side summation of these equalities yields the required condition $\rho^0 < 1$, after observing that

$$\sum_{k=1}^{K} \mathbb{E}\left(\frac{N_k \mathbf{1}_{N_k>0}}{L(\mathbf{N})}\right) = \mathbb{E}(\mathbf{1}_{\mathbf{N}\neq(\mathbf{0})}) = 1 - \mathbb{P}(\mathbf{N} = \mathbf{0}). \tag{12.17}$$

B) To address the sufficiency of (12.16), first note that $\rho_S = \rho_S^0 < 1$ obviously holds, ensuring that the queue is stable whatever the load of mobile users. Besides, writing relation (12.2) for any class k yields $\lambda_k^0 + \lambda_k^{In} = \mu_k \, \mathbb{E}\left(N_k \, \mathbf{1}_{N_k>0}/L(\mathbf{N})\right) + \lambda_k^{Out}$ which, after dividing each side by μ_k and summing over all $k \in \{1, \ldots, K\}$, gives

$$\rho^0 = \sum_{k=1}^{K} \frac{\lambda_k^0}{\mu_k} = 1 - \mathbb{P}(\mathbf{N} = \mathbf{0}) + \sum_{k=1}^{K} \frac{\lambda_k^{Out} - \lambda_k^{In}}{\mu_k}. \tag{12.18}$$

At this stage, we assume that there is only one single class M of mobile users. As $\lambda_k^{In} = \lambda_k^{Out} = 0$ for all static classes k, it follows from (12.18) that the left equilibrium equation in (12.15), $\lambda_M^{In} = \lambda_M^{Out}$, is equivalent to

$$\mathbb{P}(\mathbf{N} = \mathbf{0}) = 1 - \rho^0. \tag{12.19}$$

We use the notation of Sect. 12.2.2 to class M so that $\mathbb{P}(\mathbf{N} = \mathbf{0}) = Q(\lambda_M)$ is a function of the arrival rate $\lambda_M = \lambda_M^0 + \lambda_M^{In}$ of class-M users. Since $\rho^0 < 1$ is assumed, the existence of a unique solution λ_M to (12.19) is ensured if it is shown that

  I. $Q(\lambda_M^0) \geq 1 - \rho^0$;
  II. $Q(.)$ is strictly decreasing over $\mathbb{R}^+$;
  III. $Q(.)$ is continuous over $\mathbb{R}^+$;
  IV. $Q(\lambda) \, \searrow \, 0$ as $\lambda \to +\infty$.

Items **I, II, III,** and **IV** can be successively proved as follows.

  I.   For any impatience queuing system, summing the conservation relation (12.2) applied to each class and recalling (12.17), we get the following expression for the average number of moving users:

$$\mathbb{E}(N_M) = \frac{\mu_M}{\theta_M} \left(\rho + \mathbb{P}(\mathbf{N} = \mathbf{0}) - 1\right). \tag{12.20}$$

By identity (12.20) for the mean number $\mathbb{E}(N_M)$, its non-negativity entails that

$$\forall \, \lambda_M \geq 0, \quad Q(\lambda_M) = \mathbb{P}(\mathbf{N} = \mathbf{0}) \geq 1 - \rho. \tag{12.21}$$

Lower bound (12.21) holds, in particular, for the value $\lambda_M = \lambda_M^0$ for which $\rho = \rho^0$, hence $Q(\lambda_M^0) \geq 1 - \rho^0$ as required.

  II.   This is ensured by Proposition 2.3.
  III.   This is ensured by Proposition 2.4.
  IV.   All parameters being kept constant, we denote by N_M^* the number of mobile users in the same queuing system but in the absence of any static

users. A straightforward sample path argument enables us to show that $N_M \geq N_M^*$ almost surely, so that

$$Q(\lambda_M) = \mathbb{P}(\mathbf{N} = \mathbf{0}) \leq \mathbb{P}(N_M = 0) \leq \mathbb{P}(N_M^* = 0). \qquad (12.22)$$

Besides, the stationary distribution of N_M^* is that of the occupancy process of a PS queue with a single customer class M [8], with load $\rho_M = \lambda_M/\mu_M$ and impatience rate θ_M; the corresponding empty-queue probability is, in particular, given by

$$\mathbb{P}(N_M^* = 0) = \left(1 + \sum_{\ell \geq 1} \rho_M^\ell \left[\prod_{1 \leq j \leq \ell} (1 + j\theta_M/\mu_M)\right]^{-1}\right)^{-1}.$$

This probability is then upper-bounded by $(1 + \theta_M/\mu_M)/\rho_M$, which ensures together with inequality (12.22), that $Q(\lambda_M)$ tends to 0 as λ_M tends to infinity. □

12.4 Numerical Results

We report here some numerical experiments, focusing on the important case where users are gathered into two classes, namely one static and one mobile class. In all subsequent scenarios, we fix a cell capacity $C = 50\,\text{Mbit/s}$, a proportion of 50% mobile users and a mean flow volume $\sigma = 12.5\,\text{MB}$ (100 Mbit) for both classes.

12.4.1 Impatience Model

Regarding the generic Markovian model analyzed in Sect. 12.2, we examined in [14] the sensitivity of the performance indicators to the distributions of sojourn (impatience) time and flow volume. It was shown there that the throughput of each class is only marginally impacted by both distributions, indicating that results derived from the Markovian framework remain valid for more realistic distributions, while the handover probability is noticeably more impacted (particularly at low load), increases with the variance of T_M and decreases with the variance of Σ_M and Σ_S.

We now focus on the performance indicators provided by the (numerically solved) Impatience Model. In Fig. 12.1 are plotted the average throughputs and handover probability obtained with exponentially distributed T_M, Σ_M, and Σ_S, and considering a series of three normalized mobility rates: θ_M equals 0.2, 1, or 5 times the service capacity $\mu_M = 0.5\,\text{s}^{-1}$. Large throughput gains for static and mobile

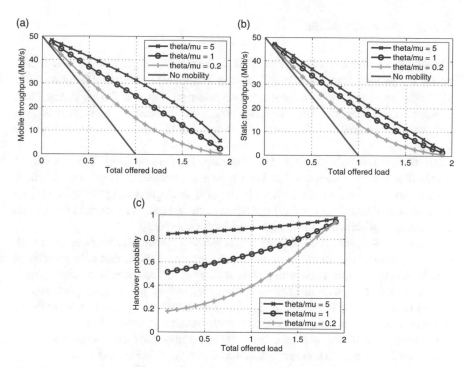

Fig. 12.1 Performance indicators: (**a**) mobile throughput, (**b**) static throughput, and (**c**) handover probability obtained from the Impatience Model, for 50% mobile users and a normalized mobility rate of 0.2, 1, or 5

users are observed, compared to a scenario where all users would be static; besides, we note a significant gain of mobile users throughput over that of static users. We have further observed that the throughput gain of mobile over static users appears to be the greatest when the mobility rate is large (say, more than twice the service rate for mobile users) and when the proportion of mobile users is small (say, 20%).

Such throughput gains are expected in this open-loop system, and were already observed in [2]: they result from the impatient nature of mobile users who may leave the system without re-entering it (hence the gain w.r.t. the all-static scenario) and tend to do it especially when local congestion occurs (hence the gain mobile/static). Very large handover rates (see Fig. 12.1c) may however counterbalance these gains.

Interpreting the above results helps us to assess the impact of cell size. Assuming a constant speed v, the mean distance the mobile user travels in the cell is $\mathbb{E}(D) = v/\theta_M$; this mean distance is typically of the order of the radius R of a circular cell. Thus if $v = 90$ km/h for example, the values of θ_M considered above, namely $5\,\mu_M$, μ_M, and $0.2\,\mu_M$ (with $\mu_M = 0.5\,\mathrm{s}^{-1}$) respectively correspond to a radius of 10 m, 50 m, and 250 m, typical of a Femto, Pico, and Micro cell. As expected, users in Femto cells experience the largest throughput since their mobility rate is the highest.

12.4.2 Mobility Model

We assess the Mobility Model proposed in Sect. 12.3, i.e., the Markovian model where the handover arrival rate λ_M^{In} exactly balances the outgoing handover rate λ_M^{Out}. We consider the homogeneous four-cell ring network shown in Fig. 12.2, where all cells have the same capacity and traffic parameters as in Sect. 12.4.1, and three normalized mobility rates: θ_M is equal to 0.1, 1, or 10 times the service capacity μ_M.

Event-driven simulations have been performed at the flow level. The accuracy of results drawn from simulation has been tightly controlled. In every configuration, ten independent simulation runs have been performed, generating around 1 million discrete events each. The obtained confidence intervals are very small in most cases and thus are not shown in the following plots for simplicity.

Figure 12.3 depicts the performance indicators for mobile users in each cell. For each value of θ_M, the four curves (each corresponding to one cell) are almost indistinguishable from each other, thus assessing the robustness of simulations. We observe that the stability region is characterized by $\rho^0 < 1$, as predicted by Proposition 3.1 and, from Fig. 12.3a, that the throughput gains due to mobility increase with the mobility rate. Other complementary results have shown that the mobile/static throughput gain is all the more important that the proportion of mobile users is weak. The latter simulation is compared in Fig. 12.4 to the Mobility Model (applied to the *representative cell*) when $\theta_M = 10\,\mu_M$. We note that the representative cell model provides slightly optimistic throughputs compared to those obtained from simulation.

The good match between model and simulation results validates our approach for reducing a homogeneous network to a single representative cell: the assumption quoted in Sect. 12.3.1, that the handover traffic flow re-enters the representative cell

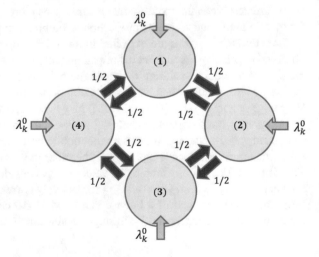

Fig. 12.2 A homogeneous ring network of four identical cells with symmetric routing

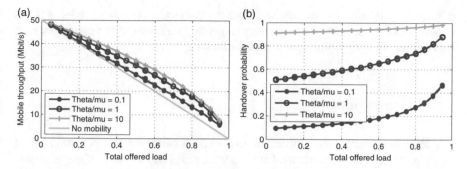

Fig. 12.3 Homogeneous ring network: performance indicators, (**a**) mobile user throughput and (**b**) handover probability, obtained from simulation versus the total offered load in each cell (proportion of 50% mobile users and $\theta_M / \mu_M = 0.1, 1, 10$)

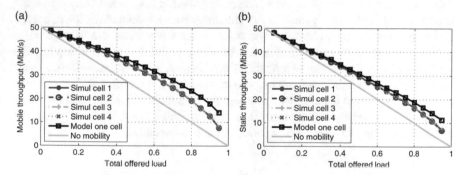

Fig. 12.4 Homogeneous ring network: (**a**) mobile throughput and (**b**) static throughput obtained from simulation and the Markovian Mobility Model ($\theta_M / \mu_M = 10$)

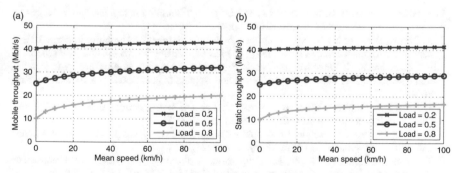

Fig. 12.5 Impact of the users speed on (**a**) mobile throughput and (**b**) static throughput for a proportion of 50% mobile users, cell radius 50 m, and total offered load 0.2, 0.5 or 0.8

as a supplementary Poisson flow, appears reasonable. The robustness of the latter assumption has also been checked in the case of a heterogeneous ring network [14].

Finally, we evaluate the impact of mobile speed for a given cell size (a cell radius of 50 m corresponding to a Pico cell). Figure 12.5 depicts the static and mobile users throughputs in terms of the speed v for different values of the total offered load (0.2,

0.5, and 0.8). Results are here derived from the Markovian Mobility Model only. As expected, all performance indicators are increasing functions of the speed; but note that the impact at very high speed is rather limited.

12.5 Conclusion

We have investigated the impact of inter-cell mobility on the performance of dense networks with small cells. Our approach relies on two main ideas: a simple performance model can be developed to capture mobility on the basis of the multi-class Processor-Sharing queue with impatience; the performance of a network of small cells can be handled by applying the generic model to each individual cell. The present paper extends the former contribution [14], notably by providing mathematical proofs for the stability of the impatience model and for the existence and uniqueness of an equilibrium regime for the handover flows in a homogeneous network. Further practical outcomes can be stated as follows: (1) as a step beyond available studies, the handover probability has been evaluated to assess the trade-off between throughput gain and signaling overhead due to mobility; (2) both classes of users are shown to benefit from a throughput gain induced by inter-cell mobility; this gain is created by the opportunistic displacement of mobile users within the network according to local load variations in individual cells.

References

1. N. Abbas, T. Bonald, B. Sayrac, Opportunistic gains of mobility in cellular data networks, in *Proceedings of WiOpt 13th International Symposium* (2015)
2. B. Baynat, R.-M. Indre, N. Nya, P. Olivier, A. Simonian, Impact of mobility in dense LTE-A networks with small cells, in *Proceedings of VTC2015- Spring Workshop*, Glasgow (2015)
3. R. Bhattacharya, E. Waymire, *Stochastic Processes with Applications*. Classics in Applied Mathematics, vol. 61 (Society for Industrial and Applied Mathematics (SIAM), Philadelphia, 2009)
4. T. Bonald, A. Proutière, Wireless downlink data channels: user performance and cell dimensioning, in *Proceedings of MobiCom'03*, San Diego (2003)
5. T. Bonald, S. Borst, N. Hegde, How mobility impacts the flow-level performance of wireless data systems, in *Proceedings of IEEE Infocom* (2004)
6. T. Bonald, S. Borst, N. Hegde, M. Jonckheere, A. Proutière, Flow-level performance and capacity of wireless networks with user mobility. Queueing Syst. **63**, 131–164 (2009)
7. S. Borst, A. Proutière, N. Hegde, Capacity of wireless data networks with intra- and inter-cell mobility, in *Proceedings of IEEE Infocom* (2006)
8. E.G. Coffman, A.A. Puhalskii, M.I. Reiman, P.E. Wright, Processor-shared buffers with reneging. Perform. Eval. **19**, 25–46 (1994)
9. E. Dahlman, S. Parkvall, J. Skold, *4G: LTE/ LTE-Advanced for Mobile Broadband*, 2nd edn. (Academic Press, London, 2013)

10. F. Guillemin, S. ElAyoubi, C. Fricker, P. Robert, B. Sericola, Controlling impatience in cellular networks using QoE-aware radio resource allocation, in *Proceedings of 27th International Teletraffic Congress* (2015)
11. M.K. Karray, User's mobility effect on the performance of wireless cellular networks serving elastic traffic. Wirel. Netw. **17**, 247–262 (2011)
12. L. Kleinrock, *Queueing Systems, Volume I: Theory* (Wiley Interscience, New York, 1975)
13. X. Lin, R.K. Ganti, P.J. Fleming, J.G. Andrews, Towards understanding the fundamentals of mobility in cellular networks. IEEE Trans. Wirel. Commun. **12**(4), 1686–1698 (2013)
14. P. Olivier, A. Simonian, Performance of data traffic in small cells networks with inter-cell mobility, in *Proceedings of ValueTools 2016*, Taormina (2016)
15. A. Osseiran et al., Scenarios for 5G mobile and wireless communications: the vision of the METIS project. IEEE Commun. Mag. **52**(5), 26–35 (2014)
16. P. Robert, *Stochastic Networks and Queues* (Springer, Berlin, 2003)
17. C. Schindelhauer, Mobility in wireless networks, in *Proceedings of SOFSEM 2006: Theory and Practice of Computer Science*, Merin (2006)

Part III
Optimization and Quantitative Evaluation Techniques Applied to Cloud Computing and the Internet of Things

Chapter 13
A Technique to Identify Data Exchange Between Cloud Virtual Machines

Nicola Bicocchi, Claudia Canali, and Riccardo Lancellotti

13.1 Introduction

Resource management in cloud data centers is a critical task to reduce energy consumption in IT infrastructures. A large corpus of literature focuses just on reducing the number of powered-on hosts [5, 7, 21]. However, a more advanced approach to the problem aims at taking into account energy expenditures for data exchanges, for example by placing highly interacting VMs on the same physical host [9, 20].

This evolution towards a network-aware VMs allocation is even more evident as certain trends, such as software-defined networks, become widespread. However, network-aware allocation requires to map network interactions between each couple of VMs in a data center. In the present study, we assume the point of view of an Infrastructure as a Service (IaaS) cloud provider. In such vision, VMs are just opaque objects that do not offer any insight on the software they are running. Hence, it is unlikely to have a data traffic matrix between the VMs, relying just on monitoring services in industry[1] or proposed in scientific literature [3]. Indeed, most monitoring systems just provide the input/output traffic rate of each VM, without a per-source/per-destination breakdown. A specialized monitoring infrastructure able to provide a complete traffic matrix among VMS may be developed and deployed over the data center, as in [22, 25]; however, in this case overhead and scalability

---

[1]https://aws.amazon.com/cloudwatch/.

N. Bicocchi · C. Canali · R. Lancellotti (✉)
Department of Engineering "Enzo Ferrari", University of Modena and Reggio Emilia, Modena, Italy
e-mail: nicola.bicocchi@unimore.it; claudia.canali@unimore.it; riccardo.lancellotti@unimore.it

© Springer International Publishing AG, part of Springer Nature 2019   201
A. Puliafito, K. S. Trivedi (eds.), *Systems Modeling: Methodologies and Tools*,
EAI/Springer Innovations in Communication and Computing,
https://doi.org/10.1007/978-3-319-92378-9_13

are likely to represent an issue. This approach, indeed, involves on one side the overhead of dedicated resources of the infrastructure devoted to the monitoring of every communication between couples of VMs, and on the other side the presence of huge amount of data to be stored and processed, that would limit the scalability of the system [22]. For this reason, in this book chapter we propose a solution that is not relying on the presence of a data traffic matrix but is able to infer communication patterns starting from the input/output network time series of each VM.

The contribution of this chapter is twofold. First, we outline a methodology that accepts as the input the time series of the number of network packets inbound and outbound for each VM, and provides information about which VMs intensively communicate with each other. The proposed methodology, which was initially proposed in [8, 17], is explicitly tailored to cope with the most challenging scenarios, that is the case where multi-tier applications are deployed over a cloud infrastructure relying on vertical replication. Second, we discuss how such methodology can be implemented adopting solutions, such as gossip protocols and distributed algorithms, that can provide high scalability even in large data centers. A qualifying point of our proposal is exploiting correlation of data traffic time series to infer the interaction between VMs. Our study compares multiple correlation metrics to identify the best solution to cope with the scenario of replicated multi-tier applications. Our experiments, carried out using a benchmark application over a real cloud infrastructure, demonstrate the viability of our approach to discover interacting VMs within a cloud data center.

The remainder of this paper is organized as follows. Section 13.2 describes the reference scenario of this paper. Section 13.3 models the problem and outlines the proposed methodology. Section 13.4 details an implementation of the methodology making use of distributed agents and a gossiping protocol. Section 13.5 describes the experimental results used to validate our proposal, while Sect. 13.6 discusses the state of the art in scientific literature. Finally, Sect. 13.7 concludes the paper and details final remarks.

13.2 Reference Scenario

We now outline the reference scenario for this chapter that is depicted in Fig. 13.1.

The upper part of Fig. 13.1, namely *Virtual view*, represents how the cloud infrastructure is perceived by the cloud customer. We assume to have a collection of VMs interacting among themselves to support a multi-tier application. The application is deployed over s replicated vertical stacks. Each vertical stack contains the r tiers of the application, hosted on separate VMs. The communication patterns in this scenario involve just VMs within the same vertical stack. No communication occurs between VMs belonging to two different vertical stacks. We assume that incoming request load is balanced across the vertical stacks. Hence, VM belonging to different vertical stacks but to the same tier are likely to show similar network traffic patterns (referring to Fig. 13.1 this may occur, for example, for $VM_{1,1}$ and

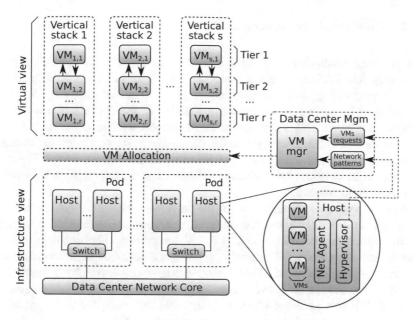

Fig. 13.1 Reference scenario

VM$_{2,1}$), the final effect is that we may have correlation in network utilization even if no communication occurs between the two VMs. Dealing with this effect is one of the main challenges of our research.

The lower part of the figure represents the physical infrastructure of the data center (namely *Infrastructure view*). The data center is composed of manageable subareas, called PODs [4]. The physical hosts are interconnected using a hierarchically structured network infrastructure (actual topologies in a data center may be more complex such as the fat-tree topology [1]) and, as a consequence, we may experience different communication costs when allocating VMs over the hosts. This heterogeneous cost of communication motivates our proposal for a network-aware VM placement [9].

Figure 13.1 provides also the detail of a physical host. Each host is equipped with a *Hypervisor*, that supports also the collection of data about the VMs resource usage. Typical resources, such as CPU, memory, I/O utilization, are made available for the data center management, while network utilization is processed by the *Network Agent* that is described in detail in Sect. 13.4. Finally, the dashed box on the right side of the figure describes the Data Center Management: such component receives the resource utilization of each VM and collects the information generated by the network agent about VMs interaction. These data are then used to take decision about the allocation of VMs over the physical infrastructure, creating a mapping between the virtual and infrastructure views.

13.3 Methodology Description

The proposed methodology to support a network-aware VMs allocation in the previously described scenario has a twofold goal. First, we aim to identify couples of VMs exchanging data: the solution of the problem of allocating VMs can be improved by adding in the objective function of the optimization problem a component to capture VMs interactions [6, 9]. Second, we aim to gain an insight on how the applications are deployed in the data center to support more sophisticated management strategies. As an example, we can consider that the availability of the services can be increased by replicating and spreading the interacting VMs over the data center infrastructure.

We now propose a formalization of the previously introduced basic principles. Let $\mathcal{N}$ be a set of VMs in a data center. P_j^{out} and P_j^{in} are the two time series of packet rate in output from and in input to VM $j \in \mathcal{N}$, respectively. For both time series let τ be the sampling interval. Our goal is to use this description of network traffic in the data center to infer which VMs exchange data among themselves. Specifically, we rely on the correlation between input and output time series. In our analysis we assume that there is no re-organization of the stacks composing the Web applications (i.e., there is no re-organization of the communication patterns during the collection of the time series). We introduce a methodology based on three steps:

1. Traces interpolation and synchronization;
2. Correlation matrix creation;
3. Interacting VMs identification.

13.3.1 Traces Interpolation and Synchronization

We will detail the mechanism to collect samples about the network data transfer in Sect. 13.4. For the goal of this analysis it is important to point out that the data for each VM will be a collection of records with the fields <timestamp, pkt_in, pkt_out>, with pkt_in and pkt_out being the number of packets received and transmitted in the last τ seconds, respectively. The last two values contain the number of packets received and transmitted in the last τ seconds. However, the gossip protocol used to distribute data and the absence of explicit synchronization between monitor for each VM results in time series from different VMs that can be not synchronized.

A set of preliminary experiments suggest that feeding not synchronized traces in the correlation analysis may lead to poor results for identifying interacting VMs. This motivates our choice to introduce a preliminary step in our methodology to synchronize the input data. To reach this result we exploit data interpolation as follows:

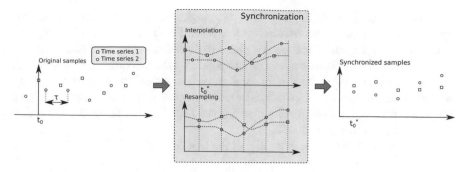

Fig. 13.2 Interpolation and synchronization

- we define a starting time t_0 and we remove every sample before that time. As a result, each time series will start in the time interval $[t_0, t_0 + \tau]$. Furthermore, we make sure that every time series contains T samples;
- we define a synchronization time t_0^* as the average value of the starting times of each trace. We define a set of new sampling times as $t_0^*, t_0^* + \tau, \ldots, t_0^* + i\tau, \ldots, t_0^* + T\tau$;
- for each time series $P_{j_1}^{out}$, we define a synchronized time series $P_{j_1}^{*out}$ using cubic interpolation. In our prototype the implementation of the cubic interpolation is provided by the Python *Pandas*[2] framework. A similar procedure is carried out also for the time series of inbound packets $P_{j_2}^{in}$.

Figure 13.2 provides an example of the above described process. We start with two time series (left part of the figure). The samples in the series are shown as squares and circles, with each sample occurring after τ seconds from the previous one. The center of the figure illustrates the spline interpolation of the samples to obtain a continuous function from the samples, that is used in a re-sampling process synchronized to start at time t_0. Finally, on the right, we have the output time series of synchronized samples.

In order to maintain the readability of the paper, in the following we will adopt a simplified notation using $P_{j_1}^{*out}(i)$ and $P_{j_2}^{*in}(i)$ to indicate the i-th sample of a synchronized time series on VMs j_1 and j_2.

13.3.2   Correlation Matrix Creation

The correlated time series are used to create a correlation matrix **C** taking into account input and output packet rates of each pair of VMs. We explicitly focus on the correlation between output packet of a VM and input packet of another

---

[2]http://pandas.pydata.org/.

VM with the goal to identify the VMs exchanging data among themselves. Our approach is consistent with existing proposals in literature, such as [25, 29], that consider not just communication but also the direction of the data flow with the goal to optimize the data center communication infrastructure. It is also worth noting that this process is inherently parallel, so it is possible to leverage the gossip protocol described in Sect. 13.4 to locally generate portions of the $\mathbf{C}$ matrix and then propagate this information to make sure that the Data Center Management function has the complete matrix available.

We define the generic element c_{j_1, j_2} of the matrix $\mathbf{C}$ as:

$$c_{j_1, j_2} = \mathrm{Cor}\left(P_{j_1}^{*out}, P_{j_2}^{*in}\right) \tag{13.1}$$

where j_1 and j_2 are two generic VMs and $\mathrm{Cor}(\cdot)$ is the correlation function between time series. The resulting matrix $\mathbf{C}$ may be not symmetric. An example to explain this case is when we have a uni-directional data transfer, for example from j_1 to j_2. In this case the correlation between $P_{j_1}^{*out}$ and $P_{j_2}^{*in}$ is high, but the opposite correlation between $P_{j_2}^{*out}$ and $P_{j_1}^{*in}$ is mainly related to ACKs and is likely to be low. Several alternatives to quantify correlation can be found in literature. Our study compares the use of two metrics: *Pearson* and *Spearman* correlation functions [23]. The first is the most popular correlation metric, while the Spearman coefficient was identified in preliminary tests as a promising alternative to operate on time series sharing long-term trends, that is common in our reference problem.

The Person Correlation coefficient ρ is defined as:

$$\rho\left(P_{j_1}^{*out}, P_{j_2}^{*in}\right) = \frac{E\left[\left(P_{j_1}^{*out} - \mu\left(P_{j_1}^{*out}\right)\right)\left(P_{j_2}^{in} - \mu\left(P_{j_2}^{*in}\right)\right)\right]}{\sigma\left(P_{j_1}^{*out}\right)\sigma\left(P_{j_2}^{*in}\right)} \tag{13.2}$$

with $\mu(\cdot)$ being the mean value and $\sigma(\cdot)$ the standard deviation. We use the notation $E[\cdot]$ for the average function, that appears in Eq. (13.2) to estimate the covariance.

The Spearman coefficient ρ_s is the correlation between two time series containing the *ranks* of the original time series values:

$$\rho_s\left(P_{j_1}^{*out}, P_{j_2}^{*in}\right) = 1 - \frac{6\sum_{i=0}^{T} r\left(P_{j_1}^{*out}(i)\right) - r\left(P_{j_2}^{*in}(i)\right)}{T(T^2 - 1)} \tag{13.3}$$

where $r(\cdot)$ is the rank of a sample in the same time series and T is the number of samples. Operating on ranks rather than on values improves the ability to identify small fluctuations in values that may distinguish two time series sharing the same long-term trends, which is common when we have a workload equally distributed among replicated vertical stacks (as in our reference scenario).

13.3.3 Identification of Interacting VMs

The last part of the proposed methodology is the identification of couples/groups of interacting VMs using the correlation matrix $\mathbf{C}$.

This step is part of the operation of the Data Center Management function and can focus on either identifying single couples of interacting VMs or on groups of VMs belonging to the same vertical stack (as in Fig. 13.1).

With respect to the first task, that is identifying couples of VMs exchanging large amount of data, a first solution is to apply a threshold on the correlation matrix to identify *correlated* and *uncorrelated* time series of network resource utilization. In a nutshell, every couple of time series with a correlation c_{j_1,j_2} higher or equal than the threshold is correlated. On the other hand, if c_{j_1,j_2} is below the threshold, we assume that the two time series are not uncorrelated.

However, if we have a broader scope and we aim at identifying the vertical stacks (that are of VMs running different tiers of the same application), we must *cluster* together VMs that show a similar behavior in terms of network traffic patterns. To this aim we consider the correlation matrix as an *affinity matrix* between VMs and a clustering algorithm is applied to group together the VMs. Multiple clustering algorithms have been proposed in literature. However, we must focus on just a subset of them that accepts an input in the form of an affinity/distance matrix (instead of a feature vector). Specifically, in the remaining of this paper we consider and compare: *spectral clustering* [19], *affinity propagation*, and *agglomerative clustering* [11].

Spectral clustering uses the Laplacian operator applied to the input similarity matrix. The eigenvalues and eigenvectors of the result are used to create a new coordinate system. The original samples projected into this new space of coordinates are then clustered using the k-means clustering algorithm [19]. Affinity propagation bases its operation on simulated message exchange between samples. The algorithm is quite fast as it identifies a representative for each cluster and assigns elements to each cluster based on the affinity with their representatives [14]. Finally, agglomerative clustering creates a *dendogram* starting with on cluster for each VM and then merging at each iteration the two most similar clusters. Depending on the cluster we want to obtain, we can then cut the dendogram at any point. In our experiments, we use the implementation of the three algorithms provided by the *SciKit-Learn* library.[3]

---

[3]http://scikit-learn.org/.

13.3.4 Scalability and Computational Complexity

We now discuss the scalability of the proposed technique with respect to the network size. In particular, we consider the computational complexity of the three steps composing the technique with respect to the number of hosts n.

The synchronization of each trace depends linearly on the network size, that is has a complexity that is $\mathcal{O}(n)$ (assuming that at most a finite number of VMs can run on each host). The creation of the correlation matrix has complexity $\mathcal{O}(n^2)$, that can be reduced to $\mathcal{O}(n)$ if we parallelize the task so that each host computes a slice of the global matrix (as described in the following of the paper). Finally, the most critical part of the technique is the clustering process that has a complexity ranging from $\mathcal{O}(n^2)$ to $\mathcal{O}(n^3)$, depending on the chosen algorithm.

We can thus conclude that the overall computational cost of the technique is polynomial. Hence, we can expect limited scalability problem with respect to the problem size.

13.4 Implementation

In this section, we discuss an actual implementation of the proposed methodology. The goal of identifying clusters of VMs communicating to each other requires, as shown in Sect. 13.3, the knowledge of communication patterns of all VMs. This knowledge can be considered as an aggregate property of the whole system. From an engineering perspective, it is possible to calculate aggregate functions with either reactive or proactive approaches [13, 15].

Reactive approaches are based on queries issued by specific nodes in the network. The answers are returned directly to the issuer while the rest of the network may or may not receive the answer. An example of this approach could be a dedicated node periodically polling all VMs and making the needed computation in a centralized fashion.

Proactive approaches, instead, provide the value of aggregate functions to all nodes in the network in an adaptive fashion. By the term adaptive we mean that, if changes due to network dynamism or to variations in the input values arise, the output of the aggregation protocol should be capable of tracking these changes. Proactive protocols are frequently used in completely decentralized solutions of complex tasks.

Given the significant increase in size and complexity of modern data centers [12], we focus on a solution that exploits a reactive protocol based on gossiping. The nature and functioning of gossip-based aggregation is detailed in the following section.

13.4.1 Gossip-Based Aggregation

Let us consider a generic network with a set of $\mathcal{M}$ nodes, where each node corresponds to a physical host. Each node interacts with a small number of other nodes (neighbors). The basic protocol is based on the *push–pull* scheme illustrated in Algorithm 1. Time is divided into time slots of length τ, corresponding to the monitoring intervals. Each node $p \in \mathcal{M}$ executes two different threads. At every time slot, the active thread starts an information exchange with one random neighbor q by sending it a message containing the local status s_p and waiting for its remote status s_q. It is possible to extend the basic scheme to have interaction with multiple neighbors for each time slot. The passive thread waits for messages sent by an initiator and always replies with the local status. The term *push–pull* refers to the fact that each information exchange is performed in a symmetric manner: both nodes send and receive their status. It is worth noticing that the local status s_p could represent both properties of the node p itself or values measured by it. In the considered application, the status information basically consists in the time series of network utilization samples of both the local node and of the known neighbors. The methods send() and receive() are used for this message exchange.

The method getNeighbors(ω) showed in Algorithm 1 can be considered as a service underlying the aggregation protocol. It returns a uniform random sample over the entire set of neighbors. Furthermore, the parameter $\omega \in [0, 1]$ can be used for specifying how many neighbors have to be returned. In particular, with $\omega = 1$ the method will return all the available neighbors while with $\omega = 1/neighbors$ only one neighbor will be returned. The method update computes a new local status based on the current local status and the remote status received during the information exchange. The output of update as well as the semantic of the node status completely depends on the aggregation function needed by the application.

This gossip scheme tends to impose a uniform load to the system [15]. Each node executes the same amount of operations. Incidentally, reducing τ or increasing ω reduces convergence times of the algorithm (at expense of more messages exchanged). Therefore, applications can constantly manage the trade-off aggregation accuracy and communication costs.

13.4.2 Network Agent

Given the above definition of gossip-based aggregation, we now provide some details on how to implement it for the considered scenario. As pointed out in Sect. 13.2, each physical host is equipped with a *Hypervisor*, that manages the locally hosted VMs, and with a *Network Agent* that is a gossiping node responsible for the data exchange about the network resource utilization. We recall that the status of each VM is represented as a time series of tuples in the form <timestamp, pkt_in, pkt_out>. Given a sampling period $\tau = 30\,$s, a time series s_j

---

Algorithm 1 Push–pull gossip protocol executed by node p

Ensure: Active thread
  for once every τ at random time in time-slot **do**
    $q[] \leftarrow \texttt{getNeighbors}(\omega)$
    for $q \in q[]$ **do**
      $\texttt{send}(s_p, q)$
      $s_q \leftarrow \texttt{receive}(q)$
      $s_p \leftarrow \texttt{update}(s_p, s_q)$
    end for
  end for
Ensure: Passive thread
  for ever **do**
    $s_q \leftarrow \texttt{receive}(*)$
    $\texttt{send}(s_p, \texttt{sender}(s_q))$
    $s_p \leftarrow \texttt{update}(s_p, s_q)$
  end for

---

comprising 240 elements would describe the latest 2 h of network usage of VM_j. The status of a node consists of the status of each VM managed by that host hypervisor that is, for node i the status is $s_i = \{s_j, j \in \mathcal{N}_i\}$, where $\mathcal{N}_i$ is the set of VMs hosted on physical host i.

The problem being addressed consists in making available at each host, in a timely fashion, the status of all the physical hosts comprising the considered data center, that is $s_i, \forall i \in \mathcal{M}$. The gossip protocol is used to spread data concerning the VMs running on one physical host to all others.

Let us consider two hosts p and q and their respective status s_p and s_q. At each gossiping iteration, as detailed above, nodes p and q exchange their statuses (i.e., the status s of all the VMs they are running). It is worth noticing that the nodes, as an example we consider node p, might already have received the status s_q from a former exchange with either q or another node. In this case, the aggregation function selects the most recent version of s_q and discards the other. As a consequence, each physical host can keep track of an extended status comprising both s_p and s_q. If executed on the hosts of a data center, this algorithm allows each host to receive the status of all VMs running within the data center. To better clarify the details of the data exchange, Fig. 13.3 provides a view of the message involved in the operations of the gossip protocol. A gossip message contains both information about the local host and propagates data about remote hosts. Each status for a host consists in the status of the hosted VMs, that, in turn, contains the samples time series.

It is also worth noticing that despite gossip-based communication schemes have proven several desirable properties over flooding, they are slower to converge [13]. Thus, for the sake of being able of successfully correlating time series, it is relevant that the time needed to spread network data across all the hosts is smaller than the network sampling period τ (i.e., in our case is 30 s). If this condition is not met, gossiping can be suitably tuned by adjusting the ω parameters.

Finally, another benefit of the proposed approach comes from a reduction of the computational complexity of the clustering process. In fact, in case of reactive

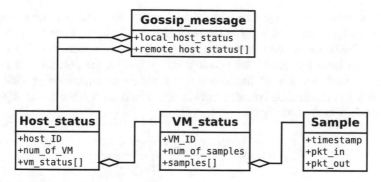

Fig. 13.3 Message format (UML notation)

approaches a dedicated node has to collect the whole set of VMs statuses and find eventual correlations. In the proposed approach, instead, each node can compute correlations concerning only the VMs it hosts. Basically, the network agent on each host computes the correlation between each hosted VM and every other VM. This creates a set of columns in the final correlation matrix **C**, that are then forwarded to the Data Center Management system for the identification of interacting VMs and/or clustering operations.

13.5   Experimental Results

13.5.1   Setup Description

The proposed approach is tested using an experimental setup exploiting the TPC-W benchmark.[4] The considered traces are referred to a TPC-W run with 12 VMs organized in 4 vertical stacks and covering a 12 h period during which the samples are collected every 30 s (however, we consider also a granularity of 1 and 2 min for samples collection in our experiments).

To define the metrics to compare the different correlation functions, we need to distinguish between two different goals: first, identifying the interacting couples of VMs; second, clustering together VMs to identify vertical stacks.

For the first goal, that is identifying interacting VMs, we rely on a classic measure for classification problems: the *F-measure*, which is the harmonic mean of *Precision* and *Recall*. Precision is defined as $P = \frac{TP}{TP+FP}$ while recall is $R = \frac{TP}{TP+FN}$, with TP and TN true positives and negatives and FP and FN being false positives and negatives, respectively. F-measure is: $F = 2\frac{P \cdot R}{P+R}$.

---

[4]www.tpc.org/tpcw/.

For the second goal, we take into account one of the most popular metrics for clustering evaluation, that is clustering *purity* [2]. Purity is the fraction of correctly identified VMs and is used to evaluate the performance of the clustering step. In order to measure the purity, we consider the output of the clustering **S** and we compare it with a vector **S*** that contains the correct classification of VMs (that, for each VM, contains the correct vertical stack it belongs to). We can thus provide a formal definition of purity as:

$$purity = \frac{|\{s^j : s^j = s^{j*}, \forall j \in \mathcal{N}\}|}{|\mathcal{N}|}$$

with s^j being the vertical stack to which VM j is assigned in by the clustering algorithm, s^{j*} being *correct* vertical stack of VM j, and $\mathcal{N}$ is the set of considered VMs.

13.5.2 Correlation Coefficients Analysis

The first experiment aims at comparing the use of Pearson and Spearman correlation coefficients, that are used to compute the correlation matrices over every couple of time series.

Heat maps are used to represent the two correlation matrices in Fig. 13.4: the two time series of input and output packets are presented for each VM (numbered 1–12 in the figure). An outline of the vertical stacks (the four white square frames on the main diagonal of the matrix) is also shown in each picture. The correlation should ideally present this behavior: high values within the four squares delimiting each vertical stack and low values outside the squares.

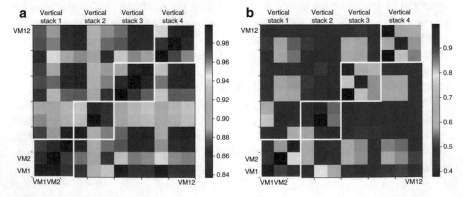

Fig. 13.4 Heatmap of the correlation matrices. (**a**) Pearson correlation coefficient. (**b**) Spearman correlation coefficient

The observation of Fig. 13.4a, b offers interesting insights. On one hand, the Pearson correlation coefficient (Fig. 13.4a) is characterized by values very high (close to 1, as shown by the predominance of orange-red hues). We also note some reddish shades located far from the main diagonal outside from the vertical stack frames: this is particularly critical because it suggests the presence of high correlation between couples of VMs not belonging to the same vertical stack. If we observe the results achieved by the Spearman correlation coefficient (Fig. 13.4b), we note that the use of this coefficient allows to better distinguish between VMs of different stacks and VMs belonging to the same vertical stack (as shown by the red areas located near to the diagonal), thus revealing a better capacity to identify VMs that are actually exchanging data.

13.5.3   Identification of Communicating VMs

In this experiment we evaluate the effectiveness of using the two correlation coefficients to identify couples of communicating VMs on the basis of a threshold. Specifically, we analyze the sensitivity of the VMs classification for different threshold values.

The comparison of the F-measure for the two correlation coefficients is shown in Fig. 13.5 for increasing values of the threshold. The observation of the graph reveals that the use of the Spearman correlation coefficient may provide a double advantage: (a) better performance with respect to the Pearson coefficient, characterized by higher maximum F-measure values; (b) higher stability of the performance, with an accuracy higher than 0.5 achieved by different threshold values (0.5–0.8), while the Pearson function achieves the best values for a single peak close to 0.96 as threshold value.

Fig. 13.5 F-measure comparison

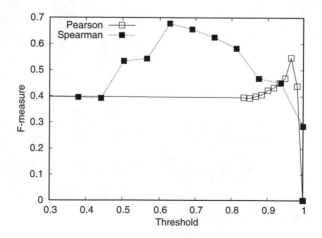

13.5.4   VMs Clustering

In this set of experiments we compare the use of the correlation coefficients to create the input affinity matrix for the three considered clustering algorithms (spectral clustering, affinity propagation, and agglomerative clustering).

Figure 13.6 represents the clustering purity of the clustering algorithms for each of the two considered correlation coefficients. From the figure, it is evident that the use of the Spearman coefficient allows the clustering algorithms to achieve significantly higher purity if compared to the alternative. Moreover, we note that the best performing algorithm of the three considered alternatives is the spectral clustering. The different performance of the algorithms may be explained by considering the example in Fig. 13.7 that represents a spectral clustering output on the top against the actual clustering (that is, the ground truth) shown on the bottom. We may observe two kinds of errors in the clustering output: first, two VMs are swapped (VMs 1 is assigned to the cluster to which VM 4 belongs, and vice versa); second, VM 12 is misplaced. On the other hand, the poor performance of the affinity propagation algorithm is worsened by an additional problem: the wrong estimation of the clusters number. Then, we note that the agglomerative clustering causes a high number of cluster swaps between VMs.

Fig. 13.6 Clustering purity comparison

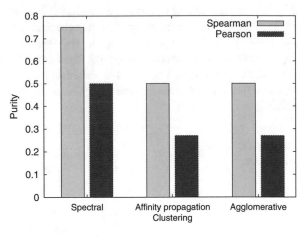

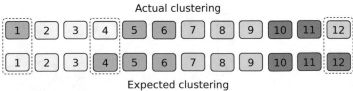

Fig. 13.7 Clustering example

13.5.5 Comparison of Time Sampling Intervals

The last experiment evaluates the impact of data sampling granularity on the quality of the classification (that identifying correlated and non-correlated time series) and of the VMs clustering. Let us start the evaluation from the classification process. Figure 13.8 compares the F-measure obtained by using the Spearman correlation coefficient as the basis for VMs classification: the considered sampling period τ of network utilization ranges from 30 s to 2 min. We note that the impact on the classification F-measure is very significant, even with the small considered change in the sampling granularity. A coarse-grained data collection has, indeed, two main effects. First, the sensitivity to the threshold value of the Spearman correlation is reduced for large sampling periods. The increased robustness is caused by the smoothing of the time series. Indeed, the time series taken into account basically only shows the most evident fluctuations: the stability to a wider range of threshold values is motivated by the ease of identifying low-frequency and highly evident fluctuations (which is basically the main way to distinguish two vertical stacks) even over a wider range of threshold values. Second, the accuracy significantly decreases for increasing sampling intervals if we observe the threshold range of values corresponding to higher F-measure ([0.6–0.8]). The motivation of this effect is quite intuitive: a smoother time series of the network resource utilization is the result of an increased sampling period. In this way, we still are able to appreciate the main effects of the correlation; however, we miss the high frequency fluctuations, thus experiencing a reduction in the classification performance.

The final analysis evaluates the effect on the purity of the clustering solution of the sampling frequency. Figure 13.9 shows the comparison among the clustering purity of the three clustering algorithms making use of the Spearman coefficient of correlation. The graph offers two validations of previous observations. First, better performance of the spectral clustering with respect to other algorithms is confirmed, with the results showing a gain in the purity values ranging from 97% to 31%

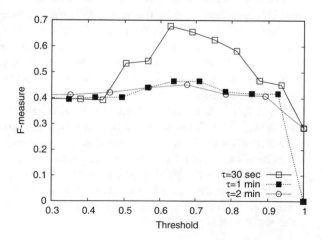

Fig. 13.8 Sensitivity to sampling period τ

Fig. 13.9 Clustering purity comparison

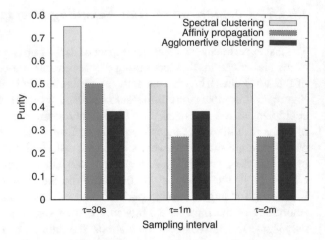

over the alternatives. Second, the increase of the sampling period τ significantly decreases the performance even for the spectral clustering algorithm, causing a reduction in the purity of 33% as τ increases from 30 s to 1 min. This effect can be explained by the same motivations that cause a decrease in the F-measure values for the detection of communicating VMs: the distinction of different vertical stacks is more challenging in case of smoother time series.

13.6 Related Work

The problem of describing network utilization patterns and using this information to improve the performance and efficiency of data transfer has been widely investigated in literature.

In particular, the problem of monitoring network traffic patterns to identify data flows has been tackled over the years, with special focus on the links over geographically distributed networks, that may become bottlenecks in the data delivery. A large corpus of literature defines this problem as *network tomography* [10]. Among these studies, [10] exploits metrics based on single links to produce a topological description of the network flows in a wide-area network. The same goal is considered in [26], where a bayesian approach is used to describe and match the characteristics of data flows to reconstruct routing between origins and destinations. Papagiannki et al. focus on extracting the routing topology of a geographically distributed network, with special attention on increasing scalability of the process thanks to a distributed approach, where computation is carried out at the point of presence and at the peering points of the network [24]. A more conservative approach is presented in [16], where Kowalski and Warfield derive from the theory on telephonic networks a probabilistic model to correlate the distance between two nodes and the traffic between the two. All these studies are related to our proposal, but the considered scenario, and the solution to address the problem, are more oriented towards wide-area networks, with complex and irregular graphs topology rather than on data center. Furthermore, these studies do not take into account the cloud scenario, where VMs may migrate over the infrastructure. Our focus on a modern data center with virtualization functions of computing and, possibly, of networking functions places our paper clearly aside from the above-mentioned studies.

Another branch on literature proposes the application of the Software-Defined Network paradigm to the problem of network monitoring. For example, [27] proposes NetFPGA, a system that exploits programmable hardware to monitor networks extracting useful statistics from the network links. A similar proposal is made in [28], that introduces a way to leverage the characteristics of SDN appliances to monitor large networks on a geographic scale, and in [18] that aims at improving the scalability of the process by selecting the most relevant flows in a network. Even if our approach could integrate and use information collected from an SDN infrastructure, we recall that our focus is more oriented to a cloud-based scenario, where the deployment of SDN appliances is still at its infancy, so we can't fully rely on it.

Other studies are complementary to our work, as they can take advantage from our effort. For example, [22] focuses on exploiting the knowledge of data exchange between VMs in a cloud environment to improve the performance of the cloud-hosted applications, while in [25] the final goal is to increase the efficiency of network resource usage.

Finally, we recall that preliminary versions of this research were published in [8, 17]. However, in this paper we provide a more detailed discussion on the issue of scalability and we propose a new software architecture to implement the considered theoretical model.

13.7   Conclusions

This paper focused on the problem of identifying groups of VMs exchanging data traffic in a cloud data center when we do not have access to detailed information among the data exchange between each specific couple of VMs. Our reference scenario is a multi-tiered application deployed over vertical stacks that are replicated horizontally. This case is a very common and challenging scenario in modern cloud systems where replication is used to achieve scalability. To discover communicating VMs we described a methodology that relies on the correlation between the time series of VMs network packet flows and exploits clustering to identify the vertical stacks of VMs. We evaluate and compare different correlation coefficients to find the best performing solution. From our experiments it is clear that solutions based on tanking techniques (such as the Spearman correlation coefficient) provide better performance than the alternatives. Furthermore, we compare three clustering algorithms for detecting the vertical stack: our experiments show that spectral clustering clearly outperforms the other alternatives. As a last analysis, we perform a sensitivity evaluation about the sampling period of network data: the results demonstrate that a fine-grained collection is more suited to find communicating VMs than a coarse-grained approach.

Acknowledgements The authors acknowledge the support of the project S^2C: *Secure Software-defined Cloud* funded by the University of Modena and Reggio Emilia.

References

1. M. Al-Fares, A. Loukissas, A. Vahdat, A scalable, commodity data center network architecture, in *Proceedings of the ACM SIGCOMM 2008 Conference on Data Communication, SIGCOMM '08*, New York, NY (ACM, New York, 2008), pp. 63–74. http://doi.acm.org/10.1145/1402958. 1402967
2. E. Amigó, J. Gonzalo, J. Artiles, F. Verdejo, A comparison of extrinsic clustering evaluation metrics based on formal constraints. J. Inf. Retr. **12**(4), 461–486 (2009)
3. M. Andreolini, M. Colajanni, M. Pietri, A scalable architecture for real-time monitoring of large information systems, in *Proceedings of IEEE Symposium on Network Cloud Computing and Applications*, London (2012)
4. H. Ballani, P. Costa, T. Karagiannis, A. Rowstron, Towards predictable datacenter networks. ACM SIGCOMM Comput. Commun. Rev. **41**(4), 242–253 (2011)
5. A. Beloglazov, J. Abawajy, R. Buyya, Energy-aware resource allocation heuristics for efficient management of data centers for cloud computing. Futur. Gener. Comput. Syst. **28**(5), 755–768 (2012)
6. D. Boru, D. Kliazovich, F. Granelli, P. Bouvry, A.Y. Zomaya, Energy-efficient data replication in cloud computing datacenters. Clust. Comput. **18**(1), 385–402 (2015)
7. C. Canali, R. Lancellotti, Exploiting classes of virtual machines for scalable IaaS cloud management, in *Proceedings of the 4th Symposium on Network Cloud Computing and Applications (NCCA)* (2015)
8. C. Canali, R. Lancellotti, Identifying communication patterns between virtual machines in software-defined data centers. SIGMETRICS Perform. Eval. Rev. **44**(4), 49–56 (2017)
9. C. Canali, R. Lancellotti, M. Shojafar, A computation- and network-aware energy optimization model for virtual machines allocation, in *Proceedings of International Conference on Cloud Computing and Services Science (CLOSER 2017)*, Porto (2017)
10. R. Castro, M. Coates, G. Liang, R. Nowak, B. Yu, Network tomography: recent developments. Stat. Sci. **19**, 499–517 (2004)
11. W.H. Day, H. Edelsbrunner, Efficient algorithms for agglomerative hierarchical clustering methods. J. Classif. **1**(1), 7–24 (1984)
12. M. Dayarathna, Y. Wen, R. Fan, Data center energy consumption modeling: a survey. IEEE Commun. Surv. Tutorials **18**(1), 732–794 (2016)
13. P.T. Eugster, R. Guerraoui, A.M. Kermarrec, L. Massoulieacute, Epidemic information dissemination in distributed systems. Computer **37**(5), 60–67 (2004). http://doi.ieeecomputersociety. org/10.1109/MC.2004.1297243
14. B.J. Frey, D. Dueck, Clustering by passing messages between data points. Science **315**(5814), 972–976 (2007)
15. M. Jelasity, A. Montresor, O. Babaoglu, Gossip-based aggregation in large dynamic networks. ACM Trans. Comput. Syst. **23**(3), 219–252 (2005)
16. J.P. Kowalski, B. Warfield, Modelling traffic demand between nodes in a telecommunications network, in *Proceedings of ATNAC'95* (1995)

17. R. Lancellotti, C. Canali, A correlation-based methodology to infer communication patterns between cloud virtual machines, in *Proceedings of the 10th EAI International Conference on Performance Evaluation Methodologies and Tools (VALUETOOLS)*, Taormina (2017), pp. 251–254
18. D. Li, N. Dai, F. Li, C. Xing, F. Dai, Estimating SDN traffic matrix based on online informative flow measurement method, in *Proceedings of 2017 Fifth International Conference on Advanced Cloud and Big Data (CBD)* (2017), pp. 75–80
19. U. Luxburg, A tutorial on spectral clustering. Stat. Comput. **17**(4), 395–416 (2007)
20. A. Marotta, S. Avallone, A simulated annealing based approach for power efficient virtual machines consolidation, in *Proceedings of 8th International Conference on Cloud Computing (CLOUD), IEEE* (2015)
21. C. Mastroianni, M. Meo, G. Papuzzo, Probabilistic consolidation of virtual machines in self-organizing cloud data centers. IEEE Trans. Cloud Comput. **1**(2), 215–228 (2013). https://doi.org/10.1109/TCC.2013.17
22. X. Meng, V. Pappas, L. Zhang, Improving the scalability of data center networks with traffic-aware virtual machine placement, in *Proceedings of the 29th Conference on Information Communications (INFOCOM)*, San Diego, CA (2010)
23. L. Myers, M.J. Sirois, *Spearman Correlation Coefficients, Differences Between* (Wiley, Hoboken, 2014). http://dx.doi.org/10.1002/9781118445112.stat02802
24. K. Papagiannaki, N. Taft, A. Lakhina, A distributed approach to measure ip traffic matrices, in *Proceedings of the 4th ACM SIGCOMM Conference on Internet Measurement* (ACM, New York, 2004), pp. 161–174
25. J. Sonnek, J. Greensky, R. Reutiman, A. Chandra, Starling: minimizing communication overhead in virtualized computing platforms using decentralized affinity-aware migration, in *Proceedings of 39th International Conference on Parallel Processing (ICPP)*, San Diego, CA (2010)
26. C. Tebaldi, M. West, Bayesian inference on network traffic using link count data. J. Am. Stat. Assoc. **93**(442), 557–573 (1998)
27. M. Yu, L. Jose, R. Miao, Software defined traffic measurement with opensketch, in *Presented as part of the 10th USENIX Symposium on Networked Systems Design and Implementation (NSDI 13), USENIX*, Lombard, IL (2013), pp. 29–42
28. L. Yuan, C.N. Chuah, P. Mohapatra, Progme: towards programmable network measurement. IEEE/ACM Trans. Netw. **19**(1), 115–128 (2011)
29. Y. Zhang, N. Ansari, Hero: hierarchical energy optimization for data center networks. IEEE Syst. J. **9**(2), 406–415 (2013)

Chapter 14
Container Orchestration: A Survey

Emiliano Casalicchio

14.1 Introduction

Nowadays, cloud architectures are moving from virtual machine (VM) centric to container centric.

Containers technologies are strongly supported by PaaS providers [5, 10], IaaS providers [40], and Internet service providers [31]. Moreover, container technologies are used to deploy large-scale applications for big data processing (e.g., [30, 32]), scientific computing (e.g., [14]), IoT (e.g., [4, 9]), and edge computing (e.g., [18]).

Containers became so popular because they potentially may solve many cloud application issues, for example: the *dependency hell* problem, typical of complex distributed applications. Containers give the possibility to separate components, wrapping up the application with all its dependencies in a self-contained piece of software that can be executed on any platform that supports the container technology. The *application portability* problem; a container can be executed on any platform supporting the container image format. The Open Container Initiative (opencontainers.org) has been created to define a standard image format and a standard runtime environment for containers. Docker, Kubernetes, and Cloudify (cf., Table 14.1) are container technologies compliant with the cloud portability frameworks TOSCA [23, 33]. The *performance overhead problem*; containers are

E. Casalicchio (✉)
Department of Computer Science and Engineering, Blekinge Institute of Technology, Sweden

Department of Computer Science, Sapienza University of Rome, Italy
e-mail: emiliano.casalicchio@bth.se

© Springer International Publishing AG, part of Springer Nature 2019   221
A. Puliafito, K. S. Trivedi (eds.), *Systems Modeling: Methodologies and Tools*,
EAI/Springer Innovations in Communication and Computing,
https://doi.org/10.1007/978-3-319-92378-9_14

Table 14.1 Container technologies considered in this work

Technology	Type	URL
Linux Container (LXC)	SC, AC	https://linuxcontainers.org/
OpenVZ	SC, CM	https://openvz.org/
Windows Hyper-V Container (WHC)	SC	https://docs.microsoft.com/en-us/virtualization/windowscontainers/about/
Docker	AC, CM	https://www.docker.com/
Windows Server Container (WSC)	AC	https://docs.microsoft.com/en-us/virtualization/windowscontainers/about/
rkt	CM	https://coreos.com/rkt
LXD	CM	https://linuxcontainers.org/
Amazon EC2 Container Service (ECS)	CM, OF	https://aws.amazon.com/ecs/
Google Container Engine (GCE)	CM, OF	https://cloud.google.com/container-engine/
Microsoft Azure Container Service (ACS)	CM, OF	https://azure.microsoft.com/en-us/services/container-service/
Kubernetes	OF	https://kubernetes.io/
Swarm	OF	https://www.docker.com/
Marathon	OF	https://mesosphere.github.io/marathon/
Cloudify	OF	http://cloudify.co/

Type meaning: *AC* application container, *SC* system container, *CM* container manager, *OF* orchestration framework

lightweight and have a small performance footprint than VMs [8, 13, 21, 28], moreover, launching, re-launching, and stopping a container is an order of magnitude faster than VMs.

The container technology landscape is developing and expanding at the speed of light, however, we are far away from the maturity stage and there are still many challenges to be solved, for example: the reduction of networking overheads compared to hypervisors; the secure resource sharing and isolation to enable multi-tenancy; the improvement of container monitoring methodologies and tools; the enhancement of runtime scale-up and migration capabilities; the enhancement of autonomic capabilities in orchestration frameworks by means of self-healing and self-optimization mechanisms.

This work is a survey of the state of the art in container orchestration technologies. To the best of our knowledge there is no similar contribution to the literature. The contribution and organization of this paper is as follows. First, Sect. 14.2 reports and classifies the state-of-the-art container technologies (system container, application container, container manager, and orchestrator). Then, Sect. 14.3 focuses on container orchestrator proposing a taxonomy based on the orchestration feature. Section 14.4 describes the reference architecture of an autonomic container orchestrator. The state-of-the-art research results in autonomic orchestration are analyzed in Sect. 14.5. Specifically the focus is on monitoring, workload characterization and performance evaluation of container technologies, self-optimization and self-healing solutions.

What emerges from our study is that: container technologies are used not only in enterprise applications but also in big data analytics, IoT, fog, and edge computing; the research community has devoted a significant effort in performance evaluation studies, and in proposing new solutions aiming at improving the self-optimization and self-healing capabilities of container orchestrators; the cloud industry offer in terms of autonomic orchestration is still at the infancy.

14.2   System Container, Application Container, and Container Manager

Container technologies can be classified into system container, application container managers, and orchestrators (cf., Fig. 14.1). In this section we will focus on the first three categories, while orchestration is discussed in the next section.

Application container is designed to run application or application components into containers. An application container built up from a series of image layers. Usually a developer can start from a base image (e.g., the operating system kernel and default libraries) then the application can be copied in a new layer (eventually in two—a bottom layer with the source code and libraries and a top layer with the executable). Application data and commands to be executed by the container will be added in upper layers. After the multilayer image is built, all the layers are read-only except the last one: read and write but not persistent. This means that the content is lost when the container is deleted.

The widely used application container technology is Docker, a multiplatform solution designed for Linux, OSX, and Windows. Docker extends LXC with a kernel and application-level API to facilitate container management and the management of containerized applications. LXC also can be used as application container. rkt is specifically designed to run application container in a cloud native environment, i.e., CoreOS (Container Linux). Windows Server Container (WSC) is the Microsoft version of application container.

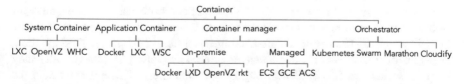

Fig. 14.1 A taxonomy that classifies container technologies as application container, system container, and container manager

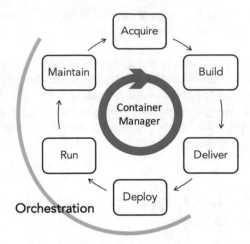

Fig. 14.2 The container life cycle [26]. Acquire is related to select and download a base image to build the containerized application. Deliver concerns to bring the application in production. Deploy is about deploying the application and maintaining its update. The run step set the management system and runtime environment (e.g., scaling policy, health check, recovery policies). The final step, Maintain, determines how you will get visibility into your application for maintenance

System containers allow to run a full OS in a container; hence, the container is like a lightweight virtual machine. The layering concept applies also to system container. Starting from a base operating system image, it could be customized adding libraries, tools, and data in new layers. Linux Container (LXC) and OpenVZ allow to share a Linux kernel among containers built from the same base image. LXC uses standard kernel while OpenVZ uses a specialized kernel. Windows Hyper-V Container (WHC) is the Microsoft version of system container; to guarantee a strong environment isolation WHCs do not share the kernel among containers.

A container manager is a framework providing a set of API to easily manage all the life cycle of the container (cf., Fig. 14.2). Container manager can be classified as on-premise solutions (to be installed, configured, and managed on private datacenters or in cloud) or managed solutions (offered by cloud providers as a service). Docker has been designed as a container management system; moreover, the ecosystem of management container is in continuous evolution. For example, Windows Server container and Hyper-V container both can be managed with Docker. Also rkt offers APIs for easy application container management. Google Container Engine, Microsoft Azzure Container Service, and Amazon ECS are three examples of managed container manager offered as cloud platform (usually they support Docker and LXC). Concerning system container, LXD is the manager for LXC. OpenVz also provides container management APIs.

14.3 Container Orchestration

Container orchestration allows cloud and application providers to define how to select, to deploy, to monitor, and to dynamically control the configuration of multi-container packaged applications in the cloud.

Container orchestration is concerned with the management at runtime to support the deploy, run, and maintain phases (cf., Fig. 14.2). Container orchestrator usually offers the following main features (cf., Fig. 14.3): resource limit control, scheduling, load balancing, health check, fault tolerance, and autoscaling.

Resource limit control allows to reserve a specific amount of CPU and memory for a container; that constraints can be used to make scheduling decisions and to limit the interference among containers. Resource limit control features leverage the equivalent mechanisms offered by the container manager. Indeed, while a container can use all the resource available in the underlining system, container managers provide APIs to limit the amount of memory and CPU used and the specific CPU used.

Scheduling defines the policy used to place the desired amount of container on desired nodes at a given time instant. Scheduling can be done on the basis of resource constraints or node affinity or both. More sophisticated scheduler can usually be integrated as external components (the custom feature in Fig. 14.3).

The load balancer does the work of distributing the load among multiple container instances. Round-robin is the default implemented policy. More complex policy can be provided by external load balancer (the custom feature in Fig. 14.3).

Health check is achieved controlling that a container is capable to answer requests. Implementations foreseen TCP/UDP/SSH ports connection open checking and HTTP request receive and answer checking.

Fault tolerance can be implemented as replica control and/or high availability controller. Replica control allows to specify and maintain a desired number of containers. Health check is used to determine when a faulty container should be destroyed and a new one lunched to maintain the target number of replicas. High

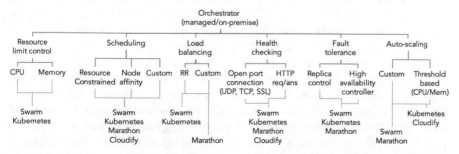

Fig. 14.3 A taxonomy that classifies container orchestration technologies. In the up part of the diagram are the features, and in the bottom part the specific technologies implementing the specific features

availability controllers allow to configure multiple orchestration manager to have always control on the application in case an orchestrator node fails or is overloaded. The same technique used to create high availability controllers can be used also to implement scalable controllers.

Autoscaling allows to automatically add and remove containers. The implemented policies are threshold based (on CPU and memory utilization) but in some cases it is possible to plug-in more sophisticated autoscaler or to define custom autoscaling policies (the custom feature in Fig. 14.3).

In the landscape of container orchestration frameworks the choice is between on-premise solutions or managed solutions (as for the container managers). The main on-premise solutions are: Docker Swarm, the native Docker orchestrator offering clustering functionality for Docker containers, which lets system administrator to turn a group of Docker engines into a single, virtual Docker engine. Kubernetes, an orchestration system for Docker containers capable to handle scheduling and to manage workloads based on user-defined parameters. Mesosphere Marathon, a container orchestration framework for Apache Mesos. It offers key features for running applications in a clustered environment. Cloudify, a cloud orchestration framework that enables modeling of applications and services and automating their entire life cycle. Cloudify is TOSCA compliant and could be used to deploy Docker container, Docker Swarm cluster, or Kubernetes clusters.

The main examples of container orchestration as a service are Google Container Engine, Amazon Elastic Container Service, and Microsoft Azure Container Service.

14.4   Reference Architecture

As described in the previous section, container orchestration can be automatized using mechanisms such as scheduling, autoscaling, health check, load balancing, and fault tolerance. However, the adaptation policies available are very simple, e.g., round-robin scheduling/load balancing and threshold-based autoscaling.

Because of the highly dynamic and complex environment in which containers are operated, the variety of service level objective stated in service level agreement, the willingness of customers to minimize costs, and the many goal a service provider could have (e.g., maximize resource usage, reduce energy consumption, and satisfy SLA in multitenant environment), there is an urgent need of complex adaptation policies to make orchestration a real autonomic process with at least the following capabilities: self-healing, to improve the application availability and resiliency; self-optimization, to trade-off costs and performance while fulfilling Service Level Agreement; and self-protection, to increase system/application security.

In the remaining the focus is mainly on self-healing and self-optimization of application container management over distributed datacenters and in multitenant environments. The attention is concentrated on why, when, where, and how to deploy, to launch, to run, to migrate, and to terminate containers. All those actions are defined in what is called an orchestration or adaptation policy.

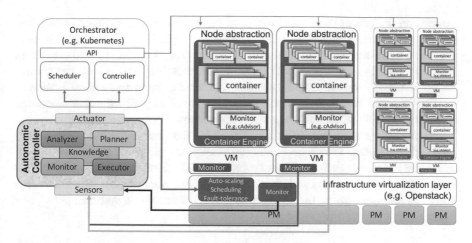

Fig. 14.4 The architecture for autonomic orchestration of containers

Figure 14.4 shows the architecture of an autonomic container orchestration system based on the classical MAPE-K cycle [20].

As shown in Fig. 14.4, the orchestration should be multilayer. This consideration has an impact on the whole orchestration problem. Adaptation at container platform layer and at infrastructure layer should be coordinated by the adaptation policy. Separating the infrastructure and the container platform adaptation could result in interfering decisions leading to suboptimal configurations.

Why orchestration actions should occur is determined by specific objectives of the customers or of the service providers. *When* the orchestration should take place depends on events such as an increase in the workload intensity or volume, change of SLA, node failures or critical states, and critical node load/health state. Such events are monitored at different layers of the platform and the infrastructure. The Planner component executes the orchestration policy to find the appropriate system configuration (*How* and *Where*). Typically, the adaptation policy finds the optimal or suboptimal system configuration that maximizes the provider utility and that satisfies non-functional and functional constraints. The challenge is to define sophisticated adaptation polices that account for the complexity of the system and of the objectives. *How* is concerned with the method used to formulate the adaptation problem, e.g., as linear or non-linear optimization problem, as ant colony problem, as a genetic algorithm, and so on. The decision taken by the orchestrator is based on the system and environment state information collected by the Monitor component and analyzed by the Analyzer component. As shown in Fig. 14.4 the data should be collected not only at container level but also at infrastructure level to coordinate the adaptation of the two layers.

Where the adaptation will take place could be: at container or cluster of container, at infrastructure level, at datacenter level, and at federation level. Compute nodes (VMs) are grouped in clusters that can be part of the same datacenter or can be geographically distributed (federation of datacenters). Each compute node in a cluster can run one or more containers, and the execution of containers is managed by the container engine (CE) on the node (cf., Fig. 14.4). Some orchestrators introduce an abstraction of the compute node that group logically related containers (e.g., that does the same work or that communicate intensively) in a cluster usually called pod. Such node abstraction can be considered a single entity of adaptation.

The Executor component implements the adaptation plan using the Container Orchestrator API and the infrastructure management API.

For scalability and availability purposes, more than one orchestrator can be created and configured. The orchestrators can simply be replicated for high availability or can be federated (each orchestrator controls a different cluster of container). In the latest case orchestrators will act independently and no solution provides methods and tools for coordinating the action of federated orchestration systems.

14.5 State of the Art

Container orchestration raises many research issues in the following research areas:

- *Monitoring and analysis*: workload characterization of containerized applications; coordinated monitoring and analysis of performance state at all the layer of the cloud stack; correlation among container performance counters, VMs performance counters, and PMs performance counters.
- *Planning*: definition of appropriate performance, scalability, and fault tolerance models of container-based systems; definition of adaptation algorithms that account for the multilayer nature of container-based systems, for the possible distributed orchestrator (choreography) and for the variety of objectives and constraints imposed by the functional and non-functional requirements.

In what follows, the state-of-the-art research is classified along the two main research area mentioned above.

14.5.1 Monitoring and Analysis

Performance monitoring and characterization is a topic of increasing interest for the containers' research community. Also if has been recognized that the container workload is different from the traditional cloud application workload, to the best of our knowledge, there are no studies on workload characterization. The majority of the literature reviewed is about performance evaluation studies except for [8] where the authors assess the different measurement methodology used to collect performance counters for CPU and disk I/O intensive Docker workload, and in

[7] where the authors conducted a preliminary study that correlates container performance counters (counters from /cgroup file system) with system performance counters (from /proc file system).

The first seminal work on container performance evaluation [13] provides an extensive comparison among a native Linux environment, Docker, and KVM. In this work are compared the three environments in the presence of CPU intensive, I/O intensive, network intensive, and NoSQL/SQL workloads. The main intention of the work is to assess the performance improvement of running workloads in containers rather than in VMs. The comparison is based on the performance metrics collected by the benchmarking tools rather than on the workload footprint. A similar study, aimed at comparing the performance of containers with hypervisors is [28]. While standard benchmarks were used in the previous studies, the scientific workload is considered in [2]. The authors shown that Docker memory configuration can be tuned to make container performance be slightly better than VMs.

In [21] the authors studied the performance of container platforms running on top of a cloud infrastructure, the NeCTAR cloud. Specifically the authors compared the performance of Docker, Flockport (LXC), and the VM using the same benchmarks as in [13]. The comparison was intended to explore the performance of CPU, memory, network, and disk.

In [24] the authors proposed a study on the interference among multiple applications sharing the same resources and running in Docker containers. The study focuses on I/O, and it proposes also a modification of the Docker's kernel to collect the maximum I/O bandwidth of the machine it is running on.

The performance of containers running on Internet-of-Things devices are investigated in [27]. Docker runs on a Single Board Computer device such as Raspberry Pi 21. As benchmark are used: system benchmarks to independently stress CPU, memory, network I/O, and disk I/O; and application benchmarks reproducing MySQL and Apache workload. The reference for comparison is the performance of the system without any virtualization.

A qualitative evaluation of Docker as a platform for Edge Computing is in [18]. The evaluation criteria were: deployment and termination, resource and service management, fault tolerance and caching. Results demonstrate that Docker provides fast deployment, small footprint, and good performance.

In [41] the authors investigates the impacts of Docker configuration and resource interference on the performance and scalability of big data workloads (four typical Spark applications). The authors propose also a performance prediction model based on the Support Vector Regression to predict the application performance with different configurations and resource competition settings.

In [39] the authors present an extensive study that unravel the multi-faceted nature of Docker storage, for wide range of file systems, and demonstrate its impact on system and workload performance. In [11] is compared the performance of Docker when the Union file system and the CoW file system are used to build image layers.

14.5.2 Self-optimization

Autonomic orchestration of containers could leverage more than 10 years of research results in the field of autonomic computing [12, 17, 22, 36].

The adoption of container technologies call for new autonomic management solutions [6, 34]. That is confirmed by the significant amount of research works, published in the last 2 years, and addressing the problem of autoscaling and self-healing in container-based systems.

An early study on container management [16] shows that Elastic Application Container-based resource management outperforms the VM-based approach in terms of feasibility and resource-efficiency. C-Port [1] is the first example of orchestrator that makes it possible to deploy and manage container across multiple clouds. The authors plan to address the issues of resource discovery and coordination, container scheduling and placement, and dynamic adaptation. However, the research is at an early stage. In terms of orchestration policy, they developed a constraint-programming model that can be used for dynamic resource discovery and selection. The constraints that they considered are availability, cost, performance, security, or power consumption. In [30] the authors provide a general formulation of the elastic provisioning of virtual machines for container deployment as an integer linear programming problem, which takes explicitly into account the heterogeneity of container requirements and virtual machine resources. Only QoS and cost are considered in the problem formulation. In [37] the authors propose an adaptive multi-instance container-based architecture targeting time-critical applications. The solution is implemented using Docker and Kubernetes. In [7] the authors analyze the behavior of the Kubernetes Horizontal Pods Auto-scaling algorithm and propose a new solution to make a more appropriate allocation of resources to fulfil application response time constraints. In [3] the authors propose ELASTICDOCKER, an autonomic controller powering vertical elasticity of Docker containers autonomously. ELASTICDOCKER scales up and down both CPU and memory assigned to each container according to the application workload, and live-migrates containers when there is no enough resources on the hosting machine. The experiments show that ELASTICDOCKER helps to reduce expenses for container customers, make better resource utilization for container providers, and improve Quality of Experience for application end-users. As compared to horizontal elasticity, ELASTICDOCKER outperforms Kubernetes Horizontal elasticity by 37.63%. In [19] the authors propose an Ant Colony Optimization (ACO) algorithm to schedule docker containers with the ultimate goal to use resources more efficiently. The ACO algorithm has been implemented in Docker SwarmKit and performance compared with the existing greedy scheduling algorithm: ACO throughput is 15% higher than the greedy algorithm. In [15] the authors design a novel application oriented Docker container (AODC)-based resource allocation framework to minimize the application deployment cost in datacenters and to support automatic scaling while the workload of cloud applications varies.

In [4] is addressed the problem of properly assigning resources to containers in a IoT scenario to reduce energy consumption. The authors propose DockerCap, a software-level power capping orchestrator for Docker containers that follows an Observe-Decide-Act loop structure: this allows to quickly react to changes that impact on the power consumption by managing resources of each container at runtime, to ensure the desired power cap. The paper shows that it is possible to obtain results comparable with the state-of-the-art power capping solution provided by Intel RAPL, still being able to tune the performances of the containers and even guarantee SLA constraints.

14.5.3 Self-healing

The self-healing problem is addressed in [38] and [29].

In [38] the authors propose Serfnode a lightweight, platform agnostic, and easy to integrate with an existing system of Docker containers. The proposed solution contains a monitoring and self-healing mechanism based on Supervisor (supervisord.org) for added resiliency.

In [29] is proposed an intuitive approach based on Computational Intelligence (CI) for enhancing the dependability of container Docker Swarm. The proposed CI-based approach predicts the possible failure of the host of a manager node by observing its abnormal behavior. Thus, this indication can automatically trigger the process of creating a new manager node or promoting an existing node as a manager for enhancing the orchestrator's dependability.

14.6 Final Remarks

As pointed out in the previous sections, the state-of-the-art mechanisms for container orchestration should be enhanced introducing models and algorithms for runtime self-adaptation. Among the many research challenges, an urgent answer is required for the followings.

Monitoring and Workload Characterization Monitoring techniques and tools used for the operating system and application levels do not allow to catch the performance behavior of containers. Moreover, there is no a commonly agreed definition of QoS metrics for container-based systems. Docker offers the docker stat command that returns CPU and memory utilization for each running container. More detailed CPU, memory, and network statistics can be accessed through the /containers/(id)/stats API. In a recent work [25], the authors modify Docker and Docker Swarm in order to monitor the I/O capacity and utilization of the containers with the goal of controlling the QoS level of a Docker cluster. Characterization of container workload is fundamental for the definition of performance models and autonomic orchestration policies. To the best of our knowledge the literature lack of such studies.

Performance Models Validated performance models and energy consumption models of container-based systems are inexistent. Performance models are widely used in autonomic computing to determine the reconfiguration actions needed to maintain the desired level of service. An alternative approach is the use of machine learning techniques to determine the more appropriate reconfiguration action (e.g., reinforcement learning [35]).

Adaptation Models for Container Orchestration As already pointed out, the container orchestration policies used until now are very simple. What the industry need is the definition of a framework for QoS-aware, energy-aware, and legislation-aware optimal adaptation for container orchestration. This framework should allow to define system models, QoS, energy, and legal constraints, to find optimal adaptation policies for container orchestration at runtime.

To conclude, the way toward the next generation of cloud computing platforms is based on application virtualization rather than on hardware virtualization and that requires a strong contribution from the research community focused not only on the above research challenges but also on different area, for example, networking and security.

Acknowledgements This work is funded by the research project *Scalable resource-efficient systems for big data analytics* and financed by the Knowledge Foundation (grant: 20140032) in Sweden.

References

1. M. Abdelbaky, J. Diaz-Montes, M. Parashar, M. Unuvar, M. Steinder, Docker containers across multiple clouds and data centers, in *2015 IEEE/ACM 8th International Conference on Utility and Cloud Computing (UCC)* (2015), pp. 368–371. https://doi.org/10.1109/UCC.2015.58
2. T. Adufu, J. Choi, Y. Kim, Is container-based technology a winner for high performance scientific applications? in *17th Asia-Pacific Network Operations and Management Symposium, APNOMS 2015*, Busan, August 19–21 (IEEE, New York, 2015), pp. 507–510. https://doi.org/10.1109/APNOMS.2015.7275379
3. Y. Al-Dhuraibi, F. Paraiso, N. Djarallah, P. Merle, Autonomic vertical elasticity of docker containers with elasticdocker, in *2017 IEEE 10th International Conference on Cloud Computing (CLOUD)* (2017), pp. 472–479. https://doi.org/10.1109/CLOUD.2017.67
4. A. Asnaghi, M. Ferroni, M.D. Santambrogio, Dockercap: a software-level power capping orchestrator for docker containers, in *2016 IEEE International Conference on Computational Science and Engineering (CSE) and IEEE International Conference on Embedded and Ubiquitous Computing (EUC) and 15th Intl Symposium on Distributed Computing and Applications for Business Engineering (DCABES)* (2016), pp. 90–97. https://doi.org/10.1109/CSE-EUC-DCABES.2016.166

5. B. Burns, B. Grant, D. Oppenheimer, E. Brewer, J. Wilkes, Borg, omega, and kubernetes. ACM Queue **14**, 70–93 (2016). http://queue.acm.org/detail.cfm?id=2898444
6. E. Casalicchio, Autonomic orchestration of containers: problem definition and research challenges, in *10th EAI International Conference on Performance Evaluation Methodologies and Tools, EAI* (2016)
7. E. Casalicchio, V. Perciballi, Auto-scaling of containers: The impact of relative and absolute metrics, in *2017 IEEE 2nd International Workshops on Foundations and Applications of Self* Systems (FAS*W)* (2017), pp. 207–214. https://doi.org/10.1109/FAS-W.2017.149
8. E. Casalicchio, V. Perciballi, Measuring docker performance: what a mess!!! in *Proceedings of the 8th ACM/SPEC on International Conference on Performance Engineering Companion*, ICPE '17 Companion (ACM, New York, 2017), pp. 11–16. https://doi.org/10.1145/3053600.3053605.
9. A. Celesti, M. Fazio, M. Giacobbe, A. Puliafito, M. Villari, Characterizing cloud federation in IoT, in *2016 30th International Conference on Advanced Information Networking and Applications Workshops (WAINA)* (2016), pp. 93–98. https://doi.org/10.1109/WAINA.2016.152
10. R. Dua, A.R. Raja, D. Kakadia, Virtualization vs containerization to support PaaS, in *Proceedings of 2014 IEEE International Conference on Cloud Engineering*, IC2E '14 (2014), pp. 610–614
11. R. Dua, V. Kohli, S. Patil, S. Patil, Performance analysis of union and cow file systems with docker, in *2016 International Conference on Computing, Analytics and Security Trends (CAST)* (2016), pp. 550–555. https://doi.org/10.1109/CAST.2016.7915029
12. F. Faniyi, R. Bahsoon, A systematic review of service level management in the cloud. ACM Comput. Surv. **48**(3), 43:1–43:27 (2015). https://doi.org/10.1145/2843890
13. W. Felter, A. Ferreira, R. Rajamony, J. Rubio, An updated performance comparison of virtual machines and Linux containers. Technical Report, RC25482(AUS1407–001), IBM, IBM Research Division, Austin Research Laboratory (2014)
14. W. Gerlach, W. Tang, K. Keegan, T. Harrison, A. Wilke, J. Bischof, M. D'Souza, S. Devoid, D. Murphy-Olson, N. Desai, F. Meyer, Skyport: container-based execution environment management for multi-cloud scientific workflows, in *Proceedings of the 5th International Workshop on Data-Intensive Computing in the Clouds*, DataCloud '14 (IEEE Press, Piscataway, NJ, 2014), pp. 25–32. http://dx.doi.org/10.1109/DataCloud.2014.6
15. X. Guan, X. Wan, B.Y. Choi, S. Song, J. Zhu, Application oriented dynamic resource allocation for data centers using docker containers. IEEE Commun. Lett. **21**(3), 504–507 (2017). https://doi.org/10.1109/LCOMM.2016.2644658
16. S. He, L. Guo, Y. Guo, C. Wu, M. Ghanem, R. Han, Elastic application container: A lightweight approach for cloud resource provisioning, in *2012 IEEE 26th International Conference on Advanced Information Networking and Applications* (2012). pp. 15–22. https://doi.org/10.1109/AINA.2012.74
17. M.C. Huebscher, J.A. McCann, A survey of autonomic computing — degrees, models, and applications. ACM Comput. Surv. **40**(3), 7:1–7:28 (2008). http://doi.acm.org/10.1145/1380584.1380585
18. B.I. Ismail, E.M. Goortani, M.B.A. Karim, W.M. Tat, S. Setapa, J.Y. Luke, O.H. Hoe, Evaluation of docker as edge computing platform, in *2015 IEEE Conference on Open Systems (ICOS)* (2015), pp. 130–135. https://doi.org/10.1109/ICOS.2015.7377291
19. C. Kaewkasi, K. Chuenmuneewong, Improvement of container scheduling for docker using ant colony optimization, in *2017 9th International Conference on Knowledge and Smart Technology (KST)* (2017), pp. 254–259. https://doi.org/10.1109/KST.2017.7886112
20. J.O. Kephart, D.M. Chess, The vision of autonomic computing. Computer **36**(1), 41–50 (2003). https://doi.org/10.1109/MC.2003.1160055

21. Z. Kozhirbayev, R.O. Sinnott, A performance comparison of container-based technologies for the cloud. Futur. Gener. Comput. Syst. **68**, 175–182 (2017). http://dx.doi.org/10.1016/j.future. 2016.08.025. http://www.sciencedirect.com/science/article/pii/S0167739X16303041
22. A.L. Lemos, F. Daniel, B. Benatallah Web service composition: a survey of techniques and tools. ACM Comput. Surv. **48**(3), 33:1–33:41 (2015). http://doi.acm.org/10.1145/2831270
23. B.D. Martino, G. Cretella, A. Esposito, Advances in applications portability and services interoperability among multiple clouds. IEEE Cloud Comput. **2**(2), 22–28 (2015)
24. S. McDaniel, S. Herbein, M. Taufer, A two-tiered approach to I/O quality of service in docker containers, in *2015 IEEE International Conference on Cluster Computing* (2015), pp. 490–491. https://doi.org/10.1109/CLUSTER.2015.77
25. S. McDaniel, S. Herbein, M. Taufer, A two-tiered approach to I/O quality of service in Docker containers, in *Proceedings of 2015 IEEE International Conference on Cluster Computing*, CLUSTER '15 (2015), pp. 490–491
26. J. McGee, The 6 steps of the container lifecycle (2016). https://www.ibm.com/blogs/cloud-computing/2016/02/the-6-steps-of-the-container-lifecycle/
27. R. Morabito, A performance evaluation of container technologies on internet of things devices, in *2016 IEEE Conference on Computer Communications Workshops (INFOCOM WKSHPS)* (2016), pp. 999–1000. https://doi.org/10.1109/INFCOMW.2016.7562228
28. R. Morabito, J. Kjällman, M. Komu, Hypervisors vs. lightweight virtualization: a performance comparison, in *2015 IEEE International Conference on Cloud Engineering* (2015), pp. 386–393. https://doi.org/10.1109/IC2E.2015.74
29. N. Naik, Applying computational intelligence for enhancing the dependability of multi-cloud systems using docker swarm, in *2016 IEEE Symposium Series on Computational Intelligence (SSCI)*, (2016), pp. 1–7. https://doi.org/10.1109/SSCI.2016.7850194
30. M. Nardelli, C. Hochreiner, S. Schulte Elastic provisioning of virtual machines for container deployment, in *Proceedings of the 8th ACM/SPEC on International Conference on Performance Engineering Companion*, ICPE '17 Companion (ACM, New York, 2017), pp. 5–10. http://doi.acm.org/10.1145/3053600.3053602
31. S. Natarajan, A. Ghanwani, D. Krishnaswamy, R. Krishnan, P. Willis, A. Chaudhary, An analysis of container-based platforms for NFV. Technical Report, IETF (2016)
32. D.T. Nguyen, C.H. Yong, X.Q. Pham, H.Q. Nguyen, T.T.K. Loan, E.N. Huh, An index scheme for similarity search on cloud computing using mapreduce over docker container, in *Proceedings of the 10th International Conference on Ubiquitous Information Management and Communication*, IMCOM '16 (ACM, New York, 2016), pp. 60:1–60:6. http://doi.acm.org/10. 1145/2857546.2857607
33. OASIS, Topology and orchestration specification for cloud applications. Technical Report Version 1.0, OASIS Standard (2013)
34. C. Pahl, Containerization and the PaaS cloud. IEEE Cloud Comput. **2**(3), 24–31 (2015). https:// doi.org/10.1109/MCC.2015.51
35. A. Pelaez, A. Quiroz, M. Parashar, Dynamic adaptation of policies using machine learning, in *2016 16th IEEE/ACM International Symposium on Cluster, Cloud and Grid Computing (CCGrid)* (2016), pp. 501–510, https://doi.org/10.1109/CCGrid.2016.64
36. S. Singh, I. Chana, Qos-aware autonomic resource management in cloud computing: a systematic review. ACM Comput. Surv. **48**(3), 42:1–42:46 (2015). http://doi.acm.org/10.1145/ 2843889
37. V. Stankovski, J. Trnkoczy, S. Taherizadeh, M. Cigale, Implementing time-critical functionalities with a distributed adaptive container architecture, in *Proceedings of the 18th International Conference on Information Integration and Web-based Applications and Services*, IIWAS '16 (ACM, New York, 2016), pp. 453–457. http://doi.acm.org/10.1145/3011141.3011202
38. J. Stubbs, W. Moreira, R. Dooley Distributed systems of microservices using docker and serfnode, in *2015 7th International Workshop on Science Gateways* (2015). pp. 34–39. https:// doi.org/10.1109/IWSG.2015.16

39. V. Tarasov, L. Rupprecht, D. Skourtis, A. Warke, D. Hildebrand, M. Mohamed, N. Mandagere, W. Li, R. Rangaswami, M. Zhao, In search of the ideal storage configuration for docker containers, in *2017 IEEE 2nd International Workshops on Foundations and Applications of Self* Systems (FAS*W)* (2017), pp. 199–206. https://doi.org/10.1109/FAS-W.2017.148
40. W. Vogels, Under the hood of Amazon EC2 container service (2015). http://www.allthingsdistributed.com/2015/07/under-the-hood-of-the-amazon-ec2-container-service.html
41. K. Ye, Y. Ji, Performance tuning and modeling for big data applications in docker containers, in *2017 International Conference on Networking, Architecture, and Storage (NAS) (2017)*, pp. 1–6. https://doi.org/10.1109/NAS.2017.8026871

Chapter 15
A Cloud-Based Overlay Networking for the Internet of Things: Quantitative Evaluation

Dario Bruneo, Salvatore Distefano, Francesco Longo, Giovanni Merlino, and Antonio Puliafito ⓘ

15.1 Introduction

In a typical Infrastructure-as-a-Service (IaaS) Cloud, users are able to create and bring up virtual machines (VMs), access the instances through ssh, VNC, or Web-based virtual console as well as to instantiate even topologically complex virtual networks among a set of VMs. Any solution meant to take advantage of IoT resources available outside the datacenter, if those are to be provided according to IaaS principles, would need to implement at least similar facilities.

In a heavily distributed ecosystem, such as IoT-related scenarios, many requirements diverge significantly in comparison to typical IaaS Cloud environments, such as the presence of nodes installed behind firewalls and/or NATs (especially when IPv6 deployments are not an option) or, more in general, the necessity to deal with any restricted environment with denied-by-default (institutional or corporate) security policies. Such constraints call for more powerful mechanisms to enable core functionalities for virtual infrastructure management, i.e., remote access to board-hosted resources and instantiation of virtual networks.

Any network virtualization [1, 2] mechanism for IoT infrastructure thus requires at least some form of reconfiguration capabilities for board-side networking facilities as well. Yet, in contrast to typically datacenter-oriented IaaS, the physical environment (cabling and media access setup, logical topologies and hierarchies,

D. Bruneo · F. Longo (✉) · G. Merlino · A. Puliafito
Università di Messina, Messina, Italy
e-mail: dbruneo@unime.it; flongo@unime.it; gmerlino@unime.it; apuliafito@unime.it

S. Distefano
Università di Messina, Messina, Italy
Kazan Federal University, Kazan, Russia
e-mail: sdistefano@unime.it; s_distefano@it.kfu.ru

© Springer International Publishing AG, part of Springer Nature 2019   237
A. Puliafito, K. S. Trivedi (eds.), *Systems Modeling: Methodologies and Tools*,
EAI/Springer Innovations in Communication and Computing,
https://doi.org/10.1007/978-3-319-92378-9_15

role allocation for equipment) in IoT scenarios is not always under control of the designer of the infrastructure, which may as well be opportunistically assembled, e.g., volunteer-contributed.

In this work we describe a rationale and some mechanisms to enable such functionalities when dealing with the unique requirements and challenges dictated by IoT environments, e.g., embedded boards and other constrained devices. In particular, overlay networking is addressed here to provide suitable facilities on top of Cloud-managed IoT resources in a technology agnostic fashion, still taking into account the limitations of smart devices, while at the same time suitable to be mapped onto an IaaS-focused solution, as earlier [3] investigated in terms of a device-centric approach for sensor-hosting nodes. A preliminary quantitative evaluation is also proposed, to assess and validate the feasibility of the approach.

15.2 IoT Infrastructure as a Service

In our approach [4] to Cloud/IoT integration at the infrastructure level, we envisioned a *virtualization* layer for boards, able to provide access to I/O pins, such as GPIO, as a Service through RESTful interfaces, and to send (predefined or custom) commands to the environment running on the board. Here, we propose our implementation of such an approach, namely the Stack4Things framework, from now on referred to as S4T, designed by extending certain *OpenStack* subsystems to smoothly integrate and leverage as much existing functionalities as possible. OpenStack is a widely known, Open Source Cloud middleware exploited in both academic and commercial contexts.

In terms of hardware options with regard to boards, we confine the discussion on purpose to relatively smart embedded devices, such as modern Arduino-based hybrid systems, hosting either kind of (low power) micro-processor (MPU) and micro-controller (MCU) units. Such a MCU-equipped board is suited to host a minimal Linux distro, e.g., OpenWRT, able then to host a number of tools and runtime environments, such as Node.js or Python, where the *S4T lightning-rod*, the node-side component of the S4T architecture, executes on the MPU side and interacts with the OS tools and services of the board, and with sensing and actuation resources through I/O pins. It acts as the point of contact with Cloud infrastructure. This is ensured by a *Web Application Messaging Protocol (WAMP)* [5] and WebSocket-based communication between the lightning-rod and the Cloud, in particular the *S4T IoTronic service*, which is characterized by the standard architecture of an OpenStack service, as depicted in Fig. 15.1.

The *S4T IoTronic conductor* represents the core of the service, managing the *S4T IoTronic database* that stores all the necessary information. The *S4T IoTronic APIs* exposes a REST interface for the end users that may interact with the service both via a custom client (*S4T IoTronic command line client*) and via a Web browser. In fact, the OpenStack Horizon dashboard has been enhanced with a *S4T dashboard* exposing all the functionalities provided by the S4T IoTronic service and other

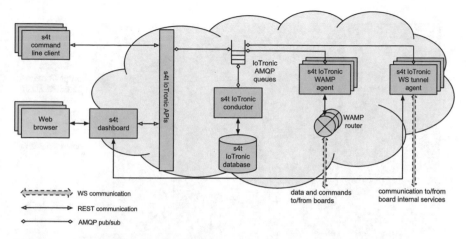

Fig. 15.1 S4T cloud-side architecture

software components. In particular, the dashboard also deals with the access to board-internal services, redirecting the user to the *S4T IoTronic WS tunnel agent*. This piece of software is a wrapper and a controller for the WS server to which the boards connect through the use of S4T *wstunnel* libraries.

Similarly, the *S4T IoTronic WAMP agent* controls the WAMP router and acts as a bridge between other subsystems and the boards assigned to the corresponding instance of the agent. The agent translates AMQP messages into WAMP messages and vice versa.

To implement this IaaS-like IoT-Cloud paradigm, specific facilities and mechanisms, such as those required to access remote (I/O) resources as well as the ones allowing to manage network topologies for things, are required. The former are addressed by a specific module running on IoT nodes, *Stack4Things lightning-rod*, adopting either pub/sub or RPC-style I/O patterns to let the users interact with board-local resources through the Cloud, while access to I/O pins is exploited for interaction with sensing and actuation resources. Figure 15.2 shows the lightning-rod architecture. It allows users to interact with boards even in the presence of NATs or firewalls by exploiting WebSocket and WAMP-based communication between a board and the Cloud. WebSocket is a network-agnostic protocol providing a full-duplex TCP communication channel over a single HTTP connection, used to overcome the restrictions imposed by middle boxes and the enforced policies therein by leveraging ubiquitous (outbound) HTTP support. The *Web Application Messaging Protocol (WAMP)* [5] is a sub-protocol of WebSocket for messaging according to both publish/subscribe and routed remote procedure call (RPC) paradigms.

This way, the S4T I/O *hardware abstraction layer* (HAL) is tasked at providing a software interface for the I/O pins by exposing them as i-nodes of a virtual filesystem. Then, the *lightning-rod engine* is at the heart of the thing-side S4T architecture, managing all the interactions with the Cloud counterpart through a

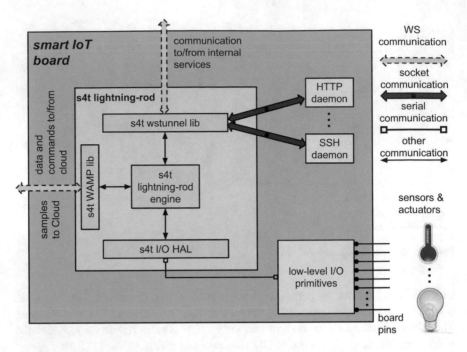

Fig. 15.2 Stack4Things thing-side stack: logical architecture

WAMP router on a WebSocket full-duplex channel. By these, specific commands to
the device sensing and actuation resources via their digital/analog I/O pins can be
remotely delivered, from either the Cloud or other devices, exploiting the WAMP
functionalities provided by the S4T *WAMP libraries*. Indeed, corresponding REST
resources get automatically generated, exposing user-defined commands Cloud-
side. Invoking these resources triggers the execution of the corresponding code on
top of the smart board. Furthermore, the S4T *wstunnel libraries* allows to establish
reverse WebSocket tunnels to a WebSocket server on the Cloud. This way, thing-
side (i.e., local) services can be directly accessed from end users by relaying traffic
coming in over WebSocket tunnels to the corresponding service, while outgoing
traffic is tunneled back up to the Cloud WebSocket server, where users connect for
this *service remoting* functionality.

15.3   Overlay Networking for IoT

Our approach to overlay networking is based on enabling mechanisms in terms
of custom layering and board-side tunneling facilities, to be coupled with the
corresponding Cloud-side adaptations. To this end such preliminary investigation
mostly focuses on virtualization architecture and patterns.

15.3.1   Tunneling

As remote infrastructure, boards are possibly going to be available over very restrictive, IPv4-only deployments. The only assumption that can (for all purposes, always) be considered true is outgoing Web traffic being permitted, i.e., board-initiated communication over standard HTTP/HTTPS ports. This actually rules out most messaging protocols, including very popular ones in IoT (e.g., MQTT), which are plainly TCP-based and by default cannot be recognized as Web traffic. We thus resorted to an HTTP-borne mechanism for bidirectional connectivity and reachability of internal services, namely WS.

WebSockets [6] as channels between a browser and a server are considered standard facilities for bidirectional communication and in particular server-pushed messaging. One of the main advantages of WS is that it is network agnostic, by just piggybacking communication onto standard HTTP interactions. This is of benefit for those environments which block Web-unrelated traffic using firewalls. Less explored is the creation of generic TCP tunnels over WS, a way to get client-initiated connectivity to any server-side local (or remote) service.

In early work [7] we devised a design and implementation of a novel *reverse* tunneling technique, as a way to provide server-initiated, e.g., Cloud-triggered, connectivity to any board-hosted service, or any other node on a contributed resource network, e.g., a WSN. In particular the latter may enable typical IoT scenarios, e.g., Machine-to-Machine (M2M) interactions, by supporting these patterns in a device-centric [3] fashion. This way, we provide a gateway to edge nodes and services, acting at a very basic level as a proxy for access to data gathered from mostly passive resources. Yet, such gateway may also act as a relay to activate remoting toward nodes in a masqueraded network, thus allowing designers to explore options beyond protocols such as Traversal Using Relays around NAT (TURN) [8].

Figure 15.3 depicts systems, flows, and interactions of such a WS reverse tunnel (abbreviated as *rtunnel*), in the case of board-provided access to a service hosted on the board itself. By leveraging the diagram, in the following we outline the sequence of operations for the setup of an rtunnel. The rtunnel client (e.g., a board) first sends a WS connection request to the rtunnel server, specifying a TCP port. When the rtunnel server receives the WS connection request, a new TCP server is brought up listening on the specified port, the WS request is then accepted, and a WS connection (depicted in the figure as "control WS") is started. When an external

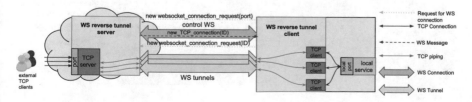

Fig. 15.3 Low-level functional diagram of WS-based reverse tunnel

TCP client tries to connect to the TCP server on the rtunnel server side, the new TCP connection is paused and, through the control WS, a WS message is sent in order to signal the request for a new TCP connection, and specify a unique ID for that connection. When the rtunnel client receives the message signaling the request for a new TCP connection, it sends a new WS connection request to the rtunnel server, specifying the ID of the connection. When the rtunnel server receives the WS connection request, it checks if the received ID does not match any of the existing TCP connections and, if so, it accepts the request and opens a new WS connection (depicted in figure as "WS tunnel"). The new TCP connection thus gets piped to the new WS connection (that acts as a WS-encapsulated tunnel for TCP segments) and then resumed. On the WS rtunnel client side, as soon as the new WS tunnel is established, a new TCP client is brought up connecting to the local service of interest, and such a new TCP connection gets piped to the new WS tunnel. TCP segments coming from the external TCP client are now able to reach the local service, and traffic thus gets to flow back and forth until the rtunnel is torn down.

15.3.2   Layering

As long as WS-based tunnels may be instantiated by the Cloud, a robust mechanism is already in place for accessing board-hosted services. What is still missing are solutions to overlay network- and datalink-level addressing and traffic forwarding on top of this facility. Some approaches for creating VPNs over WS are already available [9], but these solutions do not expose a decoupled control machinery nor the advantages of an on-demand approach.

In the following, we describe the proposed layering for WS tunnel-based layer-2 virtual networks. In Fig. 15.4 a diagram is modeled after the low-level reverse tunnel one, but focused on the instantiation of, e.g., a virtual bridge between two boards.

To better understand the description of the aforementioned diagram in the following, first it is useful to introduce a tool which is part of the solution. *Socat* is a networking "swiss army knife" available as command-line tool for Unix systems that, as its counterpart Netcat (nc), provides to network experimenters a set of facilities and shortcuts (e.g., socket piping, socket tuning, virtual TUN/TAP

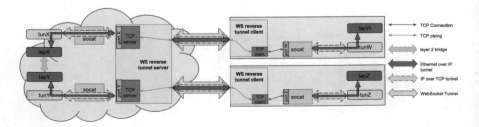

Fig. 15.4 Low-level functional diagram of bridged tunneling over WS

devices, process control). Its minimal build-time dependencies (i.e., the C library only) translate into a remarkably flexible tooling also when it comes to IoT-class, constrained devices, as long as a (possibly stripped-down) version of a POSIX-compatible system and the relevant networking stack are available.

Still sticking to the setup of a *control* WS, as a preliminary step in this workflow, in this case a rtunnel gets activated for each board to be virtually bridged. As a simplified scenario, the diagram depicts just two such boards, but no limitation is in place on the number of remote boards to be virtualized in terms of networking. As any board here, from now on referred as client, needs to go through the same set of operations, we will describe just a single instance for the sake of brevity.

Taking into consideration the uppermost board in the diagram, a first step consists in the setup of a TCP connection based on a WS-based rtunnel, obtained by exposing, on the server side, a listening socket on a local port, as soon as the rtunnel server accepts a request for a new *data* rtunnel. The TCP connection, once established, needs then to be piped to the rtunnel that encapsulates TCP segments in a WS-based stream.

On the client side of the WS rtunnel, once the rtunnel is established, a new TCP client is brought up connecting to a local (Socat-provided) listening port, piping this TCP connection back to the rtunnel.

A level-3 tunnel gets then established over such TCP-based tunnel by leveraging a Socat instance operating, on both sides of the chain, in listening mode. Upon connection, this tunnel starts exposing a virtual TUN device on each side, set up with IP addresses of user choice, as long as these belong to the same subnet.

Even if the above reported steps are to be considered logically operations to be performed early on, all steps are to be considered timing insensitive, by employing retries and listening sockets where needed.

When a level-2 encapsulation over an IP-based communication is needed, the system firstly creates a GRE tunnel to expose an Ethernet-compatible interface. Accordingly, the tunnel-hosting virtual device is set to TAP whereas endpoints are the previously configured TUN IPs. Then, in order to complete the workflow, this interface is added to a dedicated virtual bridge on the server.

In order to create IP-based tunnels, we used the *Generic Routing Encapsulation* (GRE) that is an IETF standard for a no-frills IP-in-IP tunneling protocol [10]. Besides level-3 encapsulation support, tunneling of level-2 frames (over to the corresponding virtual TAP device) is also possible.

Through the *reversed* layering devised here, we are able to provide a basic but flexible mechanism for instantiation of a virtual network, either placing boards in the same broadcast domain, making them just routable or, alternatively, reachable higher up in the networking stack, i.e., directly at the transport level. Ultimately this means being able to set up, according to user needs, either a virtual bridge, i.e., same level-2 broadcast domain, by means of GRE TAP-based tunnels, virtual NICs, socat piping, and reverse tunneling over WS, or a virtual private network, i.e., level-3 reachability, by leveraging just a subset of the aforementioned mechanisms, plus static routes configured on the server for board-to-board forwarding.

15.3.3   Server-Less Mesh Implementation

The above described scenario can also be reproduced in a fully peer-to-peer configuration, thanks to persistent WAMP control channel availability. To do that, the Cloud selects one board, from the pool of boards that have to be connected over the virtual network, that will act as a central node in a star topology. Such a selection is performed according to a ranking for suitability in taking that role, on the basis of current load and availability of on-board resources. An ad-hoc variation of the TCP *hole punching* [11] technique, in which the Cloud would preliminarily act as the required third party in the phase of connection establishment, is exploited to allow all the other devices belonging to the virtual network to connect to the chosen central node by establishing WebSocket-based reverse tunnels. While hole punching helps to overcome the natural limitations imposed by NAT traversal, especially in the case of symmetric (e.g., double-sided NAT) configurations, reverse tunneling helps actually punch holes across firewalls with (Cloud-initiated) WebSockets. On top of such a peer-based WebSocket-based infrastructure, the virtual network is built, in a star topology. At this point, network communication is fully peer-to-peer and no further actions are required from the Cloud, until virtual network teardown. Moreover, other topologies may be enabled as well, such as a full-mesh at layer 2 among all the boards selected to be interconnected over a Cloud-initiated virtual network.

In Fig. 15.5 a simple 3-node network (plus the Cloud) is depicted, where the thick double-sided blue arrows correspond to the star topology, when the node in the middle of the figure is chosen as the center of the star, while a similar arrow in red stands for the missing link among peers to establish a full mesh. The thin single-sided black arrow represents both (transient) WS-based connection establishment and just steady command streams afterwards, having the Cloud at that point already hole-punched the nodes and subsequently torn down its tunnels to those.

In that case the topology would be a tree over the set of bridges exposed by the boards, dynamically instantiated by means of, e.g., a protocol for automatic configuration of bridges belonging to the Spanning Tree family. In this case the virtual network would not feature any single point of failure, nor have a node be the bottleneck for all the generated traffic inside the network itself. On the other hand, this configuration requires a longer setup phase, due to the establishment of an higher number of tunnels, $n(n-1)/2$, in proportion to the number of nodes n, compared to the star topology, $n-1$, and more importantly a slightly higher baseline in terms of requirements, as each node would host a virtual bridge, not just the central node.

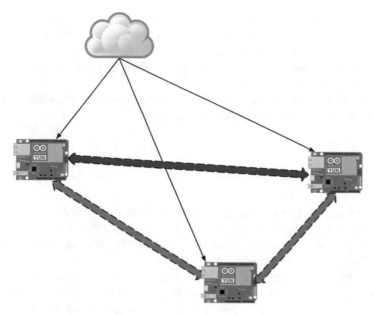

Fig. 15.5 Cloud-enabled server-less star and mesh topologies

15.4   Quantitative Evaluation

In order to test and validate the proposed solution, we developed experiments on a smart city environment. #SmartME [12] is a crowdfunding project that aims at collecting a number of field-deployed networks of sensors and actuators in Messina into a federated Smart City infrastructure, provided as-a-service [13], actually an instance of a so-called Software-Defined City [14]. Leveraging such infrastructure, it is possible to collect data and extract information to build services for citizens, possibly taking part in this city-scale network through the involvement of, e.g., smartphones, by which it is possible to interact with objects, and may even turn themselves into data producers.

15.4.1   Scenario

For our network virtualization approach, #SmartME represents an interesting testbed and case study, thanks to a development, deployment, and testing platform which is integrated with smart boxes, hosting sensors, and/or actuators, possibly already deployed in the district albeit employed for specific application domains only to this day, as such useful for a limited number of purposes.

The virtual networking subsystem has been put to test in this context by exploiting the #SmartME testbed, considering two main scenarios, one based on clients (e.g., boards) behind residential xDSL gateways, e.g., at home, and another one based on clients belonging to the (Fast) Ethernet-switched network of the university campus, respectively.

The first one is thus about connectivity scenarios over a Wide-Area Network, whereas the second one over a LAN, respectively.

15.4.2   Experimental Results

Our quantitative evaluation begins with some theoretical considerations, such as packet size and overhead estimates, that can be considered a preliminary analysis. Later we provide an evaluation focusing on specific key performance indices, namely latency and throughput.

Analyzing the overhead of the framework in terms of protocol headers and encapsulation format, the overhead introduced by WS is equal to 6 bytes (2 for the header and 4 for the mask value) while the one introduced by TCP tunneling is equal to 20 bytes. With respect to Ethernet framing, the overhead is equal to 18 when a TAP interface is needed instead of a TUN, thus reaching a total value of 44 bytes. Moreover, when security mechanisms are adopted, by using TLS, we have to add other 41 bytes, thus reaching a value of overhead equal to 85 bytes per packet. From this analysis, we can observe that the packet size is still comparable with the overhead experienced using a typical VPN (e.g., OpenVPN) that roughly is equal to 69 bytes per packet (41 bytes of overhead for security and 28 for tunneling, respectively), i.e., only slightly smaller than in our solution.

Tables 15.1 and 15.2 report on the set of experiments that have been conducted. The experiments are based on the *iperf3* [15] tool for measuring throughput and the ubiquitous *ping* tool to gauge latency, the latter by means of ICMP echo requests to obtain Round-Trip Time values as estimation of delay. In both cases, the test setup consists in having a server ready and generating traffic over TCP (iperf) or ICMP requests (ping) from the client, which collects partial and final statistics. Values in the table are averages computed over a number of 1000 samples, where each chunk of ten samples represent a single run for the tool. We did not include variance values because they are negligible.

Table 15.1 Throughput and latency measurements: xDSL

Topology/technology	iperf: throughput [Mbps]	ping: latency (RTT) [ms]
direct	0.993	85.2
vpn-p2p	0.964	88.95
vpn-srv	0.811	136.8
s4t-p2p	0.405	83.38
s4t-srv	0.335	123.6

Table 15.2 Throughput and latency measurements: campus network

Topology/technology	iperf: throughput [Mbps]	ping: latency (RTT) [ms]
direct	92.5	0.23
vpn-p2p	88.02	0.725
vpn-srv	81.75	1.501
s4t-p2p	10.65	1.813
s4t-srv	4.017	3.579

The first column indicates the kind of technology employed and under which topology: in particular, *direct* refers to tests between two hosts directly connected, over WAN or LAN, respectively. The *vpn-p2p* abbreviation refers to an OpenVPN server to which an OpenVPN client is connected, under the same roles as the two aforementioned hosts. Same happens for *s4t-p2p*, in this case with two hosts controlled by the S4T Cloud which, by first hole-punching through in the setup phase, connects one to the other directly over a WebSocket tunnel in combination with SOCAT. In the *vpn-srv* and *s4t-srv* cases, the OpenVPN server and the S4T Cloud, respectively, enable two clients to act as the two hosts, connected over a bridge which resides on the server (or the Cloud).

For the sake of comparing under the most relevant conditions, OpenVPN has been tested in TCP mode, and various parameters (e.g., MTU of the TUN interfaces, MSS for both iperf and the TCP tunnels) had already been tuned for S4T in the implementation phases, as also discussed below. As latencies are absolutely aligned in the various scenarios, or at least very predictable (e.g., natural increase for multi-hop setups, such as server-based OpenVPN or S4T Cloud-based virtual network for two clients), we will focus the discussion on the more interesting and insightful values obtained for the throughput metric, albeit it is actually latency that may be considered the most relevant metric for the use case under consideration, as it is key for near real-time (e.g., multimedia) applications, and the most reliable metric in general for embedded systems, considering that throughput is naturally more susceptible to other factors, e.g., high CPU load or RAM usage, differences in the media interface, etc.

iPerf3 works by repeatedly sending an array of *len* bytes for *time* seconds, where *len* by default is 128 kB for TCP, and the default for *time* is 10 s. Even if not reported in the table, retransmissions, equal to about 500 in case of SOCAT-based TUN-over-TCP, i.e., where the *s4t-p2p* scenario is stripped of the WS tunnel, in a 10-s period, ramp up to almost 5000 for the same time interval when also piping over WS, i.e., the full end-to-end *s4t-p2p*, thus roughly increasing retransmissions tenfold. Throughput thus decreases correspondingly to about a tenth, from almost 90 Mbps to about 10 Mbps to be precise.

This degradation in the performance with regard to throughput derives from a limited CWND (Congestion Window) on average, less than 10 kB, compared to 2–3 M under ideal conditions, i.e., corresponding to about 90 Mbps throughput as seen in the table. The problem then lies in a relatively high number of retransmissions, leading to a comparatively small congestion window. Some workarounds exist, either to lower retransmissions in this scenario, such as disabling Nagle's algorithm, or to make TCP more aggressively reactive to retransmissions, ramping up speed more quickly, such as switching to a different congestion control policy, in particular *Scalable* [16], both of which have been applied during testing. Indeed, without the workarounds figures would be even lower.

Otherwise, it is intrinsic to how TCP works (i.e., the streaming model) the inability to fully control how TUN-inbound datagrams are encapsulated and delivered as payload in the data stream. This means that there are bound to be packet losses (and thus retransmissions) due, for instance, to the splicing of two (IP) datagrams as payload of a single (TCP) segment. When that happens, the payload cannot be passed without errors to the (outbound) TUN on the other side of the tunnel, which expects the delivery of single (whole) datagrams, and either rejects or truncates what it gets due to mismatches between the MTU and the size of such segments

In this sense, the presence of a tunnel based on WebSockets exacerbates the behavior by introducing a further chance to introduce randomness in the delivery by virtue of TCP-level decapsulation and subsequent re-encapsulation, the latter as a result of the TCP channel established over WebSockets. This as a result of trading off raw performance, at the price of high application-level complexity and an (internal) ad-hoc architecture, as is the case for OpenVPN, with the simplicity and flexibility of off-the-shelf tools acting as separate subsystems and taking care of different facets of the communication model, in line with the UNIX philosophy of using one (good) tool for each job.

Fortunately results are still absolutely in the same league as, e.g., OpenVPN, and mostly decent in typical (non-real-time) edge scenarios, and ensure the viability of such a solution for, e.g., embedded systems, where the throughput of any modular solution, when fully in userspace, is typically lower anyway, since it may be capped by the CPU maxing out, as is the case for the Node.js-based reverse tunneling, possibly due to the current limitations of the V8 engine under MIPS.

When considering the case of WAN-level connectivity, e.g., featuring significantly lower bandwidth and naturally higher delay, it may be noticed that throughput for the S4T-based setups degrades less sharply, only down to about 50%, as can be seen in Table 15.1. Moreover, there is room for improvement considering that, compared to plain TCP sockets, WebSockets support delimiting payloads according to a specific *message*-oriented semantics for delivery, a facility which may indeed be exploited in this very sense.

15.5 Conclusions

Our opinion is that a novel approach is required for IoT and Cloud integration, and models and mechanisms which are agnostic to field deployments and topologies are essential to IoT infrastructure management and service provisioning.

Our approach to Cloud-enabled virtual networking for IoT tries to provide a blueprint for a combined solution, where VPN-like behavior, albeit the most easily advertised functionality and the one most easily picked up for a comparison, is actually just one out of many features. Other useful ones include always-on centralized control by means of WS-based command streams, bypassing restrictive firewall policies by piggybacking onto HTTP, relaying traffic through the Cloud for NAT traversal, Cloud-initiated hole punching to support server-less star and tree topologies for peer-to-peer networks, or even exposing internal services through reverse tunneling.

Moreover, functional requirements aside, performance has been shown to be mostly comparable and in all cases absolutely acceptable considering inherent limitations of the hardware platforms under consideration.

In terms of performance, the aforementioned considerations about the results will be the starting point for further improvements to the design, in particular resorting to facilities such as the WebSockets message-based semantics to partly overcome limitations intrinsic to the communication model, or even modifying some of the system-level tools, e.g., SOCAT, by employing low-overhead simple bytestream-oriented protocols on top of TCP, such as SLIP [17], to mark packet boundaries within the (TCP) payload.

References

1. N.M.K. Chowdhury, R. Boutaba, A survey of network virtualization. Comput. Netw. **54**(5), 862–876 (2010)
2. A. Fischer, J. Botero, M. Till Beck, H. de Meer, X. Hesselbach, Virtual network embedding: a survey. IEEE Commun. Surv. Tutorials **15**(4), 1888–1906 (Fourth 2013)
3. S. Distefano, G. Merlino, A. Puliafito, Device-centric sensing: an alternative to data-centric approaches. IEEE Syst. J. **11**, 231–241 (2015)
4. S. Distefano, G. Merlino, A. Puliafito, Sensing and actuation as a service: a new development for clouds, in *2012 11th IEEE International Symposium on Network Computing and Applications (NCA)*, August 2012, pp. 272–275
5. T. Oberstein, A. Goedde, The web application messaging protocol. Internet-Draft draft-oberstet-hybi-tavendo-wamp-02, IETF Secretariat October 2015
6. I. Fette, A. Melnikov, The websocket protocol. RFC 6455, last visited on 2/7/2018
7. G. Merlino, D. Bruneo, S. Distefano, F. Longo, A. Puliafito, Enabling mechanisms for cloud-based network virtualization in IoT (2015), pp. 268–273. https://doi.org/10.1109/WF-IoT.2015.7389064
8. R. Mahy, P. Matthews, J. Rosenberg, Traversal using relays around Nat (TURN): relay extensions to session traversal utilities for Nat (STUN). RFC 5766, last visited on 2/7/2018
9. VPN-WS. https://github.com/unbit/vpn-ws
10. S. Hanks, T. Li, D. Farinacci, P. Traina, Generic routing encapsulation over IPv4 networks. RFC 1702, RFC Editor October 1994
11. P. Srisuresh, B. Ford, D. Kegel, State of peer-to-peer (P2P) communication across network address translators (NATs). RFC 5128, last visited on 2/7/2018
12. D. Bruneo, S. Distefano, F. Longo, G. Merlino, An IoT testbed for the software defined city vision: the #SmartMe project, in *2016 IEEE International Conference on Smart Computing (SMARTCOMP)*, May 2016, pp. 1–6

13. G. Merlino, D. Bruneo, S. Distefano, F. Longo, A. Puliafito, Stack4Things: integrating IoT with openstack in a smart city context (2015)
14. G. Merlino, D. Bruneo, F. Longo, A. Puliafito, S. Distefano, Software defined cities: a novel paradigm for smart cities through IoT clouds (2015), pp. 909–916
15. A. Tirumala, F. Qin, J. Dugan, J. Ferguson, K. Gibbs, iPerf: the TCP/UDP bandwidth measurement tool (2005). http://software.es.net/iperf/
16. T. Kelly, Scalable TCP: improving performance in highspeed wide area networks. SIGCOMM Comput. Commun. Rev. 33(2), 83–91 (2003)
17. J. Romkey, Nonstandard for transmission of IP datagrams over serial lines: SLIP. STD 47, RFC Editor, last visited on 2/7/2018

Part IV
Tools Development for the Analysis of Specific Areas of Interests

Chapter 16
Markovian Performance Evaluation with BuTools

Gábor Horváth and Miklós Telek

16.1 Introduction

Most researchers have their own set of tools that they use for the everyday research activity. Collaboration between researchers can sometimes be difficult because everybody uses his/her own set of tools and everybody has his/her own preference of mathematical framework or programming language.

The authors of this paper had faced the same problem several years ago. They were working on very similar area, but the collaboration was difficult because everybody was sticking to his own set of tools. To address this issue, the first version of BuTools has been released in 2012 with the contribution of many colleges and students.[1] BuTools turned out to be very useful, but had some drawbacks: the source code quality was not homogeneous, and the feature parity between the three supported mathematical environments was only partial.

The aim of the second version was to address these issues. Almost every function has been rewritten from the ground up with efficiency and usability being the first

---

[1] This version of BuTools was available on the internet and announced through some professional mailing lists, but never got published as a tool paper. A non-exhaustive list of contributors include: Levente Bodrog, Peter Buchholz, Armin Heindl, András Horváth, István Kolossváry, András Mészáros, Zoltán Németh, János Papp Philipp Reinecke, Miklós Vécsei.

G. Horváth (✉)
Budapest University of Technology and Economics, Department of Networked Systems and Services, Budapest, Hungary
e-mail: ghorvath@hit.bme.hu

M. Telek
MTA-BME Information Systems Research Group, Budapest, Hungary
e-mail: telek@hit.bme.hu

© Springer International Publishing AG, part of Springer Nature 2019
A. Puliafito, K. S. Trivedi (eds.), *Systems Modeling: Methodologies and Tools*,
EAI/Springer Innovations in Communication and Computing,
https://doi.org/10.1007/978-3-319-92378-9_16

priorities and has been supplemented by unit tests. A special framework has been developed to generate the documentation, the examples, and the test scripts for the three supported environments automatically from a common source.

The second version[2,3] has been finalized in September, 2015, and only very small changes were made since then. BuTools V2 is being used by our research group in the everyday work with satisfaction. The goal of this paper is to introduce this toolbox and demonstrate its capabilities in the hope that others find it useful as well and make the results presented in the related literature easily accessible for practical computations.

16.2 Related Work

Many tools exist to support specific areas of Markovian performance modeling. There are tools available for MAP fitting (KPC toolbox for MATLAB [9], mapfit package for R [23], PhFit [5], and EMPHT [3] written in C), for the solution of Markov chains with special structures (SMCSolver [4], MAMSolver [28]) and for the matrix-analytic solution of single-type and multi-type queues (Q-MAM [26]), just to mention a few.

These software tools are all valuable contributions to the research community. However, they all focus on specific areas and support only specific mathematical frameworks. BuTools, on the other hand, aims to provide a more complete solution. It covers many areas related to Markovian performance modeling, and in addition to implementing several complex and unique algorithms, it also contains the basic functionality.

16.3 Installation, Basic Concepts

BuTools is portable, no installation is needed. The packages of BuTools can be loaded individually, but there are convenience functions available to load every package in a single step as well. If BuTools is located in directory <BTDir>, all BuTools packages can be loaded by

- `run('<BTDir>/Matlab/BuToolsInit.m')` in Matlab,
- `%run "<BTDir>/Python/BuToolsInit"` in an IPython console,
- `AppendTo[$Path,"<BTDir>/Mathematica"]; <<BuTools`` in Mathematica.

There are three global variables used by BuTools, summarized in Table 16.1.

---

[2]The homepage of BuTools is http://webspn.hit.bme.hu/~butools.

[3]The source code repository is located at https://github.com/ghorvath78/butools.

Table 16.1 Global variables in BuTools

Name in MATLAB	Name in mathematica	Name in python	Default value
BuTools Verbose	BuTools 'Verbose	butools. verbose	False
BuTools CheckInput	BuTools 'CheckInput	butools. checkInput	True
BuTools CheckPrecision	BuTools 'CheckPrecision	butools. checkPrecision	10^{-12}

Setting `verbose` to True allows the functions to print as many useful messages to the output console as possible. Turning it off avoids bloating the console. The default value is False, but for the examples of the reference documentation we have set it to True.

If `checkInput` is set to True, the functions of BuTools perform as many checks on the input parameters as possible. This can be very useful to recognize typos as soon as possible, but can be a waste of computational effort in case of a computationally demanding application.

The `checkPrecision` serves as the tolerance when the validity of the input parameters are checked.

16.4   Working with PH Distributions

Continuous time phase-type (PH) distributions [18] are characterized by two parameters, the initial probability vector α and the transient generator matrix of a continuous time (transient) Markov chain, denoted by $\mathbf{A}$. The PH distribution represents the absorption time of this transient Markov chain starting from α. The cumulative distribution function (cdf) is $F_{PH}(t) = 1 - \alpha e^{\mathbf{A}t}\mathbb{1}$, where $\mathbb{1}$ is the column vector of ones.

Matrix exponential (ME) distributions [2] are the generalizations of PH distributions. Formally, the cdf is $F_{ME}(t) = 1 - b e^{\mathbf{B}t} e$ and all further formulas for the statistical quantities are very similar to the ones of PH distributions, however, b, $\mathbf{B}$, and e can hold general numbers, the entries do not have to be valid probabilities, or transition rates. ME distributions therefore lack the simple stochastic interpretation that PH distributions have. Vector b is called "starting operator," matrix $\mathbf{B}$ is the "process rate operator" and vector e is the "summing operator." BuTools uses a special form of ME distributions, where the summing operator is a vector of ones, thus $e = \mathbb{1}$. This is not a restriction, as all ME distributions defined with general summing operator can be easily transformed to this representation [19]. Assume we have an ME distribution in the general form with parameters b, $\mathbf{B}$, e. The necessary similarity transform is obtained by calling the T = SimilarityMatrixForVectors(e,$\mathbb{1}$) procedure of the BuTools `reptrans` package. The parameters of the ME distribution used by all related BuTools tools can be calculated by $b' = b \cdot \mathbf{T}^{-1}$ and $\mathbf{B}' = \mathbf{TBT}^{-1}$.

BuTools provides several tools for both distribution classes in the `ph` package. Of course, functions for obtaining the cdf, the probability density function (pdf), and the moments are available as well as functions to check the validity of the representations.

With the inverse characterization tools BuTools can create APH distributions from any two moments and from any three moments by the `APHFrom2Moments` and `APHFrom3Moments` functions, the size of the necessary representation is determined automatically [6]. Furthermore, by the `PH2From3Moments` and `PH3From5Moments` functions order 2 and order 3 PH distributions can be obtained from 3 and 5 moments, respectively, if the given moments are feasible with PH(2) and PH(3) distributions [15, 29]. An interesting procedure is `MEFromMoments` [32] that returns an order N ME distribution from any $2N - 1$ moments, however, there is no guarantee that the result is a valid ME distribution. To check that the density is non-negative for all points it is possible to call the `CheckMEPositiveDensity` function afterwards (this is a very non-trivial procedure, which relies on the transformation to monocyclic representation) [27].

An other category of functions allow transformations between various PH and ME representations. `CanonicalFromPH2` creates an order 2, `CanonicalFromPH3` an order 3 canonical representation from any PH(2) and PH(3) distributions (potentially given by non-Markovian representation). The `PHFromME` function tries to find a PH(N) representation for the given ME(N) one by applying elementary similarity transformations iteratively (note, however, that this function is not able to increase the order in the hope for a Markovian representation).

One of the most valuable tools in the `ph` package is the `MonocyclicPHFromME` function, which transforms *any* ME distributions (that fulfil the eigenvalue constraint—that is, eigenvalues with maximal real part are real—and do not touch the x axis apart from the origin) to a PH distribution [22, 27]. The required size of the representation is determined automatically. The resulting PH distribution is returned in a monocyclic representation.

With a useful set of functions it is possible to analyze the redundancy of PH distributions and to obtain the minimal representation. `MEOrder` can return the order of the PH/ME based on the analysis of the parameters of the distribution, while `MEOrderFromMoments` returns the ME order necessary to realize the moments given. Several properties of various systems can be characterized through Laplace transform expressions, from which the moments are easy to obtain. From these moments, `MEOrderFromMoments` can tell if there is a matrix-exponential-like (ME-like) behavior in the background, and if the answer is yes, what is the order of that ME distribution. Function `MinimalRepFromME` gives the minimal representation of the given ME distribution; thus, the ME distribution returned is the same as the input, but can be smaller.

The `dph` package provides similar tools for the discrete counterparts of PH and ME distributions, the discrete PH (DPH) distributions, and matrix-geometric (MG) distributions. The basic set of functions to obtain the moments, the probability mass function, and the cdf are available, of course, however, the set of inverse characterization and representation transformation tools are less comprehensive than in the continuous case due to the lack of related research results. BuTools can create order-2 and order-3 DPH distributions from 3 and 5 moments (DPH2From3Moments and DPH3From5Moments based on [25] and [17]), and the moment matching method of [32] has also been adapted to the discrete case (MGFromMoments). Unfortunately, flexible order procedures like APHFrom3Moments for DPH are not available in the literature yet.

Transforming DPHs to canonical forms for the order-2 and order-3 cases are possible (CanonicalFromDPH2 and CanonicalFromDPH3), and the iterative transformation-based DPHFromMG is also included, but the discrete equivalent to monocyclic representation is unfortunately unknown; hence, we cannot transform any MG distribution to DPH yet.

Finally, both the ph and dph packages contain functions to generate random PH and DPH distributions (RandomPH and RandomDPH), and functions to generate random samples from PH and DPH distributions (SamplesFromPH and SamplesFromDPH) for simulation purposes.

The moment matching and representation transformation capabilities of the ph package are demonstrated through two examples depicted in Figs. 16.1 and 16.2. The example in Fig. 16.1 (in MATLAB) starts by creating an ME distribution based on five moments. The MEFromMoments method always returns a vector—matrix pair $(v, \mathbf{H})$ yielding the target moments, but the result is not always a valid distribution (the density function can be negative if the target moments cannot be realized by an ME distribution of the given order). The PHFromME procedure managed to transform the ME representation to a valid PH representation characterized by $(\beta, \mathbf{B})$. The transformation matrix relating $(v, \mathbf{H})$ and $(\beta, \mathbf{B})$ is calculated by SimilarityMatrix, and the last two lines show that the two rep-

Fig. 16.1 Example for the PH package, part 1

```
>> [v, H] = MEFromMoments([0.9, 2.5, 20, 500, 22000]);
>> disp(v);
        0.33333        0.33333        0.33333
>> disp(H);
       -3.4978         0.003242      -0.91912
        3.2628        -0.88868        0.3514
       -4.0036         0.14794       -1.0921
>> [beta, B] = PHFromME(v, H);
>> disp(beta);
        0.99798        0.0010399      0.00097586
>> disp(B);
       -4.0688         1.8513         0.0014997
        0.92775       -1.3039         0.00081193
        0.0056188      0.097275      -0.10593
>> T = SimilarityMatrix(H, B);
>> norm(T*B - H*T)
        1.8229e-14
>> norm(beta - v*T)
        1.0303e-15
```

```
>> v = [0.2 0.3 0.5];
>> H = [-1 0 0; 0 -3 1; 0 -1 -3];
>> [beta, B] = PHFromME(v, H);
>> disp(beta);
      -0.10542        -0.043052            1.1485
>> disp(B);
       -2.668          2.7577          -0.096015
      0.044046        -1.5887            1.5886
      0.41081         -0.10448          -2.7433
>> [beta, B] = MonocyclicPHFromME(v, H);
>> disp(beta);
0.0055089  0.0090301  0.016938  0.015216  0.0053543  0.0087356  0.052486  0.22657  0.66016
>> disp(B);
-1        1        0        0        0        0        0        0        0
 0  -2.4226   2.4226        0        0        0        0        0        0
 0        0  -2.4226   2.4226        0        0        0        0        0
 0  0.26232        0  -2.4226   2.1603        0        0        0        0
 0        0        0        0  -4.2414   4.2414        0        0        0
 0        0        0        0        0  -4.2414   4.2414        0        0
 0        0        0        0        0        0  -4.2414   4.2414        0
 0        0        0        0        0        0        0  -4.2414   4.2414
 0        0        0        0        0        0        0        0  -4.2414
>> T = SimilarityMatrix(H, B);
>> norm(T*B - H*T)
    1.0315e-15
>> norm(v*T - beta)
    4.8962e-16
```

Fig. 16.2 Example for the PH package, part 2

resentations are identical indeed. From the example in Fig. 16.2 the transformation of an ME representation to a PH one failed, PHFromME returned an invalid initial probability vector and an invalid generator matrix. MonocyclicPHFromME, however, managed to return a valid PH representation, although the order has been increased from 3 to 9. The three feedback Erlang blocks can be clearly identified in the resulting generator matrix.

16.5   Tools for MAPs

Continuous time Markovian arrival processes (MAPs, [18]) are commonly characterized by two matrices, D_0 and D_1. Arrivals by a MAP are modulated by a background continuous time Markov chain with generator $D = D_0 + D_1$. Markov chain transitions in D_0 (D_1) do not generate (generate) arrival events. As a result MAPs are capable of generating correlated arrivals.

Rational arrival processes (RAPs, also known as matrix-exponential processes, MEPs) are generalizations of MAPs [1]. Formally, all formulas for the statistical quantities are the same to the ones of MAPs. However, both D_0 and D_1 can hold negative real numbers, the entries do not have to be valid transition rates. RAPs therefore lack the simple stochastic interpretation that MAPs have.

Both MAPs and RAPs can be generalized to multi-type arrival processes. If there are K different arrival types, marked MAPs (MMAPs) and marked RAPs (MRAPs) defined by matrices $D_0, \ldots, D_K$ are able to describe the multi-type arrival process.

BuTools provides several tools for MAPs, RAPs, and their marked variants in the map package.

With the appropriate functions BuTools can return basic properties like the marginal distribution (the parameters of the corresponding PH distribution), the marginal moments, the lag auto-correlations, and the lag-k joint moments [30] of MAPs and MMAPs.

With the set of inverse characterization tools it is possible to obtain order N RAPs or MAPs from $2N - 1$ marginal moments and $(N - 1)^2$ lag-1 joint moments (RAPFromMoments and MRAPFromMoments, using the method of [30]); or from $2N - 1$ marginal moments and $2N - 3$ auto-correlations (RAPFromMomentsAndCorrelations, based on [21]). The method for creating an order-2 MAP from 3 moments and 1 correlation parameter published in [7] is implemented by the MAP2FromMoments function. The only flexible matching procedure (that can adjust the order of the result automatically, based on the input parameters) is MAPFromFewMomentsAndCorrelations, implementing [12].

As for representation transformation, BuTools is able to transform a RAP(2) to a canonical form (CanonicalFromMAP2), transform a MRAP(N) to MMAP(N) (MAPFromRAP and MMAPFromMRAP, by successive similarity transformations, achieving a MAP is not guaranteed), and can minimize a RAP representation with functions MinimalRepFromRAP and MinimalRepFromMRAP [8].

The dmap package intends to provide the same functionality for discrete time arrival processes (DMAPs and DRAPs) and their marked variants (DMMAPs and DMRAPs), however, several results present for MAPs and RAPs are not available for DMAPs and DRAPs in the literature yet.

From the inverse characterization tools only DRAPFromMoments, DMRAPFromMoments, and DMAP2FromMoments are available (see [20] for the latter one).

Both the map and dmap packages contain functions to generate random MAPs, MMAPs, DMAPs, DMMAPs and to generate random samples from these processes.

16.6 Fitting Tools

16.6.1 The trace Package

The ph, dph, map, and dmap packages provide several functions to obtain PH distributions and MAPs from statistical quantities, like moments, auto-correlations, and joint moments. The trace package has tools to obtain these kinds of quantities from empirical data traces.

The traces are vectors consisting of measurements. After loading them from a file, cdf, pdf, moments, joint moments, and lag-k auto-correlations can be computed by invoking the CdfFromTrace, PdfFromTrace, MarginalMomentsFromTrace, LagkJointMomentsFromTrace, and LagCorrelationsFromTrace functions.

Most of these functions can cope with weighted traces as well, where each measurement data is supplemented by a weight.

16.6.2 Likelihood Based Fitting

BuTools has a `fitting` package that contains two kinds of functions: procedures for likelihood (EM) based fitting and tools to evaluate the result of the fitting (distance functions).

The `PHFromTrace` is the implementation of the G-FIT procedure [31] to create a hyper-Erlang distribution by EM-algorithm. G-FIT is one of the best performing PH fitting methods at the moment. While MATLAB, Mathematica, and Python are known for not being efficient for such computationally demanding algorithms, BuTools has a reasonably fast, vectorized implementation capable of processing traces with millions of data.

The `MAPFromTrace` function implements [14], which is similar to G-FIT. The MAP it creates has Erlang components, and a switching probability matrix determining the order of these Erlang components providing the subsequent inter-arrival times. Note, however, that this fitting procedure is much slower than the one for PH fitting.

The likelihood of a PH distribution or a MAP regarding a trace can be evaluated by `LikelihoodFromTrace`.

The functions `SquaredDifference` and `RelativeEntropy` measure the difference between two vectors (e.g., probability mass functions, lag-k auto-correlations, etc.), while `EmpiricalSquaredDifference` and `EmpiricalRelativeEntropy` are the equivalents to be used for continuous quantities (e.g., for pdf or cdf of continuous time PH variables, where they are given by a number of points only).

16.6.3 Application Example

The usage of BuTools for trace fitting is demonstrated in Fig. 16.3. The first line loads a trace file consisting of 1.78 million inter-arrival time samples. The next line calculates the marginal moments of the trace. Then, a PH(3) distribution is created by matching five moments, finally a PH(5) is obtained by fitting (G-FIT). The fitting step took 113 s on a PC with a 3.4 GHz CPU and 4 GB of RAM. After obtaining the PH distributions, the approximations are evaluated. First the moments are compared, then the likelihood.

After the comparison, the density functions are obtained and plotted. (In case of the PH distributions, the `IntervalPdfFromPH` function is used, which, instead of evaluating the pdf at the given points, returns the probability of falling into intervals divided by the interval lengths. This is the correct way to compare it with the empirical pdf of the trace.) The result is depicted in Fig. 16.4.

```
>> trace = dlmread('lbltcp3_iat.txt');
>> trmoms = MarginalMomentsFromTrace(trace,5);
>> [alpha3,A3] = PH3From5Moments(trmoms(1:5));
>> [alpha5,A5] = PHFromTrace(trace, 5);
>> disp(trmoms);
            1 2.942 16.84 150.73 1876.8
>> disp(MomentsFromPH(alpha3,A3));
            1 2.942 16.84 150.73 1876.8
>> disp(MomentsFromPH(alpha5,A5,5));
            1 2.8827 15.074 112.2 1062.3
>> disp(LikelihoodFromTrace(trace,alpha3,A3));
        -0.94802
>> disp(LikelihoodFromTrace(trace,alpha5,A5));
         -0.9343
>> [xt,yt] = PdfFromTrace (trace, (1:0.1:3));
>> [xp3,yp3] = IntervalPdfFromPH(alpha3, A3, (1:0.1:3));
>> [xp5,yp5] = IntervalPdfFromPH(alpha5, A5, (1:0.1:3));
>> plot (xt,yt,xp3,yp3,xp5,yp5);
```

Fig. 16.3 Application example for fitting

Fig. 16.4 The output of the example of Fig. 16.3

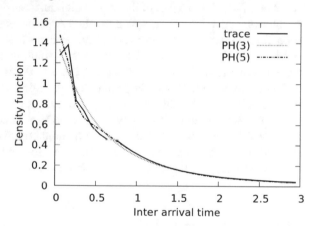

16.7 Analysis of Queues

The queues package heavily relies on the matrix-analytic methods.

16.7.1 Support for Matrix-Analytic Methods

The mam package of BuTools provides solution methods for problems based on non-linear matrix equations.

Functions solving the three-diagonal block-structured Markov chains, namely the quasi birth-death processes (QBDs), Markov chains skip-free to the left, M/G/1 type Markov chains, and Markov chains skip-free to the right, G/M/1 type Markov chains

[18] are included. The underlying algorithms are from the SMCSolver toolbox [4]. The MATLAB version of BuTools requires the presence of SMCSolver, while the Mathematica and the NumPy/IPython versions include the necessary parts of SMCSolver ported to these environments.

The fundamental matrices for QBDs are returned by the QBDFundamentalMatrices function (matrices $\mathbf{R}, \mathbf{G}$, and $\mathbf{U}$ [18]). Matrix $\mathbf{G}$, the fundamental matrix of M/G/1 type Markov chains is provided by MG1FundamentalMatrix, and the $\mathbf{R}$ matrix of G/M/1 type systems is given by GM1FundamentalMatrix. Based on the fundamental matrices the stationary solutions are provided by functions QBDStationaryDistr, MG1StationaryDistr, and GM1StationaryDistr, respectively, that return the stationary probabilities themselves. For QBDs the ingredients of the stationary matrix-geometric solution (the initial vector and the coefficient matrix) are returned by QBDSolve.

BuTools also supports the solution of continuous queueing systems, Markovian fluid flows [10]. A large part of the related literature considers *canonical* Markovian fluid flows, where the rate at which the fluid level increases or decreases is always 1. For these systems FluidFundamentalMatrices returns the most important matrices, $\mathbf{\Psi}, \mathbf{K}$, and $\mathbf{U}$ [10]. The stationary solution for the fluid level at the requested points is given by FluidStationaryDistr, and the components of the matrix-exponential solution (initial vector, matrix exponent) can be obtained by FluidSolve. For non-canonical fluid systems having non-unit fluid rates, however, it is better to use GeneralFluidSolve, which, based on the generator of the background Markov chain and the diagonal matrix of fluid rates returns the probability mass at level 0, and for positive levels the initial vector, matrix exponent and closing matrix of the matrix-exponential solution.

16.7.2 Queueing Models

Building upon the mam package, the queues package provides functions to obtain many performance measures of several queueing systems. The following queues are supported:

- The MAP/MAP/1 queue (MAPMAP1). Special cases of this queue are the PH/MAP/1, the MAP/PH/1, the PH/PH/1, etc. (see [18] and [11]).
- The QBD queue (QBDQueue). In this queue the arrival and the service process are not independent of each other, they share the same background process. Marked transitions of this background process are accompanied by a level increase, other marked transitions by a level decrease event. The MAP/MAP/1 queue is the special case of the QBD queue, however, several performance measures are more demanding to compute for the QBD queue. The implementation is based on [18] and [24].

- The MMAP[K]/PH[K]/1- FCFS queue (MMAPPH1FCFS). This is a first-come-first-served (FCFS) multi-type queue with K types of customers. Each customer type can have a different (PH) service time distribution. The solution is based on [11].
- The MMAP[K]/PH[K]/1 queue with non-preemptive and preemptive resume priority service (MMAPPH1NPPR and MMAPPH1PRPR). This is a multi-type queue with preemptive service, the efficient solution is based on a recent result [13].
- The fluid queue (FluidQueue). In this queue there is a common background Markov chain, a diagonal matrix of fluid arrival rates, and a diagonal matrix of fluid service rates in each state of the background process. The queue length and the sojourn time of the fluid drops are the two most interesting performance measures (based on [10] and [16]). This is the continuous counterpart of the QBD queues.
- The Flu/Flu queue (FluFluQueue), which is similar to the ordinary fluid queue, but the fluid input and output processes are independent, they are modulated by two separate background Markov chains. This independence is exploited in the solution, thus Flu/Flu queues are easier and faster to solve than the general fluid queues. This is the continuous counterpart of the MAP/MAP/1 queues.

The performance measures that can be obtained from these queues are summarized in Table 16.2. The abbreviation of the performance measures are

- ncMoms/flMoms: Stationary moments of the number of customers (in case of discrete queues)/fluid level (in case of fluid queues).
- ncDistr/flDistr: Stationary distribution of the number of customers/the fluid level.
- ncDistrMG/flDistrME: The parameters of the MG/ME distribution of the number of customers/fluid level.
- ncDistrDPH/flDistrPH: The parameters of the DPH/PH distribution of the number of customers/fluid level. The DPH/PH representation is obtained from the MG/ME one, which is always possible. However, in some rare cases (when a phase has a very low probability) this transformation can introduce numerical errors, hence ncDistrMG/flDistrME are safer to use.

Table 16.2 Performance measures that can be computed

Perf. meas.	QBDQueue	MAPMAP1	MMAPPH1FCFS	MMAPPH1-Prio	FluidQueue	FluFluQueue
ncMoms	✓	✓	✓	✓	✓	✓
ncDistr	✓	✓	✓	✓	✓	✓
ncDistrMG	✓	✓	–	–	✓	✓
ncDistrDPH	✓	✓	–	–	✓	✓
stMoms	✓	✓	✓	✓	✓	✓
stDistr	✓	✓	✓	✓	✓	✓
stDistrME	✓	✓	✓	–	✓	✓
stDistrPH	✓	✓	✓	–	✓	✓

– stMoms, stDistr, stDistrME, stDistrPH: the same as above for the sojourn time of the customers/fluid drops. Again, stDistrME behaves better numerically than stDistrPH.

When calling these functions, the performance measures to compute are listed in the function arguments. Several performance measures can be computed at the same function call, and BuTools will save as much computational effort as possible by avoiding repeated re-computation of some demanding steps.

16.7.3 Application Examples

The examples in Figs. 16.5 and 16.7 demonstrate how well the different packages of BuTools play nicely together. In Fig. 16.5 a two-class non-preemptive priority queue is studied. The arrival process of the low priority class is created based on three moments and the lag-1 auto-correlation of a measurement trace, while high priority customers arrive according to a Poisson process with rate 0.6. The service time distributions are obtained by matching two moments. All the performance measures are obtained by the single call of the MMAPPH1NPPR function. According to the function arguments, three moments for the number of customers, the distribution of the number of customers up to 20, and the sojourn time distribution are requested at certain points. This function returns the performance measures in the same order for both customer classes. After that, the solution is displayed either on the screen or in a plot (Fig. 16.6).

In Fig. 16.7 a fluid queue is considered. In this example both the fluid arrival and the fluid service processes are Markov modulated. After calculating the product space of the arrival and service processes by Kronecker operations, the FluidQueue function is called to compute the ME representation of the fluid

```
>> trace = dlmread('lbltcp3_iat.txt');
>> trmoms = MarginalMomentsFromTrace(trace,3);
>> tracf1 = LagCorrelationsFromTrace(trace,1);
>> [D0,D1] = MAP2FromMoments(trmoms,tracf1);
>> D2 = 0.6*eye(size(D0));
>> D0 = D0 - D2;
>> [alpha1,A1] = APHFrom2Moments([0.6,6.5]);
>> [alpha2,A2] = APHFrom2Moments([0.5,0.7]);
>> [ncm1, ncd1, std1, ncm2, ncd2, std2] = MMAPPH1NPPR({D0, D1, D2},
{alpha1, alpha2}, {A1, A2}, 'ncMoms', 3, 'ncDistr', 20, 'stDistr', 0.1:100);
>> disp(ncm1);
        57.073        8205.2    1.7778e+06
>> disp(ncm2);
        3.2657        39.861        780.53
>> plot([ncd1',ncd2']);
>> plot([std1',std2']);
```

Fig. 16.5 Example for the priority queue

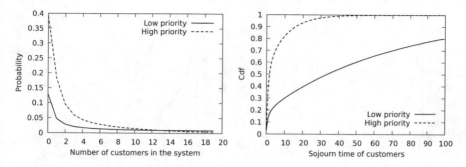

Fig. 16.6 The output of the example of Fig. 16.5

```
>> Qin = [-2 1 1; 2 -5 3; 4 0 -4];
>> Rin = diag([3 7 0]);
>> Qout = [-4 1 3; 6 -8 2; 3 7 -10];
>> Rout = diag([1 7 15]);
>> I = eye(3);
>> [alphap, Ap, betap, Bp] = FluidQueue(kron(Qin,I)+kron(I,Qout),
kron(Rin,I), kron(I,Rout), 'flDistrME', 'stDistrME');
>> disp(size(Ap));
      2      2
>> disp(size(Bp));
     18     18
>> [alphai, Ai, betai, Bi] = FluFluQueue(Qin, Rin, Qout, Rout,
false, 'flDistrME', 'stDistrME')
>> disp(size(Ai));
      2      2
>> disp(size(Bi));
      2      2
>> disp(norm(MomentsFromME(alphap,Ap,5) - MomentsFromME(alphai,Ai,5)));
     1.4325e-14
>> disp(norm(MomentsFromME(betap,Bp,5) - MomentsFromME(betai,Bi,5)));
     1.6779e-15
>> plot(PdfFromME(alphai,Ai, 0:0.01:8));
>> plot(PdfFromME(betai,Bi, 0:0.01:2));
```

Fig. 16.7 Example for the fluid queue

level and sojourn time distributions. With the parameters of the example, the order
of the fluid level distribution is 2, but the one of the sojourn time distribution
is 18. Working with such large representations can be slow and more sensitive
numerically. However, if the arrival and services are independent (as they are in
this example and in many practical cases), it is possible to use the FluFluQueue
instead of FluidQueue, since it exploits the independence and returns a much
more compact representation for the sojourn time (order 2 in this case). Note that—
despite the different order—the representations returned by FluidQueue and
FluFluQueue belong to the same distribution, as demonstrated by printing the
difference of the moments in the figure. The plots of the density functions of the
fluid level and the sojourn time are shown in Fig. 16.8.

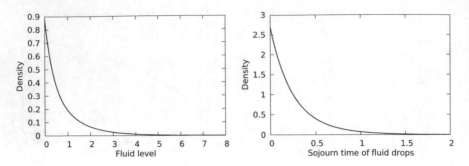

Fig. 16.8 The output of the example of Fig. 16.7

16.8 Some Further, Small Packages

16.8.1 The moments Package

Several moment expressions are being used in various publications related to PH distributions. Most works rely on the ordinary (raw) moments, while others are able to arrive to simpler formulas by introducing some alternative moment expressions, like reduced moments, normalized moments, and Hankel moments. Furthermore, in case of discrete systems, transform domain techniques make it easy to obtain the factorial moments. The moments package provides conversion routines between these moment expressions. Additionally, it provides the CheckMoments function to determine if a sequence of real numbers is a valid moment sequence (there exists a distribution with the given moments), or not.

16.8.2 The mc Package

A couple of basic functions to obtain the stationary distribution of Markov chains is mandatory in a tool devoted to Markovian performance analysis. The CTMCSolve and DTMCSolve functions return the stationary distribution of continuous and discrete time Markov chains, respectively, based on the direct (non-iterative) solution of the corresponding set of linear equations.

Two further functions, CRPSolve and DRPSolve provide the same functionality on rational processes, which are similar to Markov chains without the restrictions on the sign of the elements of the generator matrices. These functions are mostly used internally by the procedures operating on ME distributions and RAPs.

16.9 Conclusion

BuTools collects the implementations of many research results related to stochastic models with Markov background process. This way it makes several complex research results of the field easily accessible for practical application. BuTools is heavily used by our research group and found to be efficient for practical computations. The reader is encouraged to check and use BuTools, which is facilitated with online documentation and application demo.

The authors would be glad to receive any related ideas, comments, feature requests, or bug fixes.

Acknowledgements This research is supported by the ÚNKP-17-4-III New National Excellence Program of the Ministry of Human Capacities, Hungary, and by the OTKA-123914 project.

References

1. S. Asmussen, M. Bladt, Point processes with finite-dimensional probabilities. Stoch. Processes Appl. **82**(1), 127–142 (1999)
2. S. Asmussen, C. O'Cinneide, *Matrix-Exponential Distributions* (Wiley, New York, 2004)
3. S. Asmussen, O. Haggström, O. Nerman, EMPHT - a program for fitting phase-type distributions, in *Studies in Statistical Quality Control and Reliability, Mathematical Statistics) (Chalmers University and University of Göteborg*, Göteborg, 1992)
4. D. Bini, B. Meini, S. Steffé, B. Van Houdt, Structured Markov chains solver: software tools, in *Proceeding from the 2006 Workshop on Tools for Solving Structured Markov Chains* (ACM, New York, 2006), p. 14
5. A. Bobbio, A. Horváth, M. Telek, PhFit: a general phase-type fitting tool, in *International Conference on Dependable Systems and Networks, 2002. DSN 2002. Proceedings* (IEEE, New York, 2002), p. 543
6. A. Bobbio, A. Horváth, M. Telek, Matching three moments with minimal acyclic phase type distributions. Stoch. Model. **21**(2–3), 303–326 (2005)
7. L. Bodrog, A. Heindl, G. Horváth, M. Telek, A Markovian canonical form of second-order matrix-exponential processes. Eur. J. Oper. Res. **190**(2), 459–477 (2008)
8. P. Buchholz, M. Telek, On minimal representations of rational arrival processes. Ann. Oper. Res. **202**(1), 35–58 (2013)
9. G. Casale, E.Z. Zhang, E. Smirni, KPC-toolbox: simple yet effective trace fitting using Markovian arrival processes, in *Fifth International Conference on Quantitative Evaluation of Systems, 2008. QEST'08* (IEEE, New York, 2008), pp. 83–92
10. A. da Silva Soares, Fluid queues: building Upon the Analogy with QBD Processes. PhD thesis, Université Libre de Bruxelles (2005)
11. Q.-M. He, Age process, workload process, sojourn times, and waiting times in a discrete time SM[K]/PH[K]/1/FCFS queue. Queueing Syst. **49**(3–4), 363–403 (2005)
12. G. Horváth, Matching marginal moments and lag autocorrelations with MAPs, in *Proceedings of the 7th International Conference on Performance Evaluation Methodologies and Tools* (2013), pp. 59–68
13. G. Horváth, Efficient analysis of the MMAP[K]/PH[K]/1 priority queue. Eur. J. Oper. Res. **246**(1), 128–139 (2015)
14. G. Horváth, H. Okamura, A fast EM algorithm for fitting marked Markovian arrival processes with a new special structure, in *European Workshop on Performance Engineering* (Springer, Berlin, 2013), pp. 119–133

15. G. Horváth, M. Telek, On the canonical representation of phase type distributions. Perform. Eval. **66**(8), 396–409 (2009)
16. G. Horváth, M. Telek, Sojourn times in fluid queues with independent and dependent input and output processes. Perform. Eval. **79**, 160–181 (2014)
17. I. Horváth, J. Papp, M. Telek, On the canonical representation of order 3 discrete phase type distributions. Electron. Notes Theor. Comput. Sci. **318**, 143–158 (2015)
18. G. Latouche, V. Ramaswami, *Introduction to Matrix Analytic Methods in Stochastic Modeling*, vol. 5 (Siam, Philadelphia, 1999)
19. L. Lipsky, *Queueing Theory: A Linear Algebraic Approach* (Springer Science & Business Media, Berlin, 2008)
20. A. Mészáros, M. Telek, Canonical representation of discrete order 2 MAP and RAP, in *European Workshop on Performance Engineering* (Springer, Berlin, 2013), pp. 89–103
21. K. Mitchell, A. van de Liefvoort, Approximation models of feed-forward G/G/1/N queueing networks with correlated arrivals. Perform. Eval. **51**(2), 137–152 (2003)
22. Ş. Mocanu, C. Commault, Sparse representations of phase-type distributions. Stoch. Model. **15**(4), 759–778 (1999)
23. H. Okamura, T. Dohi, mapfit: an R-based tool for PH/MAP parameter estimation, in *International Conference on Quantitative Evaluation of Systems* (Springer, Berlin, 2015), pp. 105–112
24. T. Ozawa, Sojourn time distributions in the queue defined by a general QBD process. Queueing Syst. **53**(4), 203–211 (2006)
25. J. Papp, M. Telek, Canonical representation of discrete phase type distributions of order 2 and 3, in *Proceedings of UK Performance Evaluation Workshop, UKPEW*, vol. 2013 (2013)
26. J.F. Pérez, J. Van Velthoven, B. Van Houdt, Q-MAM: a tool for solving infinite queues using matrix-analytic methods, in *Proceedings of the 3rd International Conference on Performance Evaluation Methodologies and Tools* (2008), p. 16
27. P. Reinecke, M. Telek, Does a given vector-matrix pair correspond to a PH distribution? Perform. Eval. **81**, 40–51 (2014)
28. A. Riska, E. Smirni, MAMSolver: a matrix analytic methods tool, in *TOOLS '02: Proceedings of the 12th International Conference on Computer Performance Evaluation, Modelling Techniques and Tools* (Springer, London, 2002), pp. 205–211
29. M. Telek, A. Heindl, Moment bounds for acyclic discrete and continuous phase type distributions of second order, in *Proceedings of UK Performance Evaluation Workshop* (2002)
30. M. Telek, G. Horváth, A minimal representation of Markov arrival processes and a moments matching method. Perform. Eval. **64**(9), 1153–1168 (2007)
31. A. Thummler, P. Buchholz, M. Telek, A novel approach for fitting probability distributions to real trace data with the EM algorithm, in *2005 International Conference on Dependable Systems and Networks (DSN'05)* (IEEE, New York, 2005), pp. 712–721
32. A. Van de Liefvoort, The moment problem for continuous distributions. Unpublished technical report, University of Missouri, WP-CM-1990-02 (1990)

Chapter 17
J2CBROKER as a Service: A Service Broker Simulation Tool Integrated in OpenStack Environment

Riccardo Di Pietro, Maurizio Giacobbe, Carlo Puliafito, and Marco Scarpa

17.1 Introduction

Cloud computing is radically enhancing enterprises productivity, thanks to its elasticity, flexibility, efficiency, and on-demand and pay-as-you-go nature. Today, it is possible to benefit from the Cloud by deploying it in different service models, such as *Infrastructure* (IaaS), *Platform* (PaaS), and *Software* (SaaS), but many others are coming out from the market (e.g., hybrid solutions, microservices). Moreover, services may be offered by *Cloud Service Providers* (CSPs) in private Data Centers (DCs), i.e., *private Clouds*, or they can be commercially offered to customers, which is known as *public Clouds*. Yet, it is possible that public and private Clouds are combined to form *hybrid Clouds*. Despite its disruptive nature, there is the need for timely, repeatable, and controllable methodologies that evaluate the conceived Cloud algorithms and policies prior to their actual development and deployment. Simulation-based environments play a fundamental role in this direction. First of all, they allow to easily set environment variables and parameters, define models, reproduce tests, and analyze the obtained results (textual and/or graphical). More

R. Di Pietro
Università di Catania, Catania, Italy
Università di Messina, Messina, Italy
e-mail: rdipietro@unict.it,rdipietro@unime.it

M. Giacobbe (✉) · M. Scarpa
Università di Messina, Messina, Italy
e-mail: mgiacobbe@unime.it; mscarpa@unime.it

C. Puliafito
Università di Firenze, Firenze, Italy
Università di Pisa, Pisa, Italy
e-mail: carlo.puliafito@unifi.it;carlo.puliafito@ing.unipi.it

© Springer International Publishing AG, part of Springer Nature 2019   269
A. Puliafito, K. S. Trivedi (eds.), *Systems Modeling: Methodologies and Tools*,
EAI/Springer Innovations in Communication and Computing,
https://doi.org/10.1007/978-3-319-92378-9_17

importantly, the use of simulation-based approaches in Cloud environments is often a necessity, since the access to the actual infrastructure would incur payments in real currency (pay-as-you-go service model). Thus, simulation tools can significantly benefit Cloud customers by allowing them to test their services in a repeatable and controllable environment, without paying for the access to the Cloud. On the other hand, simulators can allow CSPs to evaluate, e.g., where to allocate computational resources according to varying performance, workload conditions, and monetary cost distributions. As a result, in the absence of such simulation-based environments, both Cloud customers and providers risk to make serious mistakes of assessment or to refer to non-objective evaluations, thus resulting in inefficient service performance and economic losses.

17.2 Motivations

Recent years have seen the success of the Cloud computing paradigm and the continuously increasing number of service providers and available services.

Within the Digital Single Market strategy [20] (i.e., *European Commission* priority in order to achieve better online access to digital goods and services) the Cloud computing plays a key role through the *data-driven innovation* initiatives, *data ownership, access* and *usability ownership, portability of data*, and *switching* of service providers. In this complex context, customers' discovery of the services and selection of the one which best suits their needs is not a trivial issue and might be very time-consuming and/or ineffective:*Cloud service brokerage* might help to overcome this problem. According to *MarketsandMarkets* [12], a market research firm, the Cloud service brokerage and Enablement market size is estimated to grow from USD 7.44 billion in 2016 to USD 26.71 billion by 2021.

Cloud Service Broker (CSB) is an additional computing layer which acts as an intermediary between service customers on one side and service providers on the other. *Gartner*, the world leading information technology research and advisory company, identifies three areas (i.e., aggregation, integration, and customization) in which Cloud brokerage might play an important role toward service customer, but also service providers. *Aggregation* gives the possibility to manage multiple services, possibly from different providers and present them as a unified service. This is not always easy because of the complex relationships and agreements among providers. The *integration* purpose is to make applications, which are independent at first, work nicely together and cooperate in order to fulfill the customer's needs. *Customization* consists in the tweaking of services in order to best suit users' needs.

Other applications of CSB, which still come under the concept of service selection, are the *ranking* of services according to parameters provided by users (e.g., services ordered by cost) and the *selection* of the best data center (or site), among the N available, to execute a certain job. For example, CSB plays an important role in legislation compliance and QoS management of Cloud services [4]. Some of the most important CSB companies are, in alphabetical order: *Appirio*, *ComputeNext*, and *Dell Boomi*.

17.3 Related Work

Due to the great interest and importance played by Cloud Service Brokerage, several works have been carried out in this field, surveying the possible approaches and algorithms for the service selection [1, 15, 17, 18] but also the Cloud simulators that can be used to evaluate CSBs performance [9, 16].

Probably, **CloudSim** [3] is the most popular and complete framework for modeling and simulating Cloud environments. It was developed at the *CLOUd computing and Distributed Systems (CLOUDS) Laboratory*, in the *University of Melbourne, Australia*. It is open source, entirely written in Java and provides basic classes for modeling data centers, users, brokers, computational resources, policies, and virtual machines.

CloudSim is built on top of another open source framework, namely **GridSim** [2], which was also developed at the CLOUDS Laboratory. GridSim is written in Java and essentially presents the same functionalities as CloudSim but with the difference that it is used for large-scale Grid systems and P2P networks.

Thanks to its success, CloudSim has been extended by researchers, and thus other products using it as their core have been developed. The most important example of these is **CloudAnalyst** [24], which is a Java-based simulation tool. The main feature of CloudAnalyst is the presence of an intuitive Graphical User Interface (GUI), which makes it easy to set up and run the simulation. The results of the simulation are then returned in the form of charts and tables, which is very important considering their complexity and variety.

GreenCloud [11] is an open source simulation environment built as an extension of the Ns2 network simulator. What distinguishes this environment from all the others is the focus on the energy consumed by all the components of a DC in a simulated Cloud environment. Indeed, DCs require a great amount of energy, which greatly impacts the overall operational costs. GreenCloud is a packet-level simulator, meaning that protocol processing is performed whenever a packet is to be transmitted. On the other hand, CloudSim and CloudAnalyst are event-based simulators; hence, they do not individually process packets but capture the overall effect of interactions instead. The result is that GreenCloud is slower in simulating, but it is more accurate as well.

iCanCloud [14] is an open source simulation platform entirely written in C++ and developed as an extension of the *OMNET* network simulator. The main purpose of iCanCloud is to estimate the trade-off between costs and performance, thus to help users make their decisions in order to optimize it. Besides, also iCanCloud provides a very complete and user-friendly GUI. Finally, iCanCloud has a feature that the aforementioned environments do not have: if there is a cluster of nodes available to carry out an experiment, it is possible to perform a parallel simulation among them. The only requirement is for the nodes to have MPI installed.

However, even if many simulator tools that can be used for studying Cloud systems, we cannot establish what is the best simulator to use because the evaluation depends upon actual requirement.

17.4   The J2CBROKER Simulation Tool

Based on the above considerations, we created a tool capable of simulating different cooperative Cloud Brokerage scenarios across different metrics and evaluation criteria included in *models* by using "multi-criteria" strategies. As the name implies, *"Java Json Cloud BROKER"* [6] is totally written by using JAVA language and JSON documents. More specifically, the main goal is to provide a simulation-based tool that: (1) dynamically manages JSON documents as inputs simulating both requests and offers by CSPs; (2) calculates the best choices (i.e., offers) on the basis of specific parameters through different multi-criteria engines (i.e., multi-criteria algorithms implemented in JAVA language); (3) provides the resulting best offers of its calculation as outputs, both in the form of JSON documents and *on-screen* (results can be also provided in many other forms, such as *CSV* files and diagrams). J2CBROKER service is based on a JAVA client–server architecture integrated in an OpenStack Cloud infrastructure. This integration allows J2CBROKER to be owned and hosted by a service provider and to be offered to consumers on-demand. More details about the integration between J2CBROKER and OpenStack services are introduced later. J2CBROKER uses a stateless RESTful approach for its communication. Moreover, as proposed in [5], the communication protocol used between client and server uses a data protection mechanism which combines both symmetric (AES256) and asymmetric encryption (RSA) in a smart way. Figure 17.1 shows the general architecture of the proposed *J2CBROKER* Simulation Tool.

In J2CBROKER, we introduced the concept of **Model**. A Model is essentially a JSON file which contains all the *input* metrics describing the basic characteristics of a given scenario.

More specifically, a Model identifies what needs to be simulated and how. Figure 17.2 shows several metrics and *Service Measurement Index (SMI)—Key Performance Indicators (KPIs)* that are possible to be considered in order to have different multi-criteria Models. For each Model, the architecture provides two specific components: the related Model Simulator at the client side, and the related Model Engine at the server side. The architecture has been developed in order to simulate several possible scenarios, each one defined by a Model. To this end, referring to Fig. 17.1, we mainly distinguish two blocks: the **Data Set Simulator** at the client side and the **Brokerage Engine** at the server side, both containing several Models (respectively Simulators and Engines). These blocks will be discussed in the following.

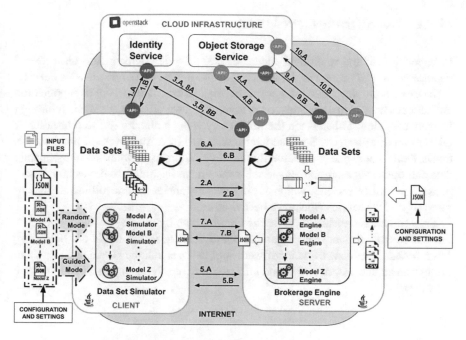

Fig. 17.1 A general architecture of the *J2CBROKER* simulation tool

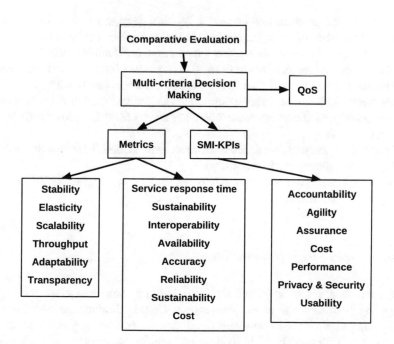

Fig. 17.2 Metrics and SMI-KPIs to realize different multi-criteria models

17.4.1   SaaS Deployment Model

In opposition to the traditional model of software deployment, the term *"SaaS deployment model"* refers to the installation and delivery of *Software as a Service*. *Software as a Service (SaaS)* is a software distribution model where the application and services are run in a centralized environment in which users access on it through the network, almost always via the Internet by using a client (e.g., web browser or GUI) as an interface. SaaS model is characterized by a **multi-tenant** architecture, that is, there is only one application that serves multiple users while keeping separate data and operating environments. SaaS deployment is similar to the establishment phase of a utility service, which is followed by metering and billing at regular intervals, for the services that have been delivered.

SaaS model is considered to be the winning one by all major vendor software. For this reason, the most important software vendors in the world are delivering *"as a service"* versions of their software applications, and they are delivering them through an ad hoc proprietary Cloud infrastructure or relying on other Cloud service providers.

17.4.2   SaaS Benefits

There are several practical and economic benefits pushing the SaaS Cloud model. From the user point of view, the big benefit is that he does not have to face a large cash outlay for software purchase, implementation, and maintenance. SaaS is used in subscription and requires lower costs at defined time intervals, and maintenance is performed directly by the software vendor. Another key benefit is that SaaS environments are based on infrastructures that can increase the amount of computing and storage offered to customers according to their needs, even momentarily and not on a regular basis.

For the above reasons, we decided to implement J2CBROKER as a Service, thus to test and to deploy it in a real scenario.

In the next subsection we explain why we decided to integrate J2CBROKER in the OpenStack environment.

17.4.3   OpenStack Integration

OpenStack is a set of software tools for building and managing open Cloud computing platforms for public and private Clouds. Maintained and supported by the biggest vendors in software development, and hosting and counting on the support of thousands of individual community members, today OpenStack represents the present and the future of open Cloud computing. OpenStack is managed by the "OpenStack Foundation" [23], a non-profit organization who deals with following, supporting and influencing both the development and the ecosystem-building around the project. At the moment, the OpenStack project consists of nine main components which represents the *"core"* of the project itself. The J2CBROKER service integrates the main components related to "Identity" and "Object Storage" functionalities, which are implemented by the projects *Keystone* and *Swift*, respectively. *Keystone* is an OpenStack service that provides API client authentication and authorization by implementing OpenStack's Identity API [10]. *Swift* is an OpenStack service that provides highly available and distributed object/blob storage by implementing OpenStack's Object Storage API [19].

17.4.4   J2CBROKER Description

17.4.4.1   The Client

In order to work, the client needs the presence of a mandatory JSON configuration file called *json-conf-file*. For smooth functioning of J2CBROKER service, the *json-conf-file* file must be filled in the proper way. As the name implies, the *json-conf-file* file contains the information about the configuration of the client, in particular its internal settings (including the symmetric encryption key used); the metrics of the Model used for that particular simulation; the configuration and setting of the communication with the server (including the public key of the server); the configuration and setting of the communication with the Identity Service of the Cloud infrastracture. As shown in Fig. 17.3, J2CBROKER can work in two different modes: the *Random Simulation Mode* and the *Guided Simulation Mode*. Both the above simulation modes are part of the setting at client side.

Fig. 17.3 The *J2CBROKER* simulation modes

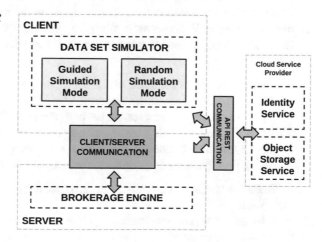

17.4.4.2   The Data Set Simulator

The Data Set Simulator represents the "core" of the client application. It consists
of a modular structure that contains different Model Simulators. This is important
because it makes this *service* a general purpose tool, which allows anyone to create
and connect his own Model Simulator. However, each Model Simulator implements
different specific behaviors. Those latter depend on both the directives received from
the *json-conf-file* file and the characteristic dedicated for the simulation scenario. If
the user uses the *service* according to the directives of the *"Random Simulation
Mode,"* the Data Set Simulator creates specific Data Sets according to a dedicated
Model [6]. Otherwise, if the user uses the *service* according to the directives of the
Guided Simulation Mode, the Data Set Simulator gathers and parses the information
from the JSON absolute paths listed inside the *json-input-list-file* file [6].

In any case, the expected behavior at the server side will be the same. It will store
all the Data Sets and, when it will receive the `active request`, it will elaborate
the Data Sets according to the Model Engine predetermined in the simulation.
Finally, the server will send a JSON file to the client with the result of the calculation
at the Brokerage Engine. In such a context, each Data Set represents a simulated
offer by a CSP at a specific Cloud site.

17.4.4.3   The Client/Server Communication

The communication phase between the client and the server is done in six different
steps (see Fig. 17.1):

1. First of all, in order to be authenticated by the Cloud, the client sends an
   `authentication request` to the Identity Service of the Cloud Infrastructure. If the Identity Service returns the token (**1.B**), which means that it is
   successfully authenticated, the client can move to the next step (**2.A**).
2. In order to verify if the server is alive, the client sends to it a `test request`
   (**2.A**). Therefore, the server needs to verify if the referenced Cloud services are
   active:

   (a) Firstly, the server sends an `authentication request` to the Identity
       Service of the Cloud Infrastructure (**3.A**). If the Identity Service returns the
       token (**3.B**), then the server can move to the next step (**4.A**).
   (b) Then, the server sends a `test request` to the Object Storage Service of
       the Cloud Infrastructure (**4.A**). If the `test request` is successfully done
       (**4.B**), then the server can move to the next step (**2.B**).

   When the test is successfully done (**2.B**), then the client can move to the next
   step (**5.A**).
3. In order to set several environment parameters at server side, the client sends
   a `set-environment request` with some encrypted parameters inside its

Headers (**5.A**). If `set-environment request` is successfully done, then the client can move to the next step (**6.A**).

4. In order to transfer all the Data Sets to the server, the client uses one `dataset request` for each Data Set. Each request will contain all the information about a particular Data Set that the client wants to transfer at that moment. All this information represents encrypted parameters stored as Headers inside the `dataset request`. If the `dataset request` was successfully done (**6.B**), then the client will send another `dataset request`, and so on, until the end. Then the client can move to the next step (**7.A**).

5. To start the server side processing phase, the client sends to the server an `active request` (**7.A**). When the server will complete the processing with success, it will return a response with the JSON file containing the output of the processing phase (**7.B**).

6. At the end of the processing phase, the server permanently store all the data created during the processing phase:

   (a) Firstly, the server sends an `authentication request` to the Identity Service (**8.A**). If the Identity Service returns the token (**8.B**), then the server can move to the next step (**9.A**).

   (b) Then, the server sends a `container-creation request` to the Object Storage Service (**9.A**). If the container creation is successfully done (**9.B**), then the server can move to the next step (**10.A**).

   (c) In order to put on the Cloud all the CSV files created during the processing phase, the server sends to the Object Storage a `create-object request` for each file (**10.A**). If the `create-object request` is successfully done (**10.B**), then the server will send another `create-object request`, and so on, until the end.

17.4.4.4   The Server

In order to work, the server needs the presence of a mandatory JSON configuration file called *json-conf-file*. For smooth functioning of J2CBROKER service, the *json-conf-file* file must be filled in the proper way. As the name implies, the *json-conf-file* file contains the information about the configuration of the server, in particular its internal settings; the configuration and setting of the communication with the client; the configuration and setting of the communication with the Identity service and the Object Storage of the Cloud infrastracture. When the server starts a communication with a client, it receives `set-environment request`. From this latter, the server acquires guidelines about the Model Simulator that characterizes the Simulation Scenario, and the successive actions to do. After the server receives the Data Sets from the client and stores them, it keeps listening for an `active request` in order to begin the processing phase through the related Brokerage Engine. This processing phase will be different depending on which Model Engine will be used during the simulation scenario. Regardless of the type of the Model Engine used, the processing result of any simulation scenario is formalized in the form of an output JSON file which is forwarded back to the client.

17.4.4.5   The Brokerage Engine

The "Brokerage Engine" represents the "core" of the server application. It consists of a modular structure that contains different Model Engines. This is important because it makes this *service* a general purpose tool, which allows anyone to create and connect his own Model Engine. However, each Model Engine implements different specific behaviors. Those latter depend on both the directives received from the client through the `set-environment request` and the characteristic of the dedicated brokerage scenario.

17.5   Case Study: Sustainability-Cost Model

In this section, we present a *case study* to prove the goodness of the proposed methodology. By referring to Fig. 17.4, we introduce a **Sustainability-Cost Model** to make the best choice in *resource allocation*. We express sustainability through several sub-metrics which are generally used to define *"how green is a datacenter"*:

Fig. 17.4 The sustainability-cost model simulator data set [6]

Sustainability-oriented Model Data Set	
Service Data Set	
Parameter	Range
Instance workload (watts)	200-300
Power basic (watts)	100
Running time (hours)	10, 24, 360, 750
Number of instances in each request	1, 10, 20, 50
Number of instances in each offer	12, 14, 16, 18, 20
Availability (%)	99.90-99.99
Service price ($/h)	0.007-0.112
Sustainability Data Set	
Parameter	Range
ITEU	0.3-0.6
ITEE	0.1-3.9
PUE	1.4-2.3
GEC	0.0-0.003
CDIE (kgCO2/kWh)	*source:* https://www.ipcc.ch

- the *Information Technology Equipment Utilization* (*ITEU*);
- the Information Technology Equipment Efficiency (*ITEE*);
- the Power Usage Effectiveness metric (*PUE*);
- the Green Energy Coefficient (*GEC*);
- the Data center Performance Per Energy (*DPPE*).

Apart from the sustainability metric, we include **availability** and **monetary cost** criteria in a *multi-criteria approach*. We remark that the goal is to prove the goodness of the methodology used in the sustainability model and not of the only multi-criteria approach that is well known in literature. Availability is the degree to which a system, product or component, is operational and accessible when required for use. It is usually expressed as percentage quota. The product quality model defined in ISO/IEC 25010 [21] comprises availability as a quality characteristic. As already reported in Fig. 17.2, it is also an important *Key Performance Indicator (KPI)*. It is generally computed as a function of the total service time, the Mean Time Between Failure (MTBF), and the Mean Time to Repair (MTTR) as follows:

$$av = (MTBF/(MTBF + MTTR)) * 100 \qquad (17.1)$$

Physical interpretation of availability is the percentage of time during which a system correctly operates. Monetary Cost is a quantifiable criterion that addresses customers and organizations in their business. By specifically referring to the IT services, it is generally expressed in $/h$ (i.e., dollars-per-hour) or $/GB$ (i.e., dollars-per-GigaByte). Usually, providers offer instance placement services with a fixed price in the service maintenance time at a site. Therefore, we can express cost for an instance i at a site node s, as follows:

$$cost_{s,i} = service\_price_{s,i} * \Delta_t \qquad (17.2)$$

where Δ_t is the *running time* of the i-th instance at the s-th site node.

We deployed the J2CBROKER Service in an OpenStack infrastructure hosted on a *IBM BladeCenter LS21* at the *Cloud Laboratory Data Center—University of Messina* [13]. We tested the service by running several application clients, hosted in several machines at the *High Performance Computing and Application—University of Messina* [8].

17.5.1 Scenario

J2CBROKER simulates a scenario where the main goal is to reduce *carbon dioxide emissions* (i.e., the *CO2*) through a Cloud brokerage ecosystem, where *Cloud Service Providers (CSPs)* cooperate in a *centralized brokerage environment* to run *instance workloads* at the most convenient Cloud sites. An instance is a temporary virtual server that needs to be allocated in order to run services. More specifically, we developed a Sustainability-Cost Model Data Set at client side and a Sustainability-Cost Model Engine at server side.

17.5.1.1 The Sustainability-Cost Model Simulator

The proposed simulation environment presents a modeling of both service and Cloud site, i.e., the **Model Data Set**, thus to provide *input* data for the related Model Engine. Each offer is modeled by a $JSON$ document, i.e., the Model Data Set, which includes two main collections: the first one defines a *service data set* describing the service parameters and among these availability and cost; the second one defines the sustainability metrics and factors at each Cloud site. The service data set, in particular, is obtained from a survey on several "top" providers of IT technologies (e.g., Dell blade servers), Cloud services and solutions (e.g., Amazon Web Services (AWS)). The second one, instead, is based on the measurement results of a real scenario, from the METI Japan project [22] on enhancing the energy efficiency and the use of green energy in data centers. Figure 17.4 shows the Model Data Set: the simulator selects a random value between the range set for each metric in order to characterize each offer by its sustainability, availability, and monetary cost values. The detailed description of the tabulated parameters and the multi-criteria algorithm implemented at the related Model Engine is part of our previous work [7].

Figure 17.5 shows an example of Data Set created by the Data Set Simulator used for the proposed case study.

In particular, it identifies an *offer* in terms of:

1. "simulationName": the parameterized name of the simulation itself;
2. "providerName": the name of the simulated provider;
3. "providerNumber": the id of the simulated provider;
4. "datasetName": the name of the simulated Data Set which represents the *offer* (according to the **Sustainability-Cost Model**);
5. "datasetNumber": the id of the simulated Data Set (each provider can present different offers);
6. "Site": the information about the metrics that describe the site of the *offer* [7];
7. "Service": the information about the metrics that describe the service of the *offer* [7].

17.5.1.2 The Sustainability-Cost Model Engine

The Brokerage Engine consists of a modular structure that contains different Model Engines and among these the Sustainability-Cost Model Engine. This one is mainly based on a decision-making algorithm in the perspective of an energy-

Fig. 17.5 An example of data set created by the data set simulator

```
{
"Simulation": {
         "simulationName": "J2CBROKER-Sim"
         },
"Provider": {
         "providerName": "PROVIDER#1",
         "providerNumber": "1"
         },
"Dataset": {
         "datasetName": "R5KL3",
         "datasetNumber": "1"
         },
"Site": {
         "ITEU": "0.4",
         "ITEE": "3.0",
         "GEC": "0.0",
         "PUE": "2.1"
         },
"Service": {
         "instanceNum": "1",
         "running_time": "10",
         "t_start": "03-05-2017 16:46",
         "availability": "99.22",
         "workLoad": "212",
         "servicePrice": "0.099"
         }
}
```

and cost-aware instance allocation in a centralized brokerage Cloud environment. It results in the *opt* value, which is an "optimum" index computed by the algorithm to weigh each offer in terms of carbon dioxide emission (sustainability in gCO2), cost ($-per-hour), and availability (%) as follows:

$$opt_{n\text{Provider},n\text{Dataset}}$$

$$= A * (gCO2_{n\text{Provider},n\text{Dataset}}/gCO2_{\text{worst}}$$

$$+ B * (cost_{n\text{Provider},n\text{Dataset}}/cost_{\text{worst}}); \qquad (17.3)$$

where $A + B = 1$. More specifically, A and B represent the *weights* respectively assigned to "sustainability" and "monetary cost saving" in the simulation. In that formula, *n*Provider is a unique number which identifies the CSP site; Dataset is a unique number identifying the offer of that CSP; gCO2 quantifies the related CO2 emission to run an instance workload; cost is the service price in $-per-hour. Both CO2 emission and cost are normalized to the respective worst case calculated at each iteration on all the offers by all the CSPs (Fig. 17.6).

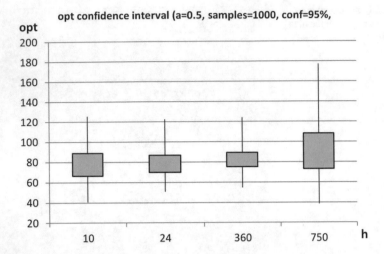

Fig. 17.6 Confidence interval of the *opt* index for the allocation of ten instances

Fig. 17.7 Experimental
results

opt range (a=0.5)				
h	10	24	360	750
MIN	40,483	50,777	54,660	38,549
MAX	125,38	122,195	124,219	177,089
AVG	77,53	78,275	82,064	90,456
CONF	22,686	16,943	14,009	35,095
AVG−(CONF/2)	66,194	69,803	75,060	72,909
AVG+(CONF/2)	88,880	86,747	89,0697	108,004

17.5.2   Experimental Results

Figure 17.7 shows several *opt* values calculated on the basis of Formula 17.3.

More specifically, the experimental results are distinguished by *running time* "h" (10, 24, 360, 750 h), as reported in the Data Set at Fig. 17.4, and we consider a number of 1000 samples (iterations) for each h with a 95% confidence. This latter is an observed interval, in principle different from sample to sample, that frequently includes the value of an unobservable parameter of interest if the experiment is repeated. The desired level of confidence is set (i.e., not determined by data). If a corresponding hypothesis test is performed, the confidence level is the complement of the respective level of significance, i.e., a 95% confidence interval reflects a significance level of 0.05. For each *running time* we calculate the *minimum (MIN)*, the *maximum (MAX)*, the *average (AVG)*, the *confidence (CONF)*, and the last two confidence interval limits. Figure 17.6 provides the reader with a quick visual feedback about the confidence interval of the above-mentioned *opt* index to allocate a number N of ten instances, with $A = 0.5$, $B = 0.5$, a number of 1000 samples,

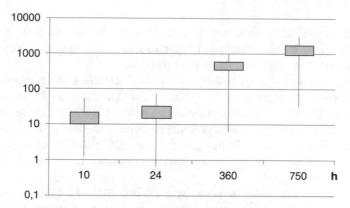

Fig. 17.8 Confidence interval of the sustainability (kgCO2/DPPE) for the allocation of ten instances

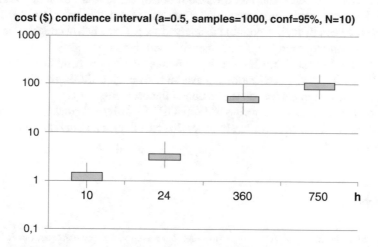

Fig. 17.9 Confidence interval of the cost saving for the allocation of ten instances

and a 95% *confidence*. The values reported in Fig. 17.6 are the result of a post-processing phase, by getting as input all the best *opt* values calculated at each run step. If we consider that for each run in our simulation the worst case results in an *opt* index close to 1000, the energy-aware algorithm at Broker is able to select offers with an *opt* index very low when compared to the worst case. The result is a good compromise between sustainability and cost since it is as better as it is closer to zero.

Figures 17.8 and 17.9 show a graphical representation of the results that are obtained in Sect. 17.5.1.2, respectively, in terms of *sustainability* and *cost saving* metrics. This latter are indicative of the *goodness* of the CSP offers.

17.6 Conclusions and Future Work

Performance evaluation in a real Cloud environment is too cost- and time-consuming. This is why simulation tools can help researchers, Cloud Service Providers, and customers to evaluate their proposals.

In this paper, we presented and discussed *J2CBROKER*, a Brokering Simulation Tool deployable as a Service and integrated in the OpenStack environment. From a technological point of view, we decided to integrate J2CBROKER in the OpenStack environment because OpenStack represents the present and the future of open Cloud computing. This integration in the Cloud allows J2CBROKER to be owned and hosted by service providers and to be offered to consumers on-demand by following the *"utility model."*

From a simulation point of view, by modeling different Cloud sites and the related economic offers on several criteria, we demonstrated how the proposed approach can accommodate different scenarios characterized by a different number of instances to allocate and based on both performance and business parameters. These last may also come from real Data Sets, and thanks to an "optimum" balance between them, it is possible to analyze scenarios where a cooperative Cloud ecosystem can reduce the gap in competition with larger providers.

Currently, the J2CBROKER service is managed via command line. The results provided by the Brokerage Engine are available through a JSON file, CSV file, and displayed via command line on the screen at the user side.

In future, we plan to develop a Web GUI in order to graphically show the simulation results, thus providing both service customers and providers with a better user experience.

References

1. E. Badidi, A framework for software-as-a-service selection and provisioning. Int. J. Comput. Netw. Commun. **5**, 189–200 (2013)
2. R. Buyya, M. Manzur, GridSim: a toolkit for the modeling and simulation of distributed resource management and scheduling for grid computing. Concurrency Comput. Pract. Exp. **14**, 1175–1220 (2002)
3. R.N. Calheiros, R. Ranjan, A. Beloglazov, C.A.F. De Rose, R. Buyya, CloudSim: a toolkit for modeling and simulation of cloud computing environments and evaluation of resource provisioning algorithms. Softw. Pract. Exper. **41**, 23–50 (2011)
4. E. Casalicchio, V. Cardellini, G. Interino, M. Palmirani, Research challenges in legal-rule and QoS-aware cloud service brokerage, in *Future Generations Computer Systems*, vol. 78(Part 1), (Elsevier, Amsterdam, 2018), pp. 211–223. ISSN 0167-739X. EISSN 1872-7115
5. R. Di Pietro, M. Scarpa, M. Giacobbe, A. Puliafito, Secure storage as a service in multi-cloud environment, in *Ad-hoc, Mobile, and Wireless Networks: 16th International Conference on Ad Hoc Networks and Wireless, ADHOC-NOW 2017, Messina, September 20–22, 2017, Proceedings* (Springer, Berlin, 2017), pp. 328–341
6. M. Giacobbe, R. Di Pietro, C. Puliafito, M. Scarpa, J2CBROKER: a service broker simulation tool for cooperative clouds, in *10th EAI International Conference on Performance Evaluation Methodologies and Tools (Valuetools 2016)* (2016), pp. 107–123

7. M. Giacobbe, M. Scarpa, R. Di Pietro, A. Puliafito, An energy-aware brokering algorithm to improve sustainability in community cloud, in *Proceedings of the 6th International Conference on Smart Cities and Green ICT Systems* (2017), pp. 166–173
8. HPCALab, High performance computing and application laboratory at University of Messina. http://hpca.unime.it/
9. F. Jrad, A service broker for Intercloud computing. Karlsruhe, Karlsruher Institut für Technologie (KIT), Dissertation (2014)
10. Keystone, the OpenStack Identity Service. Updated: 2017-10-05 00:49. https://docs.openstack.org/keystone/latest/
11. D. Kliazovich, P. Bouvry, S. Khan, Ullah: GreenCloud: a packet-level simulator of energy-aware cloud computing data centers. J. Supercomput. **62**, 1263–1283 (2012)
12. MarketsandMarkets. http://www.marketsandmarkets.com/
13. MDSLab, Mobile and distributed systems laboratory at University of Messina. http://mdslab.unime.it/
14. A. Núñez, J.L. Vázquez-Poletti, A.C. Caminero, G.G. Castañé, J. Carretero, I.M. Llorente, iCanCloud: a flexible and scalable cloud infrastructure simulator. J. Grid Comput. **10**, 185–209 (2012)
15. I. Patiniotakis, Y. Verginadis, G. Mentzas, PuLSaR: preference-based cloud service selection for cloud service brokers. J. Internet Serv. Appl. **6**, 1–14 (2015)
16. M. Radi, Efficient service broker policy for large-scale cloud environments. Int. J. Comput. Sci. Issues. **12**(1), 85–90 (2015)
17. L. Sun, H. Dong, F.K. Hussain, O.K. Hussain, E. Chang, Cloud service selection: State-of-the-art and future research directions. J. Netw. Comput. Appl. **45**, 134–150 (2014)
18. S. Sundareswaran, A. Squicciarini, D. Lin, A brokerage-based approach for cloud service selection, in *2012 IEEE 5th International Conference on Cloud Computing (CLOUD)* (2012), pp. 558–565
19. Swift, the OpenStack object storage service. Updated: 2017-10-05 13:27. https://docs.openstack.org/swift/latest/
20. The digital single market. https://ec.europa.eu/digital-single-market/en/digital-single-market
21. The ISO/IEC 25010. http://iso25000.com/index.php/en/iso-25000-standards/iso-25010
22. The ministry of economy trade and industry (METI) Japan project - enhancing the energy efficiency and use of green energy in data centers. http://home.jeita.or.jp/greenit-pc/sd/pdf/ds2.pdf
23. The OpenStack foundation. https://www.openstack.org/foundation/
24. B. Wickremasinghe, R. Buyya, Cloudanalyst: a cloudsim-based tool for modelling and analysis of large scale cloud computing environments (2009)

Chapter 18
A Software Tool for the Evaluation of Transient Removal Methods in Discrete Event Stochastic Simulations

Sushma Nagaraj and Armin Zimmermann

18.1 Introduction

Performance evaluation of complex system models helps system designers to base architectural design decisions on relevant information and better understanding, and thus reduces risk significantly. Such evaluations have to be done at model level, as there is no prototype or finished system available yet. Numerical analysis methods are often impossible for the computations when the state space becomes too large or the underlying stochastic process is too complex (non-Markovian, etc.). In such cases simulation is the only choice, and there is a vast amount of algorithms and tools available for specific application areas and model classes [1, 12, 19]. This paper concentrates on discrete event models [3, 25], where states of the system can be enumerated and there are atomic state transitions when certain events occur randomly.

The simulation in question, on which performance evaluation is to be performed, may either be in a warm-up phase, related to the system start-up, or may have steady-state properties which apply to the stationary phase. The latter include average performance under normal use and are arguably more important for systems design, as they ensure that the long-term behavior of the system will be optimal or as expected. They help in designing system parameters or decide about architectural choices.

It is well known that simulating system behavior, based on a model with well-defined semantics, is usually a technical issue. However, estimating the values of performance measures defined by the underlying stochastic discrete event model and deciding when to stop the simulation (usually similar to estimating

S. Nagaraj · A. Zimmermann (✉)
Systems and Software Engineering, Technische Universität, Ilmenau, Germany
e-mail: sushma.nagaraj@tu-ilmenau.de; armin.zimmermann@tu-ilmenau.de

© Springer International Publishing AG, part of Springer Nature 2019
A. Puliafito, K. S. Trivedi (eds.), *Systems Modeling: Methodologies and Tools*,
EAI/Springer Innovations in Communication and Computing,
https://doi.org/10.1007/978-3-319-92378-9_18

the achieved accuracy) leads to several issues [1, 12]. One is the fact that the probability distribution over the set of states and thus initial state variable values for the corresponding simulation are not known in advance, because of which the simulation models are often started empty or with default state assumptions. This leads to an initial bias of the performance measures before the simulated process has reached a (more) stationary state. Although a larger sample set leads to a better estimation quality, this is not the case here as the initial samples are biased and thus decrease result quality.

R.W. Conway through his work in [4] emphasized on the empty state of the system during the simulation start and its gradual approach towards steady-state. There are various solutions that have already been proposed for addressing this issue of the warm-up bias [10], and more research is still active in the area, owing to the unavailability of a universal solution. A commonly known method to avoid this effect is to detect this initial transient phase and ignore it in the result estimation [8, 9, 14, 15, 18]. This is obviously much easier if the full set of samples is available from a simulation log (i.e., offline); however, for practical simulation algorithm implementations, online and automatic methods for the transient phase estimation are necessary to decide on-the-fly (or online) from when samples should be taken into account.

Estimation and removal of the initial transient phase has been the most popular and successful solution for the problem at hand. There is then a trade-off about where to cut off the initial phase: the more samples are discarded, the lower is the influence of the initial bias. However, fewer usable samples will result in a higher variance of the estimated result. Taking into account simulation costs and uncertainty of truncation quality in addition to this, the necessity of the truncation may even be questioned in general [15]. Also, when the transient is detected using confidence or spectral intervals, the effectiveness of such a detection depends largely on the stationarity of the sample distribution. Hence, when a group of samples has to be considered for discarding, the accuracy of the determined truncation point is important.

There are various methods that have been suggested in diverse scenarios and applications. Methods such as the MSER-5 [24], in combination with Schruben's test [20] and Fishman's rule [5], are known to be successful and are currently being used in various simulation software tools. Apart from the tried and tested methods, there are newer algorithms such as the sequential steady-state detection (online) [7] and the MSER-5Y (offline) [24] that have been suggested to perform better than the existing methods. There also are methods like [15] that question the necessity of the transient removal itself, especially for long-running simulations.

The suitability or performance of the transient removal methods cannot be fairly judged, unless all the methods under consideration are subjected to similar testing conditions. Schruben [20] saw the advantages of combining more than one bias-removal method, and with this as an inspiration, we present a software *Framework for the Detection and Removal of Initial Transients* (fDRIT [17]) providing a possibility for a systematic and fair comparison between the different algorithms. fDRIT is a platform for a comparative analysis between various transient detection and removal methods subjected to a set of benchmark test cases. This not only helps in the qualitative and quantitative evaluation of multiple algorithms, but also eliminates the need for separate implementations.

A similar framework, AutoSimOA [11], has been designed to evaluate the simulation output of a discrete event simulation. However, in AutoSimOA, the overall quality of the discrete event simulation itself is measured, after running the output through a standard transient removal algorithm. The simulation tool Akaroa 2 [16] offers an environment for controlling simulations and implements output analysis that also contains transient removal methods, but without evaluating them explicitly.

The main intention of fDRIT is to provide an accessible platform for the understanding and comparison of the performance of any discrete event simulation algorithm in a modern computing environment, with a standard, predefined test suite, and configurable and comparable transient removal techniques. fDRIT also provides a possibility for implementation of newer algorithms and to be incorporated into or interfaced with performance evaluation tools such as TimeNET [26].

The remainder of this paper is divided as follows: Section 18.2 briefly explains the background of steady-state simulation and its transient removal problem, illustrating an overview of some of the transient removal methods from the literature. Section 18.3 explains the design and software architecture of the framework. It provides a detailed explanation of how the performance metrics are currently calculated. Section 18.4 provides an example case and the corresponding results, with a comparison between the various implemented algorithms and their corresponding quality ratings.

18.2   Methods for Transient Removal

This section briefly covers different types of transient removal methods; more details can be found in [17].

Forward and Backward Data-Interval Rule (DIC and DIG) Conway proposed truncation as an efficient method to overcome the effect of the initial transient [4]. The idea was to truncate the series of measurements up to a point where the value was neither the maximum nor the minimum of the remaining data set. This test was suggested to be made on a data set that was not collected during the initial run, but instead on a set that was collected after a few pilot runs, such that the uncertainty of the initial run could be eliminated and the stabilization period decision would be more accurate than the initial run. The recognized initial transient period could then be deleted from the result of each run. Conway's rule specifies selection of an a-priori number of replications of the simulation run with a fixed number of observations in the output per exploratory run. Following this, the output sequence is scanned using a forward pass, starting at the beginning, moving on to determine the earliest observation which is neither the maximum nor the minimum of all the later observations, which is considered as the truncation point for the corresponding simulation run. However, any system with high frequency variations and low-amplitudes resulted in premature truncations, which could be corrected by using suitable smoothing filters.

As an alternative to this rule, Gafarian et al. suggested a similar method but with a backward pass called the Backward Data-Interval Rule [8]. The only change in this method from Conway's Forward Data-Interval rule is that DIG scans the output sequence using a backward pass, starting at the last observation, to find the earliest observation which is neither the maximum nor the minimum of the observations examined earlier. This observation is considered the truncation point of the corresponding simulation run.

When DIC and DIG methods were considered together, they had the effect of defining a range of output values which contain the most appropriate truncation point. In effect, the actual truncation point lies between the truncation points specified by both of the methods. These methods can be used for specifying a range in which the truncation value could be present in any application. However, they are suitable only for systems that prefer an offline analysis.

Fishman's Rule Fishman [5] proposed a first-order autoregulatory method for investigation of the effects of the initial conditions in a simulation. The bias and variance were used to measure the effects. The truncation heuristic proposed was referred to as the Crossing-of-the-Truncation-Mean rule, which defined the cut-off point as the observation at which the sample path crossed the cumulative mean a pre-determined number of times. The cumulative mean of all the observations, starting at the initial point, was computed at the arrival of each new observation in the output sequence, maintaining a count of the number of times the output sequence crossed the mean-sequence. Following this, after a pre-specified number of such crossings, the output data was truncated.

A prominent problem in this method was pointed out by White [22]: the inability of dynamic determination of the appropriate number of crossings across different applications. Such a choice is sensitive to frequency and magnitude of oscillations in the truncated sequence, independent of the character of the reserved sequence. White suggested a revision of Fishman's rule by proposing truncation after the first crossing of the reserved sequence. Fishman, however, also discussed the possible inflation of the variance of the truncated mean led by the truncation of data from the output sequence, which was not desired. There must be a balance between elimination of bias against the corresponding increase in the variance. This idea was used as a basis by several researchers for developing various other methods, working towards achieving the mentioned balance.

Schruben's Tests Schruben [20] proposed the concept of confidence interval estimation for the mean of any simulated time-series, by standardizing the entire series, instead of using a single estimator or a scaling constant. Instead, the use of a limiting stochastic process to model the limiting behavior of the standardized time-series which is conditioned to start and end at zero is considered. There are four assumptions made in this algorithm: The first is that the process is strictly stationary. Second, the process has a zero mean and a finite variance that can be closely approximated as a real-valued function. The third assumption is that the unknown parameter to be estimated is assigned the existing value until further change. Fourth, the mixing function for the process converges with the respective process going down to the value zero at a suitable speed. In the test procedure, the simulation output is converted to a standardized time-series using the time-series analysis. Then, the converted standardized series is tested against any bias presence using null hypothesis. Test results from diverse simulation models are provided in the paper, claiming that the test is generic in its approach, suitable to a wide range of applications.

This algorithm is to date one of the most used tests to check the stationarity of a system model. Many tools use this method as one of the hybrid methods in the implemented transient removal algorithms.

In later work, Schruben et al. [21] also suggested an alternative hypothesis testing framework with a family of tests for detecting initialization bias in the mean of the simulation output series. Here a general transient mean function is used, in place of the null hypothesis with no change in the output mean. Knowledge of the type of bias contained in the time-series is, however, a prerequisite for using this method. It is also required to know if there is a positive or negative bias present in the system.

Euclidean Distance Method This algorithm was proposed by Young-Hae Lee et al. [13], as an online algorithm for the determination of steady-state in simulation outputs. The Euclidean distance method estimates the required results within a single run and in combination with the batch means method, without the need for a pilot run. Although the process of finding of the truncation point was more robust, the algorithm tended to delete more data than the other methods. Also, the generality parameter of this method was left unchecked by the authors. For achieving complete automation in simulation analysis, it was suggested that the batch size and the number of batches needed to be determined online.

MSER Rules The Marginal Standard Error Rules aim at finding an optimal truncation point based on minimizing the confidence interval estimate. This wields Fishman's suggestion of formulating a balance between the bias reduction and decreased precision or increased variance [6]. Such a balance is suggested to be achieved by minimizing the mean square error (MSE) of the mean from the truncated sample. White and Minnox [23] tested this method on four queueing simulation models and deduced that the resultant match with the expected output is based on the match of the obtained truncation point with the actual value. Later, there were further improvements made to the MCR model by comparison with the following four tests: no truncation, Fishman's test, DIC, and DIG, and it was concluded that the MCR heuristic significantly improved the value of the steady-state mean value μ. However, MCR was not represented as a fail-proof method. It was described as being "moderately superior" to the methods that were the state-of-the-art of that day. The failure of the accurate truncation point detection in case of rare-event occurrence was also mentioned and use of methods such as batch means in such circumstances was suggested.

Spratt, who used the term Marginal Standard Error Rules (MSER) instead of Marginal Confidence Rule (MCR), defined different methods from literature and compared their effectiveness of reducing the bias effect [18]. He made two significant modifications to the original MCR algorithm in order to avoid two cases: analysis on insufficient data and oversensitivity of the algorithm. One of the modifications was done on the basis of Schruben's suggestion to aggregate the data into non-overlapping fixed size batches. The recommended batch size from Spratt is 5; hence, this modified MSER rule is also referred to as the MSER-5 algorithm. However, a general form of batched MSER, proposed by White, is what is used widely today. It was known as MSER-m, where m denotes the batch size.

In 2011, Saeideh Yousefi [24] elaborated further on the quality of the MSER-5 algorithms. One of the main problems pointed out in this work was the estimation of the truncation point and its corresponding Confidence Interval estimator and estimated steady-state mean value being incorrect in simulation cases which contained pronounced initial transients. In case of such a failure, the algorithm provided no alternative way for continuing with or restarting the simulation. Hence, Yousefi proposed a modified version of the MSER-5 algorithm, called MSER-5Y, which not only delivered a truncation point always, but also provided a "nearly-unbiased" point estimator and an approximately valid CI estimator for the steady-state mean. This algorithm was found to work well in many test scenarios, with the limitation of working with only precollected data, like MSER-5.

Pawlikowski's Heuristics In 1990, Krzysztof Pawlikowski [19] carried out extensive research in estimating the main factors that can affect the accuracy of stochastic simulations designed to give insight into the steady-state behavior of queueing processes. The main aim of the research was to address the problems of estimating the optimal start and stop points for simulation runs, in order to obtain the statistical accuracy in the results. He suggested a two-step solution for the transient problem. In the first step, the different heuristics R1–R10 presented in the literature could be used to get a first approximation of the truncation point. For example, Fishman's rule or the Heuristic R5 was used in his simulation study of communication protocols. After obtaining the approximate value of the truncation point, the second step was to carry out the estimation a second time, on the sequence truncated using the approximated truncation point value. This was done in order to test the sequence for stationarity and to get a robust estimate of the onset of the steady-state phase in the simulation. The key for using the R5 heuristic is the selection of the mean-crossing value, in order to avoid the truncation point value being too high or too low.

Several tools use Fishman's rule or the R5 heuristic in Pawlikowski's heuristics, including TimeNET (Timed Net Evaluation Tool, [26]) and JMT (Java Modelling Tools [2]).

TimeNET is a software tool for the modelling and performability evaluation using various classes of stochastic Petri nets (SPNs). It supports the creation, testing, simulation, and analysis of SPN(s), which are well suited for the model-based performance and dependability evaluation. The simulation component of TimeNET can perform the transient as well as the stationary evaluation of SPNs with statistical accuracy control achieved by establishing confidence intervals and relative errors.

JMT is a suite of applications developed by Politecnico di Milano and Imperial College London. The project aims at offering a comprehensive framework for performance evaluation, system modeling with analytical and simulation techniques, capacity planning and workload characterization studies. The JSIM module within the JMT tool is a simulation module for Queueing Networks. JSIM has an automated transient detection procedure which is based on spectral analysis. This tool computes and plots the online estimated values which are within the calculated confidence intervals. The tool also supports what-if analyses, where a sequence of simulations can be run with custom control parameter values. The JMT tool also credits the transient detection as a critical statistical decision where the detection and removal, variance estimation and simulation length control need to be completely automated and have claimed to have achieved the same. The simulation is automatically stopped when the performance measures of the simulation model are estimated with the desired accuracy.

Transient Removal in Akaora This is a recent online transient phase detection algorithm for the Akaroa simulation tool [16] by Freeth [7]. This method is based on the convergence of cumulative means of the system to its steady-state value and uses forecasting techniques to determine this convergence. The truncation point is estimated as the point where the cumulative means became sufficiently horizontal and flat. The proposed method was tested against the MSER-5 statistic and the former detected the onset of the steady-state more effectively and consistently for almost all simulation models that were used for testing and hence proved to be a good consideration in sequential truncation. However, the performance of this method was tested in the *Akaroa* environment and its behavior in diverse applications still remains untested.

18.3   fDRIT: A Framework to Detect and Remove Initial Transients

The idea of implementing a framework through which multiple transient detection algorithms can be implemented and tested against a set of standard tests was inspired by the modified proposal by Schruben [20]. This work suggested the use of a testing framework, comprising of a set of standard tests with which the initialization bias and the mean of the simulation output series can be evaluated. This was found to be a useful approach, especially for comparing two or more methods under consideration. It was also observed that an efficient transient removal can be often achieved with a combination of multiple algorithms that are already known, and that different algorithms are suitable for different problems at hand. However, the comparative suitability of the available methods is unknown due to the lack of knowledge of the presence of the methods, or limited testing, proving it difficult to analyze the applicability of the algorithms to the situation at hand. The framework proposed here provides a platform for understanding and comparing the performance of any deployed methods, and aims at obtaining the performance evaluation of any discrete event simulation, with diverse transient removal methods, online. The current (academic) version of the framework may be found at https://github.com/tuiSSE/fDRIT.

The framework comprises three modular parts:

1. Algorithm implementation framework
2. Test framework
3. Evaluation tool

The algorithm implementation part consists of an extendable number of transient removal algorithms, which are connected to the remaining software by a standardized interface for easier implementation of future algorithms. At the time of writing, we have implemented eight algorithms from the literature:

1. Euclidean Distance Method by Young-Hae Lee et al. [13].
2. Schruben's stationarity Rule [20].
3. Method of cumulative means by Adam Freeth used in Akaora [7].
4. Fishman's rule or R5 Heuristic by Pawlikowski [19].
5. MSER-5 by Spratt [18, 24].
6. MSER-5Y by Saeideh Yousefi [24].
7. Initial transient detection approach used in the TimeNET tool [26].
8. Initial transient detection approach used in the JMT simulator [2].

The test framework holds the definition of various dynamic test algorithms that generate random data for the input data generation of the algorithms. Its modular structure allows easy extension by additional examples. This part is planned to be extended towards a set of benchmark data generators. The current implementation uses three M/M/1 queuing systems as first examples, which are:

- M/M/1 queue-waiting-time process with empty-and-idle initial condition and 90% server utilization.
- M/M/1 queue-waiting-time process with 113 initial customers and 90% server utilization.
- M/M/1/LIFO queue-waiting-time process with empty and idle initial condition and 80% server utilization.

The evaluation tool defines a set of methods that calculate the quality of the executed methods. There is also a part responsible for the user interaction. The user is required to provide two inputs: an algorithm or a combination of algorithms that are to be analyzed and compared, and a test method that needs to be executed, for generating the necessary test data.

Software Architecture   The software architecture of the framework has three components: The part which contains the algorithm implementation, the test data generation framework, and the part where the evaluation of the selected methods

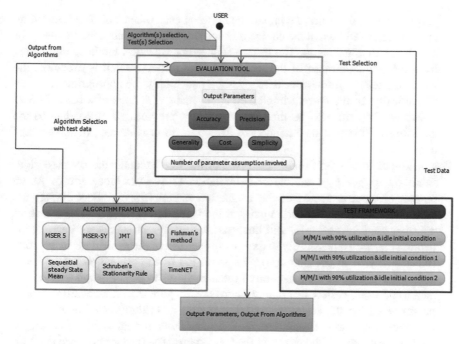

USER

Algorithm(s) selection,
Test(s) Selection

Output from
Algorithms

EVALUATION TOOL

Test Selection

Output Parameters

Accuracy   Precision

Algorithm Selection
with test data

Generality   Cost   Simplicity

Number of parameter assumption involved

Test Data

ALGORITHM FRAMEWORK

MSER 5   MSER-5Y   JMT   ED   Fishman's
method

Sequential
steady State
Mean

Schruben's
Stationarity Rule

TimeNET

TEST FRAMEWORK

M/M/1 with 90% utilization & idle initial condition

M/M/1 with 90% utilization & idle initial condition 1

M/M/1 with 90% utilization & idle initial condition 2

Output Parameters, Output From Algorithms

Fig. 18.1 fDRIT software architecture

are carried out based on the selected tests. A brief overview of the architecture is
shown in Fig. 18.1.

Algorithm Implementation   The implementation of the transient detec-
tion/removal algorithm to be evaluated is done in this module. The
implementation of the algorithm is independent of any test data, isolating it from
the testing environment. The main intention of separating the test framework
from the algorithm implementation was to enable a generic evaluation of the
implemented methods. A call is made from the evaluation framework to the
implementation framework, to fetch the required methods under scrutiny. The
job of the implementation framework is then to run the implemented algorithm,
with the test data provided by the evaluation part, and to return the calculated
output values like the steady-state mean, output or error values, or any other
output calculated for the system. The only responsibility of the algorithm
framework is to execute the implemented method with the provided test inputs
and returning the result values. This part may be extended to accommodate
as many new algorithms as necessary, with complete abstraction. Section 18.4
contains comparative results obtained from the tool for an M/M/1 queue.

Test Data Generation   The quality of algorithms implemented cannot be guaranteed with limited testing. This led to the creation of a separate module for
the purpose of testing and input data generation. The implemented algorithms
have to be subjected to different test scenarios/benchmarks, to observe their

performance and to draw a comparison between each other. This is an important process, especially when the choice of a suitable algorithm needs to be made for a certain individual setup. The testing framework is also instantiated and called from the evaluation part of the tool. When the respective call is executed, the job of the testing framework is to generate and supply the appropriate test data dynamically. In the current implementation, three variations of a basic M/M/1 queue are implemented as first prototypes, for providing the test data to the algorithms. This will be extended in the future to provide an extensive set of tests.

Evaluation of the Algorithms   This part of the tool executes the selected algorithm by triggering a sample test-simulation (Test Data Generation). As an output, the metric scores of the evaluation parameters (c.f. Sect. 18.3.1) are calculated, when the truncation point in the algorithm is reached. These metrics are then used in a Cost-Benefit analysis, with different scores for each of the parameters, with a possibility to draw a comparison of the algorithms in question (individually or in combination). The main job of the evaluation section is to gather the input, i.e., the selection of the algorithms to be compared and the test data to be used, collect test data (online) and acquire the corresponding output parameters from the selected transient removal algorithm(s) by initiating their execution with the collected test data. The evaluation engine basically acts as an interface between the user and the framework. The different criteria used for the calculation of the performance metrics are: Accuracy, Precision, Cost, and Parameter estimation, as detailed in Sect. 18.3.1.

18.3.1   Evaluation Parameters for the Quality of the Algorithms

There is no single performance metric available which quantifies the performance of the bias-removal algorithms, and guide us towards the best suitable algorithm. A few data scientists have already attempted to compare the bias-removal algorithms against one another in the past. In [14], a set of performance measures like the Mean Square Error, variance and their corresponding percentage changes and cost calculations are explained. Gafarian et al., in [8], specify five criteria to evaluate the goodness of the algorithms under consideration.

Using these as a guide, we have incorporated a set of criteria, to be able to quantify the performance of the bias-removal algorithm in question, and have devised a method for scoring the respective metrics. We believe that with such a score, the comparison of the algorithms becomes easier. The current method is only an example of the benefits of using such a scoring method and uses parameters which are believed to act as a deciding factor in the selection of a suitable algorithm. The choice of these metrics can be tailored to use, and their weighting altered according to the requirements. The chosen metrics in the current example benefit analysis are elaborated below.

18.3.1.1   Accuracy

This parameter defines a score of the accuracy of the evaluated truncation point with respect to the theoretically exact truncation point. The accuracy can be given any score between 0 and 10. This score is assigned by calculating the accuracy percentage using Eq. (18.1) given below and then dividing this percentage value by 10, to deduce a score between 1 and 10. The accuracy parameter has an overall weighting of 25% in the goodness measure of the algorithm. The weighting along with the score is used in the benefit analysis, which is used for the comparison of the goodness of multiple algorithms.

$$a = \frac{\text{Estimated Truncation Point}}{\text{Theoretical Truncation Point}} \times 100 \qquad (18.1)$$

In the case of the truncation point estimate being greater than the theoretical truncation point, i.e., if the estimated percentage is greater than 100, the score is calculated by subtracting the difference from 100 and then dividing by 10, and is also assigned a negative sign to signify the loss of useful data.

18.3.1.2   Precision

The considered algorithm can be said to be precise if the estimated truncation point is not a significantly varying value, i.e., the estimated value should be similar across runs. This value is obtained from the coefficient of variation which is the ratio of the standard deviation to the mean of the truncation point obtained across a pre-specified number of times, as given in Eq. (18.2) below. Typically, any method is expected to be precise, which means that it should have a negligible variance. This means that the closer the value p from Eq. (18.2) is to zero, the more precise is the respective value.

$$p = \frac{\text{Standard Deviation of the Estimated Truncation Point}}{\text{Mean of the Estimated Truncation Point}} \times 100 \qquad (18.2)$$

Similar to the accuracy score, the precision is also scored between 0 and 10, which is obtained by subtracting the tenth of p from Eq. (18.2) from 10. If the precision value is negative, then the precision score is considered to be 0.

The precision parameter also weighs 25% of the overall goodness measure for our sample evaluation. This weight, in combination with the precision score of the respective algorithm, and along with the other parameters is used in the beneficial analysis of the corresponding algorithms.

18.3.1.3 Computational Cost

The cost parameter is evaluated by calculating the amount of time required for the following three operations:

1. Computation time for the algorithm itself, i.e., the computational efficiency.
2. Computation time for the collection of the output data before estimating the truncation point and the subsequent discarding of the biased data.
3. The computation time associated with the determination of the truncation point.

The total computing time is calculated from the implementation code, but for the beneficial analysis, we require a score between 1 and 10. This is done by assuming a grade of 10 initially and reducing 1 point for every 30 s that the algorithm uses in our sample case. The 30 s benchmark here is model- and simulation-setting dependent, and has been chosen for the example benefit analysis based on the longest running model taking $30 \times 10 = 300$ s. To avoid the case in which an algorithm stops too early and is deemed beneficial w.r.t. cost, the weighting allocated to the cost is much lower than for accuracy and precision, for instance.

18.3.1.4 Parameter Estimation

The parameter estimation score is based on the total number of unknown configuration parameters to be set in the algorithm. This value is important because if more parameters are necessary, the riskier the algorithm may be in the case of wrong selections. This value is calculated by finding the number of parameters that are estimated in the execution of the algorithm and assigning a tenth of that value as the parameter estimation score.

The parameter estimation score can range between 1 and 10, and it has a weighting of 15% in the benefit analysis of the algorithms.

18.3.1.5 Overall Benefit Analysis

The joint performance of the various implemented algorithms is evaluated using a standard benefit analysis method. The different parameters explained in Sect. 18.3.1 have been used for the analysis. The benefit rating for the individual algorithms is computed by first calculating the parameter scores for the respective heuristics. These scores are then multiplied by the corresponding parameter weights and finally summed up for each algorithm. The performance of the algorithm under consideration is directly proportional to its benefit rating. fDRIT provides this detailed analysis to the user and can thus support an informed decision between algorithms dealing with the initial transient problem.

18.4   Example and Results

The framework was executed with the test data from a dynamic M/M/1 queue-waiting-time process with empty and idle initial condition and 90% server utilization. The results for some selected algorithms obtained are shown in Table 18.1.

In Table 18.1, the values for three of the implemented methods are depicted in detail. The benefit rating values of the other methods are given in Table 18.2. It should be noted that this criteria selection and weighting is just one possible setup that should be chosen following the intentions of a future user. If, for instance, the algorithm complexity is not significant as only the numerical results are considered important, the corresponding value can be set to 0, possibly resulting in a different algorithm to be chosen.

The framework acts as a base evaluation and testing platform for all implemented methods. This work may be reused in any other project as a dynamic library, without the need to re-evaluate and re-implement the initial transient problem for every requirement from the scratch. The possibility of drawing a comparison and conclusion to the algorithm decision is the main idea here. When the implemented algorithms is taken into consideration, it can be seen from the benefit ratings that the rating of the combination of the cumulative batch means method and the MSER-5Y algorithms has the highest rating, showing that it is the best of the compared lot. In a similar way, based on the requirement and the testing method, the best suitable algorithm(s) for the particular situation can be decided upon.

Table 18.1 Benefit rating matrix for evaluation of three truncation methods

Evaluation statistic	Weight(w)	Truncation algorithms					
		ED method		Fishman's rule		Cumulative batch means	
		Score(S)	$W * S$	S	$W * S$	S	$W * S$
Accuracy	35	3	75	5	125	7	175
Precision	35	5	125	5	1275	7	175
Cost	5	3	15	4	20	6	30
Parameter estimation	25	9	135	9	135	9	135
Benefit rating	$\sum = 100$	510		565		710	

Table 18.2 Benefit ratings for the experiments carried out on the framework

Truncation algorithms	Benefit rating
Batch means by Euclidean distance method	510.0
Fishman's rule	565.0
Cumulative batch means	710.0
MSER-5	687.5
MSER-5Y	695.0
Schruben's rule	565.0
TimeNET	687.5
JMT	730.0
Cumulative batch means + MSER-5Y	765.0

18.5   Conclusion

This paper presented the prototype software framework fDRIT for the evaluation and comparison of algorithms that solve the problem of the initial transient for stochastic discrete event simulations. Several available methods from the literature have been collected, implemented, and compared with a benchmark test. Criteria for the user-specific weighted benefit comparison are proposed and the set of algorithms is evaluated as an example. The framework should act as a decision help and standardized comparison environment to evaluate new and existing methods for certain types of application models or input data. Its software architecture is modular and extendable for future implementation of more benchmark models/data sources, and transient detection algorithms. A secondary use may be the use of the framework or parts of it inside other tools, which then do not have to re-implement a certain transient removal algorithm for themselves.

In the future, we plan to use the tool to support the development of new online algorithms for the initial transient detection of stochastic Petri net models.

References

1. S. Asmussen, P. Glynn, *Stochastic Simulation: Algorithms and Analysis. Stochastic Modelling and Applied Probability*, vol. 57 (Springer, Berlin, 2007)
2. G.S.M. Bertoli, G. Casale, An overview of the JMT queueing network simulator. Tr 2007.2, Politecnico di Milano, DEI (2007)
3. C.G. Cassandras, S. Lafortune, *Introduction to Discrete Event Systems* (Kluwer, Boston, 1999)
4. R.W. Conway, Some tactical problems in digital simulation. Manag. Sci. **10**(1), 47–61 (1963)
5. G.S. Fishman, Bias considerations in simulation experiments. Oper. Res. **20**(4), 785–790 (1972)
6. G.S. Fishman, *Monte Carlo: Concepts, Algorithms and Applications*. (Springer, New York, 1996)
7. A. Freeth, A sequential steady-state detection method for quantitative discrete-event simulation. PhD thesis, University of Canterbury (2012)
8. A.V. Gafarian, C.J. Ancker, T. Morisaku, Evaluation of commonly used rules for detecting "steady state" in computer simulation. Nav. Res. Logist. Q. **25**(3), 511–529 (1978)
9. W.K. Grassmann, Factors affecting warm-up periods in discrete event simulation. Simulation **90**(1), 11–23 (2014)
10. K. Hoad, S. Robinson, R. Davies, Automating warm-up length estimation. J. Oper. Res. Soc. **61**, 1389–1403 (2009)
11. K. Hoad, S. Robinson, R. Davies, AutoSimOA: a framework for automated analysis of simulation output. J. Simul. **5**(1), 9–24 (2011)
12. A.M. Law, D.M. Kelton, *Simulation Modeling and Analysis*, 3rd edn. (McGraw-Hill Higher Education, New York, 1999)
13. Y.H. Lee, K.H. Kyung, C.S. Jung, On-line determination of steady state in simulation outputs. Comput. Ind. Eng. **33**(3), 805–808 (1997)
14. P.S. Mahajan, R.G. Ingalls, Evaluation of methods used to detect warm-up period in steady state simulation, in *Proceedings of the 36th Conference on Winter Simulation*, WSC '04, Winter Simulation Conference (2004), pp. 663–671

15. D. McNickle, G.C. Ewing, K. Pawlikowski, Some effects of transient deletion on sequential steady-state simulation. Simul. Modell. Pract. Theory **18**(2), 177–189 (2010)
16. D. Mcnickle, K. Pawlikowski, G. Ewing, AKAROA2: a controller of discrete-event simulation which exploits the distributed computing resources of networks, in *Proceedings of European Conference on Modelling and Simulation (ECMS 2010)* (2010)
17. S. Nagaraj, A. Zimmermann. fDRIT - an evaluation tool for transient removal methods in discrete event stochastic simulations, in *Proceedings of 10th International Conference on Performance Evaluation Methodologies and Tools (VALUETOOLS 2016)*, Taormina, October 2016
18. R. Pasupathy, B. Schmeiser, The initial transient in steady-state point estimation: contexts, a bibliography, the MSE criterion, and the MSER statistic, in *Simulation Conference (WSC), Proceedings of the 2010 Winter*, December 2010, pp. 184–197
19. K. Pawlikowski, Steady-state simulation of queueing processes: survey of problems and solutions. ACM Comput. Surv. **22**(2), 123–170 (1990)
20. L. Schruben, Confidence interval estimation using standardized time series. Oper. Res. **31**(6), 1090–1108 (1983)
21. L. Schruben, H. Singh, L. Tierney, Optimal tests for initialization bias in simulation output. Oper. Res. **31**(6), 1167–1178 (1983)
22. K.P. White, A simple rule for mitigating initialization bias in simulation output: comparative results, in *IEEE International Conference on Systems, Man and Cybernetics, 1995. Intelligent Systems for the 21st Century*, October 1995, vol. 1, pp. 206–211
23. K.P. White, M.A. Minnox, Minimizing initialization bias in simulation output using a simple heuristic, in *Proceedings of IEEE International Conference on Systems, Man and Cybernetics*, October 1994, vol. 1, pp. 215–220
24. S. Yousefi, MSER-5Y: an improved version of MSER-5 with automatic confidence interval estimation. Master's thesis, North Carolina State University (2011)
25. A. Zimmermann, *Stochastic Discrete Event Systems* (Springer, Berlin, 2007)
26. A. Zimmermann, Modelling and performance evaluation with TimeNET 4.4, in *Quantitative Evaluation of Systems - 14th International Conference, QEST 2017*, pp. 300–303, Berlin, September 2017

Chapter 19
A House Appliances-Level Co-simulation Framework for Smart Grid Applications

Abdalkarim Awad, Peter Bazan, and Reinhard German

19.1 Introduction

Renewable energy sources cover a large part of the worldwide energy supply. In 2014 the share was approximately 19.1% [23]. Because of the continued expansion of renewable energy sources, the energy system is moving away from its traditional centralized structure with large producers towards a structure with many distributed generators. While the share of renewable energy in electricity production in 2005 excluding hydropower was 2.6% [24], it was already 4% at the end of 2014.

Due to the expansion of wind energy and photovoltaics (PV), fluctuating energy sources must be integrated increasingly into the system. In 2004, the global installed capacity of wind power was 48 GW, whereas with 370 GW it increased by almost eight times in the year 2014. At the same time, the installed PV capacity increased almost 48-fold from $3.7\,\mathrm{GW}_p$ to $177\,\mathrm{GW}_p$ [23]. Thus, the classic roles of producers and consumers in the energy system are supplemented by consumers who generate energy, the so-called prosumers. The integration of variable renewables increases the need for centralized and decentralized energy storage. Prosumers equipped with storage systems are also referred to as prostumers.

The increasing use of electric vehicles (EV) leads to a further increase in consumers and storage systems in the grid. Such mobile storage systems can be charged at home and also at charging stations, which intensifies the complexity of the entire system. In 2015 there were 1.26 million EVs worldwide, compared to

A. Awad (✉)
Birzeit University, Birzeit, West Bank, Palestine
e-mail: akarim@birzeit.edu; akarim@ieee.org

P. Bazan · R. German
University of Erlangen-Nuremberg, Erlangen, Germany
e-mail: peter.bazan@fau.de; reinhard.german@fau.de

© Springer International Publishing AG, part of Springer Nature 2019　　　303
A. Puliafito, K. S. Trivedi (eds.), *Systems Modeling: Methodologies and Tools*,
EAI/Springer Innovations in Communication and Computing,
https://doi.org/10.1007/978-3-319-92378-9_19

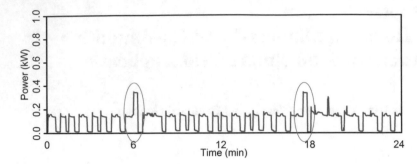

Fig. 19.1 Power consumption of a refrigerator

several hundred EVs in the year 2005 [16]. Therefore, the integration of millions of EVs will be an additional problem in the future. The increasing integration of information and communication technology into the system of consumers, prosumers, and prostumers enables the coordination of such systems and ensuring of grid stability.

Smart grid enables new applications for households such as Demand Response (DR) and Advanced Metering Infrastructure (AMI). Simulation models at appliance level can help understanding the benefits and risks of employing smart grid applications. Figure 19.1 shows the power consumption of a refrigerator. As can be seen, in the morning as well as in the afternoon, there are small increases in the power consumption (small spikes). It is the bulb in the fridge that is causing these spikes which is being captured by these readings. Such readings can be exploited to extract information about the behavior of the household. In [3, 4, 28] we used a co-simulation approach to study CVR and Volt/VAR control. In [5] we presented a short tutorial on using SGsim in electricity distributed networks. In [2] we have explored different methods to preserve privacy. In this work, we present an appliance-level co-simulation framework that enables exploring house-level smart grid applications.

The rest of the paper is organized as follows. At the beginning we present some related works. Then in Sect. 19.3 we introduce SGsim-Home. In Sect. 19.4 we present a case study, and in Sect. 19.4.2, we evaluate the proposed approaches. Finally, Sect. 19.5 concludes the paper.

19.2 Related Work

The use of simulation tools for the evaluation of new technologies and applications is a widely adopted method, and therefore, there is a wide range of tools. Some of these simulation tools have been combined into simulation environments which allow the co-simulation of various domains of a complex system. An overview of the requested requirements of such tools is given in [27]. There, the integration of of-the-shelf simulators for communication systems and electrical power systems into a co-simulation simulation framework is suggested, together with the ability to model control strategies for smart grid applications.

An example of a co-simulation framework is the modular platform mosaik for the evaluation of agent-based smart grid controls [26]. It combines different simulators and simulations and controls the data-flow between them. For this purpose, it defines its own modeling and specification language. It enables the simulation of large-scale smart grid scenarios but lacks the integration of a communication simulator. This problem is addressed with the presentation of a preliminary system architecture of integrating OMNeT++ [10]. Unlike SGsim-Home, this integration is not yet implemented.

Because communication is a key part of smart grid applications, several co-simulation tools are using the discrete-event simulator OMNeT++ [31] for modeling and simulation of communication systems. An example is the co-simulation approach of power systems, communication, and controls presented in [29]. This framework combines the commercial power system analysis software PowerFactory [12] with OMNeT++, whereas SGsim combines the electric power distribution system simulator OpenDSS [13] with OMNeT++, both of which can be used in a non-commercial environment without license fees. Another example is the communication network and power distribution network co-simulation tool for smart grids presented in [18]. There the discrete-event-based simulation of communication systems framework OMNeT++ is coupled with the continuous simulation of power systems tool OpenDSS using a Hypertext Transfer Protocol (HTTP) connection. SGsim-Home, on the other hand, couples the two simulation tools via a more runtime efficient Component Object Model (COM) interface. In addition to the two co-simulation examples, the controller component of SGsim-Home allows the connection to powerful optimizers over the internet.

The agent-based simulation engine of the co-simulation tool GridLAB-D [9], unlike SGsim-Home, has only simple network characteristics integrated like latency, bandwidth, buffer size, or congestion. Instead of using OpenDSS, it is coupled with the power system simulation and optimal power flow tool MatPOWER [33].

The approach of [20] combines three simulation tools for validating flexible-demand EV charging management. GridLAB-D controls the simulation and the charging management of the EV, the battery is modeled with OpenModelica [14], and the distribution grid with PowerFactory [12]. Because of GridLAB-D, the approach can only use simple network characteristics.

All these tools cover different aspects of the smart grid by using co-simulation. SGsim-Home integrates these aspects in one framework combining simulation tools of the power grid and the communication with a connection to an optimizer. It provides models for PV, Battery, EV, and home appliances like refrigerator, Air Condition (AC), and TV.

With SGsim-Home it is possible to analyze and minimize the privacy risk introduced by smart meters. It is shown in [17] that the detection of steady state changes from loads with an on/off switching behavior like refrigerators can identify the appliance. Even from smart meter data with a resolution of 30 min measured over 1.5 years, information about the personal circumstances of the residents can be extracted with a high probability [7].

SGsim-Home allows the analysis of integrated privacy protection and demand response techniques. The work in [22] presents a pre-processing approach to enhance user privacy. The authors have used quantization, down-sampling, and averaging to prevent successful classification of household appliance. An empirical and analytical model to study adding noise to mask smart meter readings has been presented in [6]. Additionally, they used correlation to evaluate the approach. Both methods are focused only on privacy.

Another approach introducing privacy, but this time considering several smart meters, is the homomorphic encryption of aggregated smart grid information presented in [19]. The data aggregation is performed at all smart meters involved in routing the data from the source meter to the collection unit.

In [30] privacy in smart metering systems has been studied from an information theoretical perspective in the presence of renewable energy systems and storage units. The authors describe the system as a finite state model and analyze the impact of a renewable energy system on the privacy. They also investigate the privacy and energy efficiency trade-off, but do not consider power-tariff dependent demand response and optimization.

19.3   SGsim-Home

The framework SGsim-Home is based on the co-simulation framework SGsim [1, 3–5, 28] which is based on two main simulators: OpenDSS [13] and OMNeT++ [31]. The focus of [1, 3, 5] was on transmission and distribution networks. SGsim-Home focuses on simulating home appliances. Two attractive characteristics of OpenDSS make it a suitable candidate for co-simulation. In addition to a stand-alone executable program, OpenDSS provides an in-process Component Object Model (COM) server DLL designed to be driven from an external program. The COM interface makes integrating OpenDSS into other simulators relatively easy. The second reason is the fact that OpenDSS is an open source simulator, and hence, providing this framework as open source for education and research community is possible. OMNeT++ has been selected to implement SGsim. In addition to the basic simulation tools, several frameworks have been developed for OMNeT++. For instance, INET framework has been developed with well-tuned data communication components such as TCP/IP, 802.11, and Ethernet. In order to enable the use of the framework in the field of smart grid applications, we have integrated new components for the electricity distribution network. Figure 19.2 shows the different components of the simulator. Through the COM interface, it is possible to control the execution of the circuit and to change/add/remove different components. Different approaches have been used to simulate the different devices. We have used real data to simulate some devices such as TV and washing machines. Figure 19.4 presents a 10 s resolution power consumption of a TV [15, 25]. At each time step,

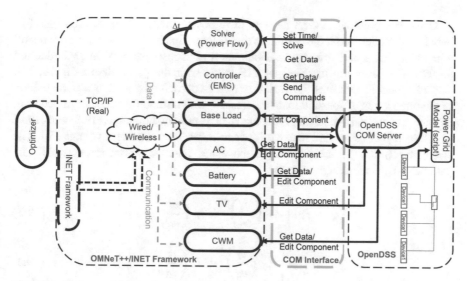

Fig. 19.2 Structure of SGsim-Home with the connections between the different components

the *Edit* command is used to change the parameter of these devices. The database provides data only for 1 h, therefore, the data will be repeated as long as the device is on. A mathematical model has been used to model thermal devices such as AC and refrigerator. The following set of equations represents this mathematical model of a refrigerator, a freezer, or an AC:

$$T(t+1) = \epsilon T(t) + (1-\epsilon)\left(\text{To}(t) - \frac{\text{COP} \times P_{\text{ref}}}{A}\right) \tag{19.1}$$

$$\epsilon = \exp\left(\frac{\delta}{\delta_c}\right), \quad \delta_c = \frac{mc}{A}, \quad A = \frac{k \times G}{\text{Th}}$$

where $T(t+1)$ is the temperature inside the room at control period $t+1$, ϵ is the factor of inertia, δ is the duration of the control period, δ_c is the time constant, mc is the total thermal mass in Wh/C, A is the overall thermal conductivity in W/C, $\text{To}(t)$ is the ambient temperature in C at control period t, COP is the coefficient of performance, and P_{ref} is the electrical power demand of the AC in W at control period t. k is the thermal conductivity coefficient, G denotes the area of the room, and Th denotes the thickness of the wall. If the refrigerator door is open, the bulb in the fridge will cause an additional power consumption P_{bulb}

$$P_{\text{Ref}}^{\text{Total}} = P_{\text{ref}} + P_{\text{bulb}} \tag{19.2}$$

OpenDSS provides several models to represent loads. We have used the ZIP-based load model *model 8* to simulate the different loads. This model is very useful when studying smart grid applications such as Conservation Voltage Reduction (CVR). The loads are modeled as ZIP loads with the parameters as in [8, 11]. The ZIP model represents the variation (with voltage) of a load as a composition of the three types of constant loads Z, I, and P which stand for constant impedance, constant current, and constant power loads, respectively. Equations (19.3) and (19.4) give the current active and reactive loads as a function of current voltage (V). The constants P_0 and Q_0 are the design active and reactive power, respectively. The parameter v_0 is the design voltage.

$$P_{Li} = P_{0i} \left[Z_P \left(\frac{v_i}{v_0} \right)^2 + I_P \left(\frac{v_i}{v_0} \right) + P_P \right] \tag{19.3}$$

$$Q_{Li} = Q_{0i} \left[Z_q \left(\frac{v_i}{v_0} \right)^2 + I_q \left(\frac{v_i}{v_0} \right) + P_q \right] \tag{19.4}$$

All devices are equipped with communication capability so that it is possible to control these devices. The INET framework provides the necessary components to simulate several kinds of communication networks such as WiFi and Ethernet.

Figure 19.3 shows a screenshot of the simulator. The devices are connected through a wireless LAN. The Smart Meter (SM) sends the energy usage at a specific frequency (e.g., 1 reading/min). The Home Energy Management System (HEMS)

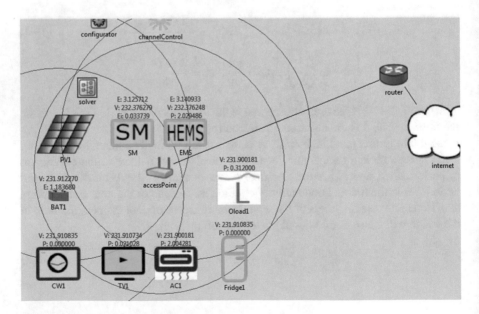

Fig. 19.3 Screenshot of the simulator

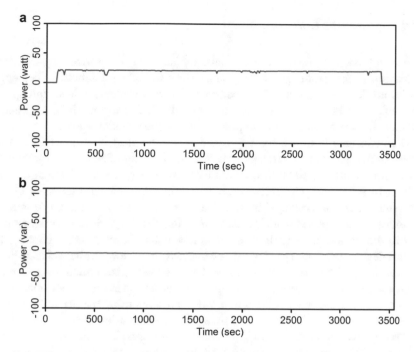

Fig. 19.4 TV active (**a**) and reactive power consumption (**b**)

coordinates the operation of the different devices. For instance, it can find the optimal operation strategy of the devices in order to minimize the electricity costs. The HEMS measures the energy usage at a higher frequency than the SM (e.g., at 1 Hz). The Oload1 represents the basic load and it is non-elastic. V is the voltage value, P represents the power consumption, E is the energy usage, Ei denotes to the last reading from the smart meter. The air condition is considered as an elastic load and the temperature should be maintained within a specific range. The clothes washing machine (CW) is considered also as an elastic load. It consists of several phases which should be run sequentially without interruption. It is possible to control the operation of the battery (charging and discharging periods) by the HEMS (Fig. 19.4).

19.4   Case Study: Integrated Privacy Protection and Demand Response

In this section, we present a case study on integrating privacy protection inside demand response. The HEMS uses the day-ahead price, storage, and load elasticity to minimize the costs. At the same time, it tries to hide load characteristics through a coordinated operation of a battery and elastic loads.

19.4.1 Smooth Consumption

In this approach, we exploit load elasticity and storage device (e.g., battery) to maximize the profit and at the same time to hide household information. The controller tries to maintain a constant power consumption level throughout the whole day through coordination between the different household appliances. Additionally, the controller tries to prevent power consumption spikes.

The main idea is to use the day-ahead price, the electricity demand, and the battery to find the optimal strategy to be followed to minimize the electricity costs and minimize privacy risks. The controller uses the day-ahead price and demand forecast to solve an optimization problem to find the optimal amount of energy to be sold, charged/discharged in/from the energy storage unit, and the amount of electricity to be imported from the main grid. Additionally, it finds the optimal time slots to run elastic loads such as washing machines. Furthermore, it controls the thermal devices (e.g., AC) to hide load characteristics. The controller solves a linear optimization problem for 1 day (i.e., $T = 1440$ min with a resolution of $\delta = 1$ min). Then, according to the results, it changes the current operating parameters of the system. We have formulated the optimization problem using the general algebraic modeling system (GAMS) and then solving the problem using the solver CPLEX.

The objective of the controller is to minimize the costs $C(t)$ and the privacy risks $PR(t)$, therefore, the objective function can be written as below:

$$\min \left\{ \sum_{t=1}^{T} \lambda 1 C(t) + \lambda 2 PR(t) \right\} \tag{19.5}$$

$\lambda 1$ and $\lambda 2$ are constants that emphasize the importance of costs or privacy, respectively. The costs come from importing energy from the grid.

$$C(t) = EP(t)\delta P_b(t) \tag{19.6}$$

$EP(t)$ is the electricity price, $P_b(t)$ is the power imported from the grid.

The above maximization problem is subject to system constraints. We considered the electrical balance constraints which can be written as:

$$P_d(t) + P_b(t) - P_l(t) - P_e(t) - P_c(t) = 0 \tag{19.7}$$

where P_d denotes power discharged from the battery, P_l represents the base load (non-elastic), P_c is the power charged in the battery, and $P_e(t)$ denotes the amount of allocated power in this time slot from the elastic energy.

The energy balance in the battery can be modeled as:

$$E(t+1) = (1 - \alpha)E(t) + \delta \eta_c P_c(t) - \delta \frac{P_d(t)}{\eta_d} \tag{19.8}$$

$$E^{\max} \geq E(t) \tag{19.9}$$

E is the state of charge of the battery, α represents the self-discharge rate from the battery, and η_c and η_d are the charge and discharge efficiencies of the battery, respectively. $E^{\max}$ is the capacity of the battery. We have also considered the following limitations in the system:

$$P_d^{\max} \geq P_d(t), \quad P_c^{\max} \geq P_c(t), \quad P_e^{\max} \geq P_e(t) \tag{19.10}$$

$P_c^{\max}$, $P_d^{\max}$, $P_e^{\max}$ denote the maximum amount of power allowed to charge, to discharge, and to allocate an extra load at each time step, respectively.

The next set of equations guarantees that the battery is either in charge or discharge state.

$$P_d^{\max} x(t) \geq P_d(t) \tag{19.11}$$

$$P_c^{\max} (1 - x(t)) \geq P_c(t) \tag{19.12}$$

$$P_e^{\max} \geq P_e(t) \tag{19.13}$$

$$x(t) \in \{0, 1\} \tag{19.14}$$

The elastic load EL should be served in a specific period, which can be written as:

$$\sum_{t=T_1}^{T_2} \delta P_e(t) = \text{EL} \tag{19.15}$$

where $[T_1, T_2]$ is the period where the elastic load should be run. If the load should be carried out continuously and it consists of several phases (e.g., washing machine), the following constraints should be added:

$$P_e(t) = w(t) P_{\text{phases}}(k) \quad \forall k \tag{19.16}$$

$$y(t) + w(t) \geq w(t - 1) \tag{19.17}$$

$$y(t) \geq y(t - 1) \tag{19.18}$$

$$w(t) \in \{0, 1\} \tag{19.19}$$

Equation (19.16) models whether an energy phase (k) is being processed during time slot t. Equation (19.17) ensures that the process will not be interrupted after it starts. Equation (19.18) ensures sequential processing of the phases.

We define the following function for privacy. The first term tries to maintain a constant consumption throughout the whole day, while the second term tries to minimize the changes of the power consumption.

$$PR(t) = |P_b(t) - P_{\text{Avg}}| + |P_b(t) - P_b(t - 1)| \tag{19.20}$$

Additionally, the following constraint prevents sudden changes in the power consumption.

$$|P_b(t) - P_b(t - 1)| \leq \Delta P \tag{19.21}$$

19.4.2 Evaluation

In order to explore the capability of the approach to preserve the privacy, we have used the constant consumption approach to hide an EV charging signal. Hiding such a signal is more challenging than hiding refrigerator cooling cycle or turning on a bulb. We evaluated the proposed methods by examining the capability of the algorithm proposed in [32] to disaggregate EV charging signals from aggregated real power signals. The methods presented in [32] can effectively mitigate the interference coming from an AC, enabling accurate EV charging detection and energy estimation under the presence of AC power signals. It is a non-intrusive energy disaggregation algorithm of EV charging signals. It has five steps. In the first step, a threshold is applied to obtain a rough estimate of the EV charging load signal. Then in the second step, it filters the AC spikes. Then it removes the so-called residual noise. Then, in the fourth step, it classifies the type of each filtered segment. In the last step, it performs the energy disaggregation based on the effective width and the effective height of a segment. We have used the same data set that has been used in [32], which came from the Pecan Street Database [21]. This database collects raw power signals recorded from hundreds of residual houses in Austin, Texas. Ten houses using EV were randomly chosen from the database. Each aggregated power signal is generally a combination of about twenty power signals of various appliances, such as EV, AC, furnace, dryer, oven, range, dishwasher, cloth-washer, refrigerator, microwave, bedroom-lighting, and bathroom-lighting. The ground-truth power signals of these appliances are also available in the database. Thus, the database is very suitable to test algorithms' performance in practice. Table 19.1 summarizes the simulation parameters. Figure 19.5a shows the non-elastic power consumption of the house and electricity price. The EV charging process occurs in the afternoon. The house tries to minimize the power usage costs through optimal allocation of an elastic load and storing energy in a battery when it is cheap (e.g., at early morning) for future usage when the electricity is expensive (at afternoon). We assumed that the house owns a 1 kWh battery and a 2 kW AC. Figure 19.5b shows the power usage of the house when coordinating the usage to maintain a constant electricity usage. The house gets the day-ahead price and calls the optimizer. Using this price signal, the optimizer finds the optimal allocation of the elastic loads and the battery charging and discharging period to minimize the costs. At the same time, it tries to maintain a constant power consumption during

Table 19.1 Parameters

Parameter	Value
Battery	1, 5 kWh
ΔP	50 W
$\lambda 1 = \lambda 2$	1
δ	1 min
T	1440 min
$P_d^{max} = P_c^{max} = P_e^{max}$	3 kW
$\eta_c = \eta_d$	90%
ϵ	0.99
A	300 Wh/C
COP	3.5

the day. The controller can adapt the operation to react to new load signals. We have repeated the same experiment for ten houses with EV charging signal. Only in one case it was possible to detect the charging time. Increasing the battery size makes it possible for the controller to further flattening of the power consumption as can be seen in Fig. 19.5c, where we tested a 5 kWh battery. Using the available components, it is possible to produce a misleading charging signal. As depicted in Fig. 19.5d, the controller has produced a consumption profile that looks similar to an EV charging signal at midday.

Based on the price signal, charging the EV in the afternoon is not the optimal charging time. In fact, the EV charging process can be considered as an elastic load which should be done in a specific period (e.g., before 8 AM). If we consider only the electricity costs, i.e., $\lambda 2 = 0$, the controller will select time slots in the early morning as an optimal charging period as can be seen in Fig. 19.6.

19.5   Conclusion

In this work, we have presented a home-appliance co-simulation framework. The simulator is able to capture the electricity as well as the ICT capabilities of smart appliances. Different components have been implemented and simulated. Additionally, the operation of the components can be adapted during the simulation (e.g., the operating parameters can be changed). This way, it is possible to simulate smart grid applications at home-appliances level. Through a case study, we have presented the possibility to integrate privacy protection into an important smart grid application, namely into the demand response. The results have shown the ability to hide load information through coordination between different components. Similarly as with SGsim, we are planning to provide this framework as open source for the academic community.

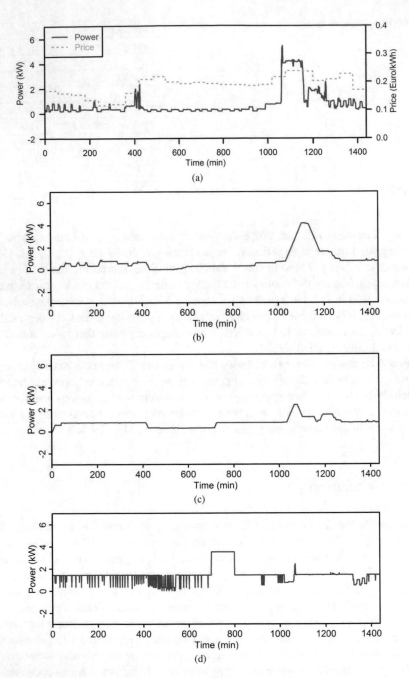

Fig. 19.5 Non-elastic load (**a**), load after smoothing with a 1 kWh battery (**b**), load after smoothing with a 5 kWh battery (**c**), and load after smoothing and adding a misleading charging signal (**d**)

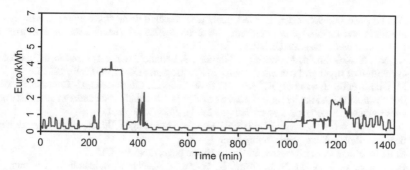

Fig. 19.6 Optimal allocation of EV charging

Acknowledgements Peter Bazan is also a member of "Energie Campus Nürnberg," Fürther Str. 250, 90429 Nürnberg. His research was performed as part of the "Energie Campus Nürnberg" and supported by funding through the "Aufbruch Bayern (Bavaria on the move)" initiative of the state of Bavaria.

References

1. A. Awad, P. Bazan, R. German, SGsim: a simulation framework for smart grid applications, in *Proceedings of the IEEE International Energy Conference (ENERGYCON 2014)*, Dubrovnik, Croatia, May 2014, pp. 730–736
2. A. Awad, P. Bazan, R. German, Privacy aware demand response and smart metering, in *Proceedings of the IEEE 81st Vehicular Technology Conference: VTC2015-Spring, First International Workshop on Integrating Communications, Control, Computing Technologies for Smart Grid (ICT4SG)*, Glasgow, May 2015, pp. 1–15
3. A. Awad, P. Bazan, R. German, SGsim: co-simulation framework for ICT-enabled power distribution grids, in *18th International GI/ITG Conference on Measurement, Modelling and Evaluation of Computing Systems and Dependability and Fault-Tolerance (MMB & DFT 2016)*, Münster, April 2016
4. A. Awad, P. Bazan, R. Kassem, R. German, Co-simulation-based evaluation of volt-VAR control. In *2016 IEEE PES Innovative Smart Grid Technologies Conference Europe (ISGT-Europe)*, October 2016, pp. 1–6
5. A. Awad, P. Bazan, R. German, A short tutorial on using SGsim framework for smart grid applications, in *Proceedings of the 10th EAI International Conference on Performance Evaluation Methodologies and Tools on 10th EAI International Conference on Performance Evaluation Methodologies and Tools*, Valuetools16 (2017), pp. 143–148
6. P. Barbosa, A. Brito, H. Almeida, S. Clauß, Lightweight privacy for smart metering data by adding noise, in *Proceedings of the 29th Annual ACM Symposium on Applied Computing*, SAC '14, New York, NY, 2014, pp. 531–538
7. C. Beckel, L. Sadamori, T. Staake, S. Santini, Revealing household characteristics from smart meter data. Energy **78**, 397–410 (2014)
8. A. Bokhari, A. Alkan, R. Dogan, M. Diaz-Aguilo, F. de Leon, D. Czarkowski, Z. Zabar, L. Birenbaum, A. Noel, R. Uosef, Experimental determination of the ZIP coefficients for modern residential, commercial, and industrial loads. IEEE Trans. Power Delivery **29**(3), 1372–1381 (2014)

9. D.P. Chassin, K. Schneider, C. Gerkensmeyer Gridlab-d: an open-source power systems modeling and simulation environment, in *2008 IEEE/PES Transmission and Distribution Conference and Exposition* (2008), pp. 1–5
10. J. Dede, K. Kuladinithi, A. Förster, O. Nannen, S. Lehnhoff, Omnet++ and mosaik: Enabling simulation of smart grid communications. arXiv preprint arXiv:1509.03067 (2015)
11. M. Diaz-Aguiló, J. Sandraz, R. Macwan, F. de León, D. Czarkowski, C. Comack, D. Wang, Field-validated load model for the analysis of CVR in distribution secondary networks: energy conservation. IEEE Trans. Power Delivery **28**(4), 2428–2436 (2013)
12. DIgSILENT power factory, DIgSILENT GmbH, Gomaringen (2016). http://www.digsilent. de/index.php/products-powerfactory.html
13. EPRI Electrical Power Research Institute, Home page, October 2015
14. P. Fritzson, P. Aronsson, H. Lundvall, K. Nyström, A. Pop, L. Saldamli, D. Broman, The openmodelica modeling, simulation, and development environment. in *46th Conference on Simulation and Modelling of the Scandinavian Simulation Society (SIMS2005)*, Trondheim, October 13–14, 2005
15. C. Gisler, A. Ridi, D. Zufferey, O.A. Khaled, J. Hennebert, Appliance consumption signature database and recognition test protocols, in *2013 8th International Workshop on Systems, Signal Processing and their Applications (WoSSPA)*, pp. 336–341, May 2013
16. International Energy Agency. Global EV Outlook 2016: Beyond one million electric cars. OECD/IEA, Frankreich, 2016
17. C. Laughman, K. Lee, R. Cox, S. Shaw, S. Leeb, L. Norford, P. Armstrong, Power signature analysis. IEEE Power Energ. Mag. **1**(2), 56–63 (2003)
18. M. Lévesque, D.Q. Xu, G. Joós, M. Maier, Communications and power distribution network co-simulation for multidisciplinary smart grid experimentations, in *Proceedings of the 45th Annual Simulation Symposium* (Society for Computer Simulation International, San Diego, 2012), p. 2
19. F. Li, B. Luo, P. Liu, Secure information aggregation for smart grids using homomorphic encryption, in *2010 First IEEE International Conference on Smart Grid Communications (SmartGridComm)*, October 2010, pp. 327–332
20. P. Palensky, E. Widl, M. Stifter, A. Elsheikh, Modeling intelligent energy systems: co-simulation platform for validating flexible-demand EV charging management. IEEE Trans. Smart Grid **4**(4), 1939–1947 (2013)
21. Pecan street database, Home page, October 2015
22. A. Reinhardt, F. Englert, D. Christin, Enhancing user privacy by preprocessing distributed smart meter data. In *Sustainable Internet and ICT for Sustainability (SustainIT), 2013*, October 2013, pp. 1–7
23. REN21. 2015, *Renewables 2015 Global Status Report*. REN21 Secretariat Paris, 2015
24. REN21 renewable energy policy network, Renewables 2005 global status report. Worldwatch Institute, Washington, DC (2005)
25. A. Ridi, C. Gisler, J. Hennebert, ACS-F2- a new database of appliance consumption signatures, in *2014 6th International Conference of Soft Computing and Pattern Recognition (SoCPaR)*, August 2014, pp. 145–150
26. S. Rohjans, S. Lehnhoff, S. Schutte, S. Scherfke, S. Hussain, mosaik-A modular platform for the evaluation of agent-based smart grid control, in *Innovative Smart Grid Technologies Europe (ISGT EUROPE), 2013 4th IEEE/PES* (IEEE, New York, 2013), pp. 1–5
27. S. Rohjans, S. Lehnhoff, S. Schu tte, F. Andrén, T. Strasser, Requirements for smart grid simulation tools, in *2014 IEEE 23rd International Symposium on Industrial Electronics (ISIE)* (IEEE, New York, 2014), pp. 1730–1736
28. SGsim, Home page (2016). https://sourceforge.net/projects/sgsim
29. M. Stifter, J.H. Kazmi, F. Andrén, T. Strasser, Co-simulation of power systems, communication and controls, in *2014 Workshop on Modeling and Simulation of Cyber-Physical Energy Systems (MSCPES)* (IEEE, New York, 2014), pp. 1–6
30. O. Tan, D. Gunduz, H. Poor, Increasing smart meter privacy through energy harvesting and storage devices. IEEE J. Sel. Areas Commun. **31**(7), 1331–1341 (2013)

31. A. Varga,   The OMNeT++ discrete event simulation system,   in *European Simulation Multiconference (ESM 2001)*, Prague, June 2001
32. Z. Zhang, J.H. Son, Y. Li, M. Trayer, Z. Pi, D.Y. Hwang, J.K. Moon,   Training-free non-intrusive load monitoring of electric vehicle charging with low sampling rate,   in *The 40th Annual Conference of the IEEE Industrial Electronics Society (IECON 2014)*, Dallas, TX, October 2014, pp. 1–6
33. R.D. Zimmerman, C.E. Murillo-Sánchez, R.J. Thomas,   Matpower: steady-state operations, planning, and analysis tools for power systems research and education. IEEE Trans. Power Syst. **26**(1), 12–19 (2011)

Index

© Springer International Publishing AG, part of Springer Nature 2019
A. Puliafito, K. S. Trivedi (eds.), *Systems Modeling: Methodologies and Tools*,
EAI/Springer Innovations in Communication and Computing,
https://doi.org/10.1007/978-3-319-92378-9

Printed in the United States
By Bookmasters